Sub-Half-Micron Lithography for ULSIs

In semiconductor-device fabrication processes, lithography technology is used to print circuit patterns on semiconductor wafers. The remarkable miniaturization of semiconductor devices has been made possible only because of the continuous progress in lithography technology. However, for the trend of ever-increasing miniaturization to continue a breakthrough in lithography technology is now needed. This book describes advanced techniques under development in Japan and elsewhere that represent the key to future semiconductor-device fabrication.

The background to developments in lithography technology, trends in ULSI technology and future prospects are reviewed, and the requirements that future lithography technology must meet are described. Several important lithography methods, such as deep-UV lithography, X-ray lithography, electron-beam lithography, and focused-ion-beam lithography are described in detail by experts in each area. The principles underlying each of these methods are illustrated at the beginnings of each chapter to help the reader understand the basis of the different approaches. Other relevant technologies, such as those that concern resist materials, metrology, and defect inspection and repair are also described. Original figures and tables are presented to highlight the key issues and recent developments.

This book will be of value to graduate students studying semiconductor-device fabrication, to engineers engaged in such fabrication and to designers of ULSI devices.

KATSUMI SUZUKI was born on April 28, 1950. He received an M.S. degree in electrical engineering from Tokyo University of Agriculture and Technology, Tokyo, Japan in 1975, and a Ph.D. in engineering from Osaka University, Osaka, Japan in 1991. He joined the NEC Corporation in 1975, and has been engaged in research and development of X-ray lithography. Now, he is managing an X-ray lithography group of NEC. Since 1996, he has also been working as a research fellow in the Super-fine SR Lithography Laboratory of the Association of Super-Advanced Electronics Technologies (ASET).

SHINJI MATSUI received M.S. and Ph.D. degrees in electronic engineering from Osaka University, Osaka, Japan, in 1977 and 1981, respectively. He joined the NEC Corporation in 1981, and engaged in nano-scale engineering using electron-beam and focused-ion-beam. He was seconded to Semiconductor Leading Edge Technologies, Inc. from 1997 to 1998 and managed the Electron-Beam Direct Writing Group. He transferred from NEC to Himeji Institute of Technology in 1998. Dr Matsui now manages the Nano-scale Science & Technology Group at the Institute.

YUKINORI OCHIAI was born on May 12, 1955. He graduated from Hiroshima University in 1977 and received a Ph.D. in engineering from Osaka University in 1986. He joined NEC Corporation in 1986, and engaged in nanofabrication technology using focused-ion-beam and electron-beam. Dr Ochiai was a research associate at Cambridge University from 1988 to 1989.

Sub-Half-Micron Lithography for ULSIs

Edited by
Katsumi Suzuki
Shinji Matsui and
Yukinori Ochiai

CAMBRIDGE UNIVERSITY PRESS
Cambridge, New York, Melbourne, Madrid, Cape Town, Singapore, São Paulo

Cambridge University Press
The Edinburgh Building, Cambridge CB2 2RU, UK

Published in the United States of America by Cambridge University Press, New York

www.cambridge.org
Information on this title: www.cambridge.org/9780521570800

First published 2000
This digitally printed first paperback version 2005

A catalogue record for this publication is available from the British Library

Library of Congress Cataloguing in Publication data
Sub-half-micron lithography for ULSIs/Katsumi Suzuki, Shinji Matsui,
and Yukinori Ochiai.
p. cm.
Includes bibliographical references.
ISBN 0 521 57080 8
1. Integrated circuits – Ultra large scale integration – Design and
construction. 2. Microlithography. 3. Printed circuits – Design and
construction. I. Suzuki, Katsumi, 1950– . II. Matsui, Shinji, 1950–
III. Ochiai, Yukinori, 1955– .
TK7874.76.S83 1999
621.39′5 – dc21 99–13558 CIP

ISBN-13 978-0-521-57080-0 hardback
ISBN-10 0-521-57080-8 hardback

ISBN-13 978-0-521-02234-7 paperback
ISBN-10 0-521-02234-7 paperback

Contents

Contributors

Dr Takayuki Abe
Senior Research Scientist
Advanced LSI Technology Laboratory
Research & Development Center
Toshiba Corporation
1, Komukai Toshiba-cho, Saiwai-ku, Kawasaki
210-8582, Japan
Tel: +81-44-549-2188, Fax: +81-44-520-1804
e-mail: takayukil_abe@toshiba.co.jp

Dr Hiroshi Ban
Senior Research Engineer, Supervisor,
Multimedia Electronics Laboratory
NTT Lifestyle and Environmental Technology
Nippon Telegraph and Telephone Corporation
3-1, Morinosato, Wakamiya, Atsugi,
Kanagawa 243-0198, Japan
Tel: +81-462-40-2545, Fax: +81-462-40-2679
e-mail: hban@aecl.ntt.co.jp

Dr Kimiyoshi Deguchi
Senior Research Engineer, Supervisor
Microfabrication Technology Laboratory
NTT Telecommunications Energy Laboratories
Nippon Telegraph and Telephone Corporation
3-1, Morinosato, Wakamiya, Atsugi,
Kanagawa 243-0198, Japan
Tel: +81-462-40-2545, Fax: +81-462-70-2372
e-mail: deguchi@aecl.ntt.co.jp

Dr Masao Fukuma
Director
Silicon Systems Research Laboratories
NEC Corporation
1120, Shimokuzawa, Sagamihara, Kanagawa
229-11, Japan
Tel: +81-42-771-0621, Fax: +81-42-771-0897
e-mail: fukuma@mel.cl.nec.co.jp

Dr Teruo Hosokawa
Senior Research Engineer, Supervisor
SOR Application Technology Project
NTT Telecommunications Energy Laboratories
Nippon Telegraph and Telephone Corporation
3-1, Morinosato, Wakamiya, Atsugi,
Kanagawa 243-0198, Japan
Tel: +81-462-40-2588, Fax: +81-462-40-4324
e-mail: hosokawa@aecl.ntt.co.jp

Dr Sunao Ishihara
Director
NTT Basic Research Laboratory
Nippon Telegraph and Telephone Corporation
3-1, Morinosato, Wakamiya, Atsugi,
Kanagawa 243-0198, Japan
Tel: +81-46-240-3300 Fax: +81-46-270-2358
e-mail: ishihara@will.brl.ntt.co.jp

Dr Kunihiko Kasama
Group Manager
Microfabrication Process Development Group
ULSI Device Development Laboratories
NEC Corporation
1120 Shimokuzawa, Sagamihara, Kanagawa 229, Japan
Tel: +81-427-79-9903, Fax: +81-427-79-9928
e-mail: kasama@lsi.tmg.nec.co.jp

Tadayoshi Kokubo
General Manager
Technology Department No. 1
Fuji film Olin Co., Ltd.
Yoshida-minami factory
4000 Kawashiri, Yoshida-cho, Haibara-gun, Shizuoka 421-0302, Japan
Tel: +81-548-32-7058, Fax: +81-548-32-7091
e-mail: tadayoshi_kokubo@ffo.fujifilm.co.jp

Dr Masanori Komuro
Assistant Director
Electron Devices Division
Electro Technical Laboratory
Ministry of International Trade and Industry
1-1-4, Umezono, Tsukuba, Ibaraki 305, Japan
Tel: +81-298-58-5509, Fax: +81-298-58-5514
e-mail: komuro@etl.go.jp

Dr Tadahito Matsuda
Senior Research Engineer, Supervisor
Microfabrication Technology Laboratory
NTT System Electronics Laboratories
Nippon Telegraph and Telephone Corporation
3-1, Morinosato, Wakamiya, Atsugi, Kanagawa 243-1098, Japan
Tel: +81-462-40-2545, Fax: +81-462-40-4316
e-mail: tmatsuda@aecl.ntt.co.jp

Dr Shinji Matsui
Professor
Advanced Science and Technology for Industry
Himeji Institute of Technology
3-1-2 Koto, Kamigori, Ako, Hyogo 678-1205, Japan
Tel: +81-791-58-0473, Fax: +81-791-58-0242
e-mail: matsui@lasti.himeji-tech.ac.jp

Dr Koichi Moriizumi
Manager
Mask CAD Section
EDA Development Department
Mitsubishi Electric Corporation
Kita-itami Works
4-1, Mizuhara, Itami, Hyogo 664-8641, Japan
Tel: +81-727-84-7149, Fax: +81-727-80-2629
e-mail: moriizum@lsi.melco.co.jp

Dr Makoto Nakase
Japan Electronic Industry Development Association (JEIDA)
Kikai Shinkou Kaikan, 3-5-8, Shiba-koen, Minato-ku, Tokyo 105-0011, Japan
Tel: +81-3-3433-6861, Fax: +81-3-3433-2003
e-mail: nakase@jeida.or.jp

Dr Yukinori Ochiai
Principal Researcher
Fundamental Research Laboratories
NEC Corporation
34, Miyukigaoka, Tsukuba, 305-8501, Japan
Tel: +81-298-50-1132, Fax: +81-298-56-6139
e-mail: ochiai@frl.cl.nec.co.jp

Dr Takeshi Ohfuji
Senior Process Engineer
Mask Engineering
Intel Corp.
5-6, Tokodai, Tsukuba, Ibaraki
300-2635, Japan
Tel: +81-298-47-5131, Fax: +81-298-47-6974
e-mail: ohfujit@intel.co.jp

Dr Shinji Okazaki
Research Manager
EUV Lithography Laboratory
Association of Super-Advanced Electronics Technologies (ASET)
Atsugi Research Center; c/o NTT Atsugi Research Laboratories
3-1, Morinosato, Wakamiya, Atsugi, Kanagawa 243-0198, Japan
Tel: +81-46-270-6688, Fax: +81-46-270-6699
e-mail: okazaki@euv.aset-unet.ocnne.jp

Dr Norio Saitou
Chief Researcher
Advanced Process Reserach Department
Central Research Laboratory
Hitachi Ltd.
1-280, Higashi-Koigakubo, Kokubunji, Tokyo 185-8601, Japan
Tel: +81-423-23-1111, Fax: +81-423-27-7704
e-mail: saitou@crl.hitachi.co.jp

Dr Hisatake Sano
Department General Manager of Research
Semiconductor Components
Semiconductor Components Operations
Dai-nippon Printing Co., Ltd.
2-2-1, Fukuoka, Kamifukuoka, Saitama 356-8507, Japan
Tel: +81-492-78-1683, Fax: +81-492-78-1698
e-mail: sano_h@cc.micro.dnp.co.jp

Dr Katsumi Suzuki
Manager, Research
LSI Basic Research Laobratory
Silicon Systems Research Laboratories
NEC Corporation
34, Miyukigaoka, Tsukuba, Ibaraki 305-8501, Japan
Tel: +81-298-50-1163, Fax: +81-298-56-6138
e-mail: katsumi@lbr.cl.nec.co.jp

Dr Tadahiro Takigawa
Senior Manager
ULSI Research Laboratories
Research and Development Center
Toshiba Corporation
1, Komukai Toshiba-cho, Saiwai-ku, Kawasaki 210-8582, Japan
Tel: +81-44-549-2316, Fax: +81-44-520-1804
e-mail: tadahiro.takigawa@toshiba.co.jp

Dr Toru Tojo
Chief Research Scientist
Advanced LSI Technology Laboratory
Corporate Research & Development Center
Toshiba Corporation
1, Komukai, Toshiba-chyo, Saiwai-ku, Kawasaki 210-8582,
Tel: +81-44-549-2188, Fax: +81-44-520-1804
e-mail: tooru.tojo@toshiba.co.jp

Dr Wataru Wakamiya
Manager
ULSI Process Development Department
ULSI Development Center
Mitsubishi Electric Corporation
4-1, Mizuhara, Itami, Hyogo 664-8641, Japan
Tel: +81-727-84-7353, Fax: +81-727-80-2675
e-mail: wakamiya@lsi.melco.co.jp

Seiichi Yabumoto
Executive Staff
Instruments Division
Nikon Corporation
1-6-3, Nishi-Ohi, Shinagawa-ku, Tokyo 140-8601, Japan
Tel: +81-3-3773-9032, Fax: +81-3-3773-6734
e-mail: yabumoto.sei@nikon.co.jp

Dr Akio Yamada
Electron Beam Lithography Division
Advantest Corporation
Ohtone R&D Center
1-5, Shintone, Ohtone-machi, Kitasaitama-gun, Saitama 349-1158, Japan
Tel: +81-480-72-6300, Fax: +81-480-78-1147
e-mail: yamada@eb.advantest.co.jp

Preface

LSI (large-scale integration) is an invention that has greatly influenced our life in the latter half of the 20th century by bringing us new forms of convenience, comfort, and entertainment. It will continue to play an important role in the 21st century. Today's multimedia systems require ultra-high-density, high-speed, and low-power-consumption devices. Within a few years, practical Gb-scale dynamic random access memories (DRAMs) and GHz-scale processors will be developed, and battery-operated multimedia processors with high-density memories and analog circuits will be among the main targets of ultra-large-scale integration (ULSI). To achieve such devices, breakthroughs in materials, fabrication processes, circuits, and systems will be needed. The development of next-generation lithography technologies is especially urgent.

The purpose of this book is to describe the present status of the principal lithography technologies and the expected future developments. To cover all of the important lithography technologies, including resist materials, metrology, inspection, and repair technologies, each item has been described by experienced engineers currently active in each field in Japan.

Chapter 1 reviews ULSI technology trends, and clarifies the requirements that lithography techniques must satisfy. Chapter 2 examines optical lithography technology, which is currently the most widely used form of lithography. Chapters 3 to 5 explain advanced lithography technologies such as X-ray lithography (XRL), electron-beam lithography (EBL), and ion-beam lithography (IBL) in that order. Resist materials and technologies, metrology, and defect inspection and repair techniques, which play an essential role in practical ULSI production, are described in detail in Chapters 6 and 7. Throughout this book, recent experimental data are cited, which will help readers appreciate the current state-of-the-art for each technology.

We have written this book mainly for researchers and engineers who are engaged in electronic device fabrication, and for university students who are interested in this field.

Katsumi Suzuki

Acknowledgements

The publishers and authors listed below are gratefully acknowledged for giving their permission to use figures based on illustrations in journals and books for which they hold the copyright. Every effort has been made to obtain permission to use copyrighted materials, and sincere apologies are rendered for any errors or omissions. The publishers would welcome these being brought to their attention.

Publisher	**Figure number(s)**
American Institute of Physics	
Journal of Vacuum Science Technology	4.1, 4.2, 4.8, 4.10, 4.11, 4.12, 4.13, 4.14, 4.15, 4.17, 4.18, 4.19, 4.20, 5.18, 5.21, 7.19
Applied Physics Letters	4.7, 4.16
Elsevier Science	
Microelectronic Engineering	4.13, 4.22, 4.23, 4.24, 4.25, 4.26, 4.27, 5.20, 5.22
G. Stengle, H. Loshner and P. Wolf	5.19
The Institute of Electrical and Electronics Engineers Inc.	4.28, 4.29, 7.21, 7.22
Japan Society of Applied Physics	4.30
Journal of the Optical Society of America	7.34
Journal of Photopolymer Science and Technology	6.35, 6.40, 6.42, 6.43, 6.44, 6.46, 6.50
Kyoritu Syuppan and Dr. Tsutomu Watanabe	3.3
T. Nakasugi	4.93, 4.94
Publication office, Japanese Journal of Applied Physics	3.21, 4.3, 4.4, 4.5, 4.9, 4.21, 4.94, 4.95, 4.96, 4.100, 4.103, 6.55, Table 4.1
Press Journal and Masaki Kurihara (Japan)	2.48
Seiichi Itabashi and Takashi Kaneko	3.9
Science Forum Inc. and Naoya Hayashi	2.50
Science Forum Inc. and Yasushi Ohkubo	2.55
Society of Photo-optical Instrumentation Engineers	2.49, 2.52, 2.54, 2.59, 6.36, 6.47, 7.26, 7.27, 7.29, 7.30

The authors express appreciation to Dr Brian Watts for his helpful assistance in correcting the manuscript of this book.

Principal abbreviations

AdMA–tBuMA	adamantylmethacrylate and t-butylmethacrylate
AFM	atomic force microscopy
ALTA-3000	name of electron-beam exposure system manufactured by ETEC Co.
ALU	arithmetic logic unit
AMIS	automated mask inspection system
ARC	anti-reflective coating
ASIC	application-specific integrated circuit
AURORA	the name of synchrotron developed by Sumito Heavy Industries Ltd.
BAA	blanking aperture array
B(DMA)DS	bis(dimethylamino)dimethylsilane
B(DMA)MS	bis(dimethylamino)methylsilane
B(DMA)TMDS	bis(dimethylamino)tetramethyldisilane
BST	barium strontium titanate $(Ba,Sr)TiO_3$
CA	chemical amplification (or chemically amplified)
CAD	computer-aided design
CCD	charge-coupled device
CD	critical dimension
CF	compression filter
CMOS	complementary metal oxide semiconductor
CMP	chemical–mechanical polishing
COP	polyglycidylmethacrylate copolymer
CP	cell projection (or character projection)
CRT	cathode-ray tube
CVD	chemical vapor deposition
DAADMDS	dialkylaminodimethyldisilanes
DDD	double diffused drain
DESIRE	diffusion-enhanced-silylation resist
DF	decompression filter
DMAPMDS	dimethylaminopentamethyldisilane
DMSDMA	dimethylsilyldimethylamine
DNQ	diazonaphthoquinone
DOF	depth of focus

DRAM	dynamic random access memory
DRM	development-rate monitor
DSP	digital signal processor
DUV	deep ultraviolet
EB	electron beam
EBES™	name of electron-beam exposure system developed by AT&T
EBL	electron-beam lithography
EBM-130/40	name of electron-beam exposure system manufactured by Toshiba Machine Co.
ECR	electron cyclotron resonance
EEPROM	electrically erasable programmable read-only memory
ESCAP	environmentally stable chemical amplification positive-tone resist
ESD	electrostatic discharge
EX-8	name of electron-beam exposure system developed by Toshiba Machine Co.
FeRAM	ferroelectric random access memory
FET	field effect transistor
FIB	focused ion beam
FIR	finite impulse response
FZP	Fresnel zone plate
GHOST	name of proximity effect correction in electron-beam exposure
HELIOS	name of synchrotron developed by Oxford Instruments Ltd.
HEMT	high-electron-mobility transistor
HMCTS	hexamethylcyclotrisilazane
HMDS	hexamethyldisilazane
HSG	hemispherical grain
HS-tBuA	hydroxystyrene with t-butyl acrylate
HT	half-tone
IBL	ion-beam lithography
IC	integrated circuit
ICA	indenecarboxylic acid
I/F	interface
I/I	ion implantation
I/O	input/output
IPA	isopropyl alcohol
KLA	KLA Instruments Co.
KrF	krypton fluoride
L&S	line-and-space (*also* **L/S**)
LaB_6	lanthanum hexaboride
L_g	gate length (= width of gate electrode) (also L_G)
LDD	lightly doped drain
LIGA	Lithographie Galvanoformung Abform Technik
LMIS	liquid-metal ion source
LOCOS	local oxidation of silicon
LPCVD	low-pressure chemical vapor deposition

LSI	large-scale integration
LUNA	name of synchrotron developed by Ishikawajima-harima Corp.
MCP	multi-channel plate
MEBES™	production name of EBES manufactured by ETEC Co.
MELCO	name of synchrotron developed by Mitsubishi Electric Corp.
MIB	masked ion beam
MIBK	methylisobutyl ketone
MIPS	millions of instructions per second
MOS	metal oxide semiconductor
NA	numerical aperture
NAND	one of logical circuits, which means an inverse of AND (product of propositions)
Nd:YAG	neodymium-doped yttrium aluminum garnet
NIJI-III	name of synchrotron developed by Sumito Electric Industries Ltd
NMP	*N*-methylpyrrolidine
OAI	off-axis illumination
OPC	optical proximity correction
OPE	optical proximity effect
PαMSt	poly-α methylstyrene
PAC	photoactive compound
PAG	photoacid generator
PAT	previous analysis of distortion and transformation of coordinates (a method of distortion correction for electron-beam delineation)
PB	pre-bake
PBOCST	poly-(t-butoxycarbonyloxy) styrene
PBS	polybutenesulfone
PC	personal computer
PEB	post-exposure baking
PED	post-exposure delay
PHS	polyhydroxystyrene
PID	Proportional Integral Differential
PLL	phase-locked loop
PMGI	polydimethylglutarimide
PMMA	polymethylmetacrylate
PMSt	polymethylstyrene
PSM	phase-shift mask
PZT	lead zirconium titanate
QTAT	quick turn-around time
Qz	quartz
RET	resolution-enhancement technology
RIE	reactive ion etching
ROM	read-only memory
S&R	step-and-repeat
SABRE	silicon-added bi-layer resist

Salicide	self-aligned silicide
SBT	strontium bismuth titanate ($SrBi_2Ti_2O_9$)
SCALPEL	scattering angular limitation in projection electron-beam lithography system
SD	source drain
S/D contact	source/drain contact
SE	secondary electron
SEM	scanning electron microscope
SEMI	Semiconductor Equipment and Materials International
SEQ	sequencer
SIM	scanning ion microscope
SIMOX	separation by implanted oxygen
SN	signal-to-noise ratio (*also* **SNR**)
SNR^{TM}	product name of resist
SOG	spin-on glass
SOI	silicon on insulator
SPD	solid-phase diffusion
SPM	scanning probe microscope
SR	synchrotron radiation
SRAM	static random access memory
STAR	simultaneous transmitted and reflected
STC	stacked capacitor
STM	scanning tunneling microscope
STO	strontium titanate
Super-ALIS	name of synchrotron developed by NTT Corp.
TAR	top anti-reflector
tBOC	t-butoxycarbonyl
tBOCM	t-butoxycarbonylmethyl
TBPB	tetrabromophenol blue sodium salt
T_g	glass-transition temperature
TIS	tool-induced shift
TMAH	tetramethylammonium hydroxide
Tri-MDS	trimethyldisilazane
TPS-OTf	triphenylsulfonium triflate
TSI	top-surface imaging
TTL	through the lens
TTR	through the reticle
TV	television
ULSI	ultra-large-scale integration
VSB	variably shaped beam
WIS	wafer-induced shift
XRL	X-ray lithography

Introduction 1

Masao Fukuma

The first commercial 0.25-μm technology was distributed to users in 1997, and is now widely used in advanced ULSIs (e.g. 64-Mb and 256-Mb DRAMs, high-end microprocessors, and ASICs). 0.18-μm technology has been developed for commercial release in 1999, and 0.13-μm technology is in the research phase.

Because lithography is a key process in the fabrication of high-performance ULSIs, this book reviews ULSI technology trends and discusses lithography requirements. It then describes in detail the most widely used optical lithography technologies as well as representative next-generation lithography technologies such as X-ray lithography, electron-beam lithography, and ion-beam lithography.

1.1 Device technology trends

1.1.1 Logic devices

Because high speed is a first priority for logic ULSIs, there have been many efforts to increase MOSFET driving capability and reduce parasitic effects. Figure 1.1 shows the technology 'roadmap' for logic-oriented CMOS devices. The driving capability is increased by making the gate oxide as thin as possible and using a carefully designed impurity profile. Gate, source, and drain resistances are reduced by using a salicide process. An n–n gate structure (n^+ gates for both n- and pMOSFETs) has been chosen for design rules down to 0.35 μm, but a p–n gate structure (n^+ for nMOSFETs and p^+ for pMOSFETs) is used when the design rule is less than 0.25 μm because the p^+ gate provides a surface channel in the pMOSFET and this results in smaller short-channel effects and a higher driving capability at a lower threshold voltage.

Experimental 0.04-μm nMOSFET[1]† and 0.07-μm CMOS devices[2] have been reported, and Figure 1.2 shows a SEM cross-section of, and waveform for, a 0.07-μm CMOS ring oscillator with a delay time of 14 ps. This dynamic performance is excellent, but the oxide thickness of the MOSFETs is about 3.5 nm, which is almost the limit for suppression of the tunneling current. Further downscaling will require the development of a gate insulator with a high dielectric constant or of a circuit that tolerates a small gate leakage current. Simulations suggest that 0.01- to 0.02-μm MOSFETs will show reasonable I–V characteristics and that the CMOS gate delay time will be only a few picoseconds.[3]

On the other hand, wiring delay in a chip is becoming a serious factor limiting ULSI performance. Not only does reducing the width and height of wiring increase the wiring resistance per unit length, but average wiring length in ULSIs also becomes longer as the chip size increases. Furthermore, wiring capacitance is not easily scaled down because fringing and/or coupling capacitance with neighboring wires becomes dominant in fine wiring. As a result, the speed of downscaled ULSI logic devices is limited by wiring delay rather than by intrinsic gate delay. Overcoming this problem will require new materials both for wiring and for interlayer dielectrics. Attention is focusing on

† Numbered references to be found at the end of each chapter.

Figure 1.1 Technology 'roadmap' for CMOS devices. τ: propagation-delay time; L_g: width of gate electrode; I/I: ion implantation.

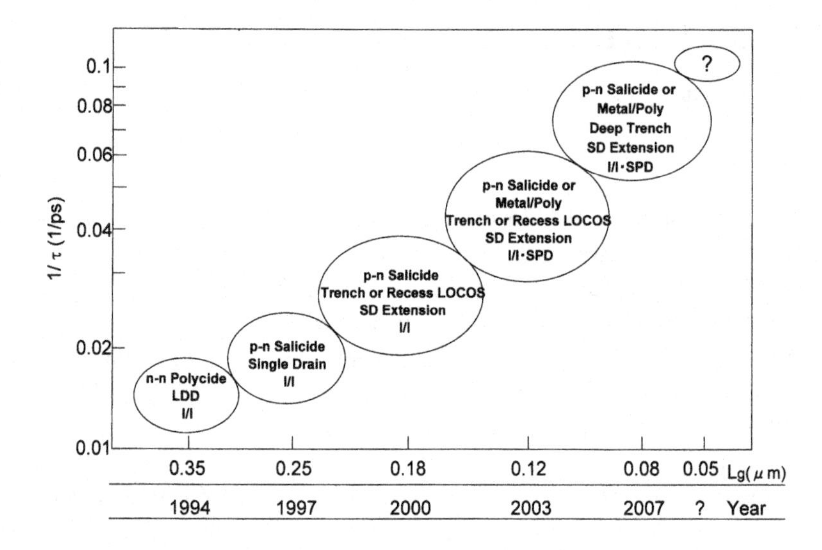

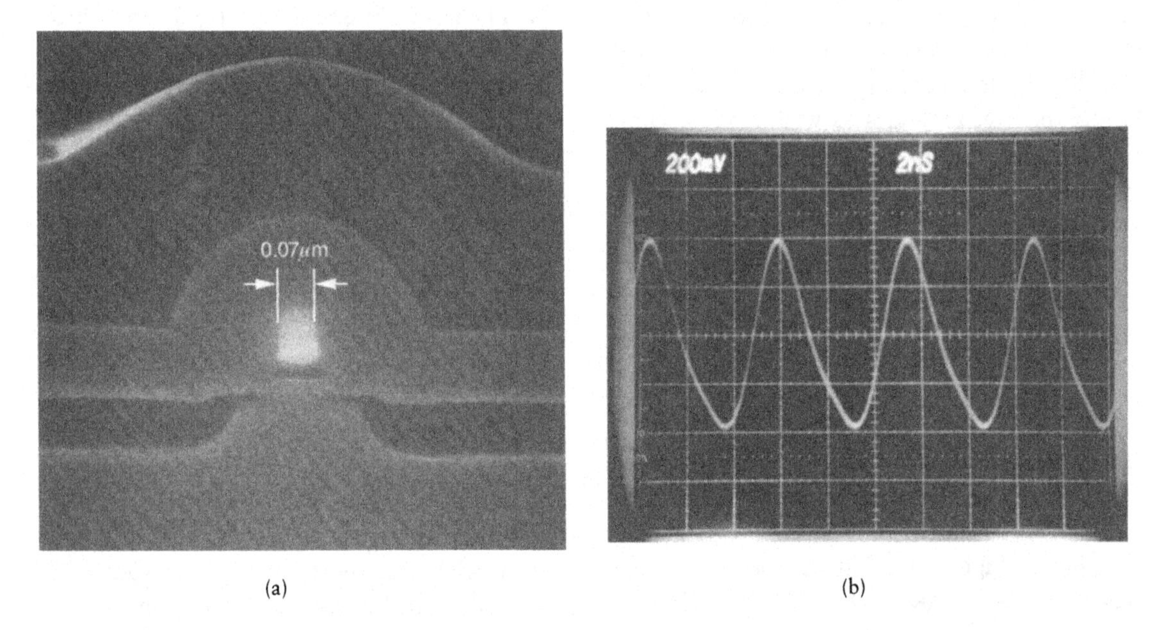

(a) (b)

Figure 1.2 (a) SEM photograph of cross-section of 0.07-μm-gate-length FET. (b) Waveform of a 201-stage CMOS ring-oscillator. Each division equals 2 ns on x-axis and 200 mV on y-axis.

copper as a low-resistance wiring material, and both SiOF and organic materials seem to be useful as interlayer insulator materials with low dielectric constants. The introduction of these materials will improve LSI performance by about 30%. This improvement, however, will not be great enough to bring us into the 0.1-μm era. Instead, circuit-oriented and architecture-oriented approaches are necessary. Figure 1.3 shows, as a typical example, a scaling

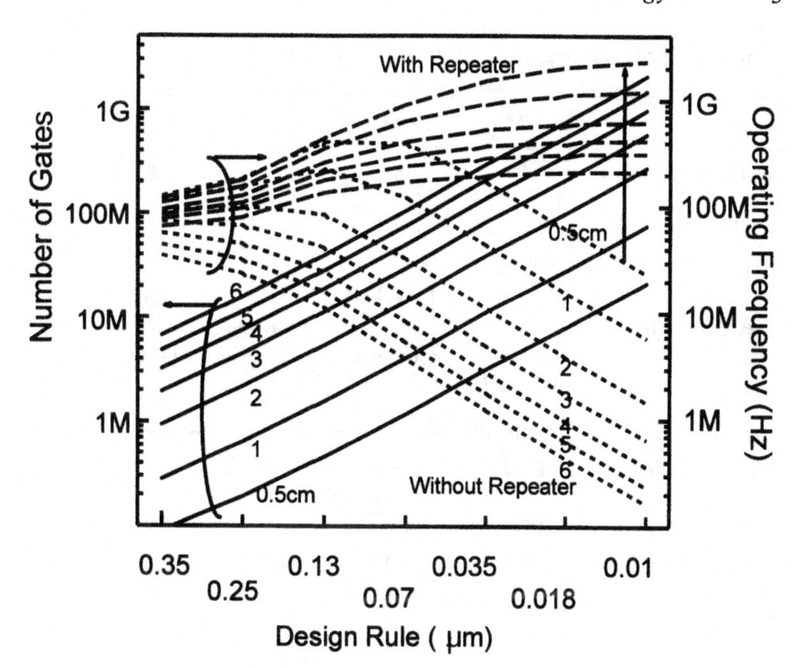

Figure 1.3 Scaling scenario for wiring. Utilizing repeaters and minimizing the size of the functional block which requires high clock frequency are essential requirements.

scenario for wiring with repeaters. Utilizing repeaters and minimizing the size of the functional block that requires a high clock frequency is expected to achieve further performance improvement even when the design rule is less than 0.1 μm.

Mobile applications require circuits that consume little power, and this leads to operating voltages below those expected from the scaling scenario or the conventional scaling 'roadmap'. A 0.9-V supply voltage, for example, is necessary even for the 0.35-μm 'generation'. High performance at such low supply voltages will require breakthroughs both in circuit design and in fabrication technology. Figure 1.4 shows an experimental low-power DSP (digital signal processor) core[4] that uses two types of MOSFET with different threshold voltages and that takes advantage of specially designed stacked CMOS circuits and busses with small signal swing. As a result, its operating frequency is as high as 100 MHz and its power dissipation at 0.9 V is only 4 mW.

In the gigahertz era, power dissipation is becoming more serious not only for mobile applications but also for general applications such as desktop PCs. Research and development on low-power architecture and circuits for low-voltage operation have recently been emphasized as much as the device scaling, and this trend will be seen even in high-end applications.

1.1.2 Memory devices

The DRAM is a typical memory LSI and has been a technology driver in Japan. Today, the capacity of the most advanced commercial DRAM is 256 Mb and that of the most advanced prototype is 4 Gb.

Table 1.1 shows a typical technology trend in DRAMs. To get a sufficient signal on a bit line, modern DRAMs require a cell charge of at least 20 fC, a value that is almost independent of the DRAM 'generation'. This requirement is a big barrier to the development of gigabit

Table 1.1 *Technology 'roadmap' for DRAMs.*

	1 Mb	4 Mb	16 Mb	64 Mb	256 Mb	1 Gb	4 Gb
Design rule (μm)	1.2	0.8	0.6	0.4	0.25	0.18	0.12
Supply voltage (V)	5		5(3.3)	3.3	3.3(2.5)	3.3(2.0)	2.0(1.5)
DRAM cell structure	Planar or Stacked or Trench	Stacked or Trench	Stacked	Stacked with HSG	HSG ring		
Gate electrode	Polycide ⟶						Metal
Drain structure	DDD	LDD ⟶			Single drain		
Isolation	LOCOS	Modified LOCOS ⟶			Trench		
Gate oxide thickness (Å)	200	180	140	110	80	65	50
Capacitor insulator	SiN					Ta_2O_5/BST/STO	
Capacitor insulator thickness (Å)		120	60	50	45	30	
Wiring metal	Al–Si–Cu	Al–Cu				Al(Cu)	Cu
Lithography	g-line		i-line		KrF	KrF + Ps	ArF/X-ray/EB
Etching	RIE		ECR/Magnetron		High-density plasma		

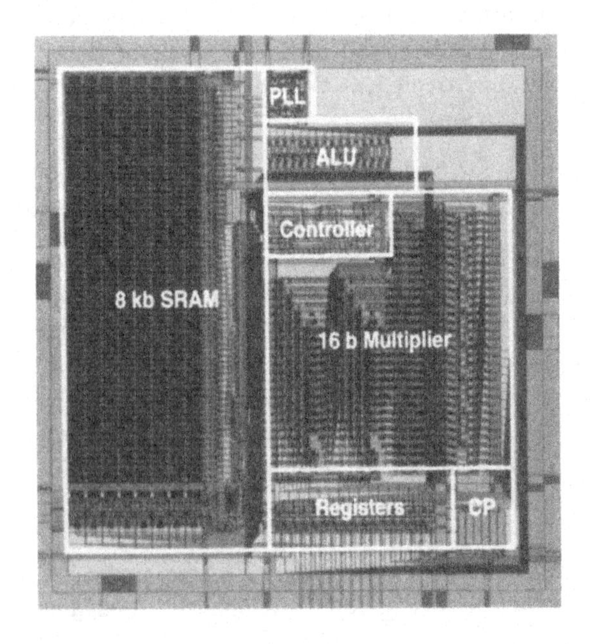

Figure 1.4 100-MHz DSP core based on 0.25-μm CMOS technology. Operating voltage is 0.9 V, and power dissipation is only 4 mW which is 1/50 that of conventional DSPs. Chip size is 1.4 × 1.4 mm^2.

DRAMs because the thickness of the cell capacitor dielectric has already reached the physical limit determined by tunneling phenomena. The memory cell in 1-Mb and smaller DRAM has a planar structure, but many types of three-dimensional structures that provide a high capacitance in a smaller structure have been proposed for 4-Mb and larger DRAMs (Figure 1.5). Many DRAM makers are now choosing the stacked trench cell because it can be fabricated more easily than a trench cell.

When the design rule becomes 0.18 μm, however, as it does for DRAMs with capacities between 256 Mb and 1 Gb, the stacked trench cell required cannot be fabricated because it is too tall or its aspect ratio is too large. Two approaches to getting sufficient cell capacitance are to increase the effective area of the capacitor cell and to use a material with a high dielectric constant ϵ. One way to increase the effective area is by making a capacitor electrode with micro-roughness. Figure 1.6 shows a typical example in which hemispherical grain (HSG) polysilicon is used as an electrode.[5] This structure provides twice the capacitor area of a flat capacitor electrode and is reasonably easy to fabricate but is not particularly suitable for further increases in the capacitor area.

The potentially more useful approach is thus to use a material with a higher dielectric constant. Some candidate materials are Ta_2O_5, BST ($Ba_{1-x}Sr_xTiO_3$), and PZT ($(Pb, Zr)TiO_3$), which have dielectric constants ranging from 20 to 5000. Figure 1.7 shows an experimental 4-Gb DRAM[6] using a BST film whose capacitance is equivalent to that of a 0.35-nm-thick S iO_2 layer. This BST film is thin enough that we can use a multi-valued architecture. That is, 2-bit data (four values) can be stored in each cell, resulting in a 4-Gb DRAM with 2-Gb memory cells.

Recently, there have been strong demands for nonvolatile memories that can be used in mobile applications. Static random access memories (SRAMs) work as quasi-nonvolatile memories, but it is difficult to get a high storage capacity with them because their memory cells are large. Because SRAMs operate at high

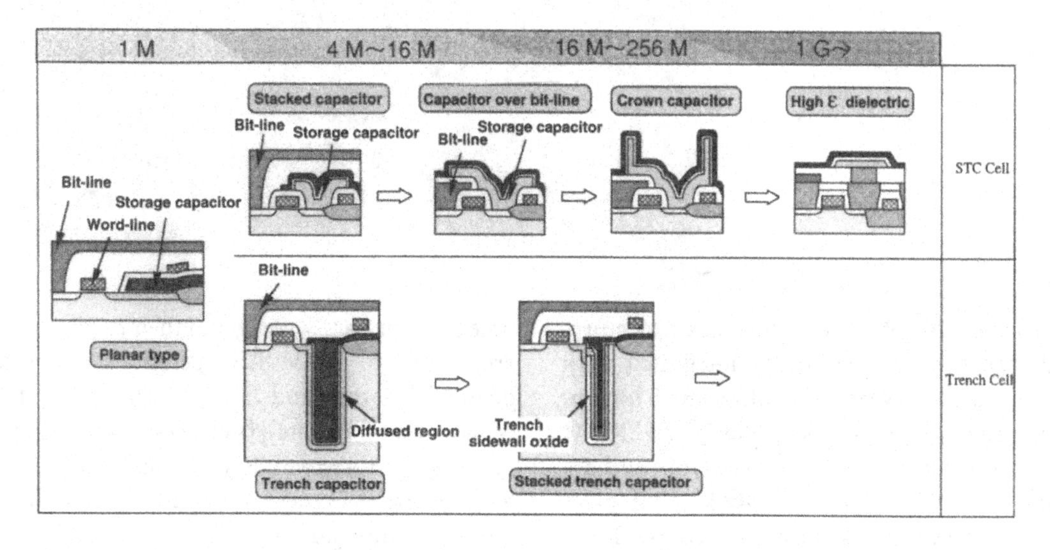

Figure 1.5 Evolution of DRAM cell structure. DRAM capacity is given in bits.

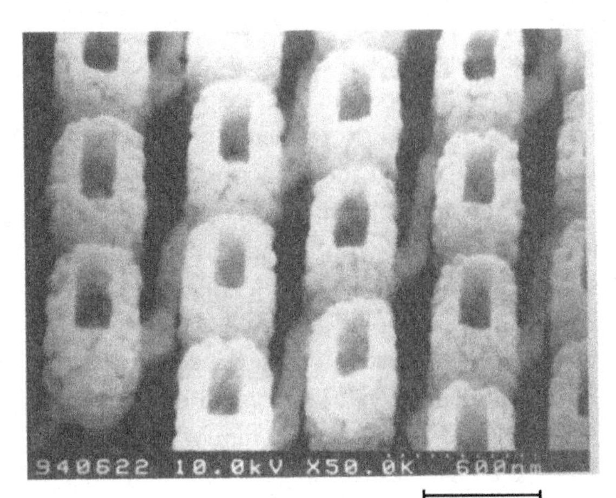

Figure 1.6 HSG capacitor cell used in 1-Gb DRAM.

Figure 1.7 Four-level storage 4-Gb DRAM where BST is utilized as cell capacitor insulator. Chip size is 33.9×29.0 mm^2.

speed, they are often used for cache memory. And, because they are easily integrated with logic circuits, they are also often used for on-chip memory. On the other hand, EEPROM and flash memory are also nonvolatile and are going to be widely used in mobile equipment such as digital cameras and IC cards. These nonvolatile memories, however, require more time (ms) for data writing and their applications are limited to ROM-like operations. Markets demand low-cost and high-density nonvolatile RAMs that operate fast. For this reason, FeRAM (ferroelectric RAM) which utilizes a ferroelectric material such as PZT (lead zirconium titanate) and SBT ($SrBi_xTi_yO_z$), is attracting attention. The ideal FeRAM cell, which consists of one transistor and one ferroelectric capacitor, would have an access time less than 100 ns and a retention time of 10 years, and would be able to be used as the nonvolatile main memory in mobile computing systems. Figure 1.8 shows an experimental 1-Mb FeRAM[7] in which SBT is used. Although the ferroelectric materials, like the high-dielectric-constant materials for DRAMs, are not easy to bring into the LSI fabrication process used in commercial manufacturing, their introduction would be a real breakthrough to the gigabit era.

We can already integrate more than 10 million transistors even in logic LSIs, and this naturally leads to the SoC 'System on a Chip.' The integration of DRAM and logic devices is advantageous in increasing memory bandwidth while reducing power consumption and assembly costs. Figure 1.9 shows a typical example of DRAM–logic integration, Compress DRAM,[8] in which a 16-Mb DRAM and a graphics data compressor/decompressor are integrated. The bandwidth is 3.2 Gb/s and the power consumption is only 1.4 W.

1.2 Demands for lithography

Lithography was not a limiting factor in scaling the design rule down to 0.35 μm, but it is now becoming a practical limiting factor. Table 1.2 lists typical lithography requirements for DRAM fabrication. There are many factors which should meet specifications for a deep-submicron design rule: resolution, depth of focus (DOF), critical dimension (CD) control, overlay accuracy, field size, throughput, and cost. Among them, CD control and overlay accuracy are the key factors for modern ULSIs. Today's most advanced processors contain more than 10 million transistors. This means that not only pure memory LSIs but also logic LSIs are now going into the six sigma (6σ) era, which requires strict control of

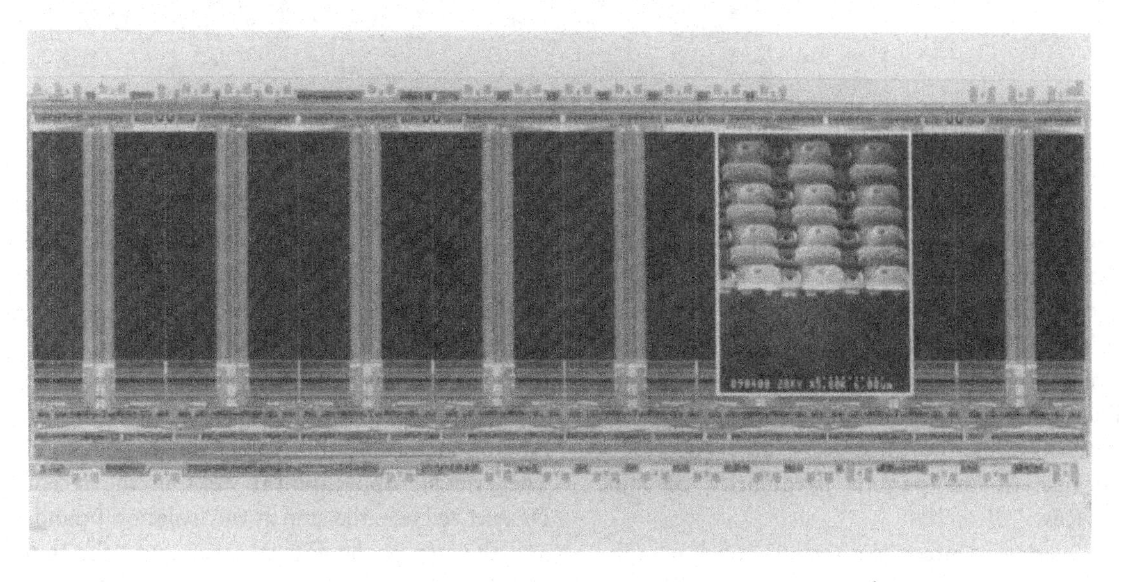

Figure 1.8 60-ns 1-Mb FeRAM. Chip size is 15.7×5.8 mm^2. Inset is a SEM photograph of the FeRAM cells.

Table 1.2 *Characteristics needed in lithography for DRAM fabrication.*

		64 Mb	256 Mb	1 Gb
Resolution (μm)	Line-and-space	0.35	0.25	0.18
	Window	0.4	0.3	0.2
Depth of focus (DOF) (μm)		1.5	1.2	1.0
Field size (mm^2) (2-chip exposure)		22 × 22 (31.1 mm ϕ)	25 × 25 (35.3 mm ϕ)	33 × 16 (1-chip exposure)
Critical dimension (CD) control (nm) (3σ)		35	25	18
Overlay accuracy (nm) ($\bar{x} + 3\sigma$)		90	60	45

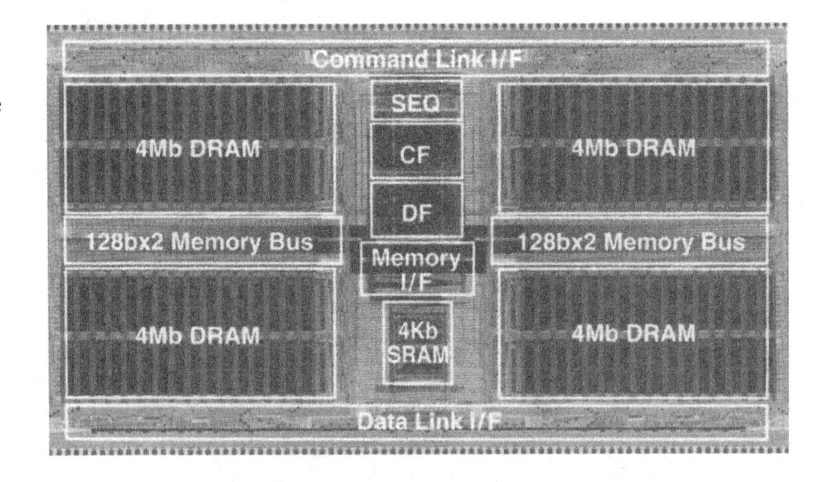

Figure 1.9 Compress DRAM; bandwidth 3.2 Gb/s, power consumption 1.4 W. Chip size is 16.5 × 8.3 mm^2.

device parameters. For example, even if the circuit allows 20% fluctuation in channel length, the standard deviation in the channel length should be less than 4–5 nm when the design rule is 0.13 μm. Beam uniformity and stable resist processes are key factors for reducing line-width fluctuation. For logic applications, pattern-size adjustment is also necessary because of the pattern irregularity of logic devices.

Overlay accuracy is a serious concern in the fabrication of high-density memories. Figure 1.10 shows, for various design rules, the calculated relationship between DRAM cell size and overlay margin. In the high-density memory, the overlay accuracy has an impact on memory size similar to that of the resolution.

Depth of focus is another key factor, especially in the case of optical lithography. The surface profile of ULSI devices depends on the type of circuit and the type of device structure. The critical topographical steps in the recent DRAM cell are: the step at the isolation boundary, the step at the gate structure, and the step at the capacitor. The last one is the most critical. To achieve the required resolution in practical

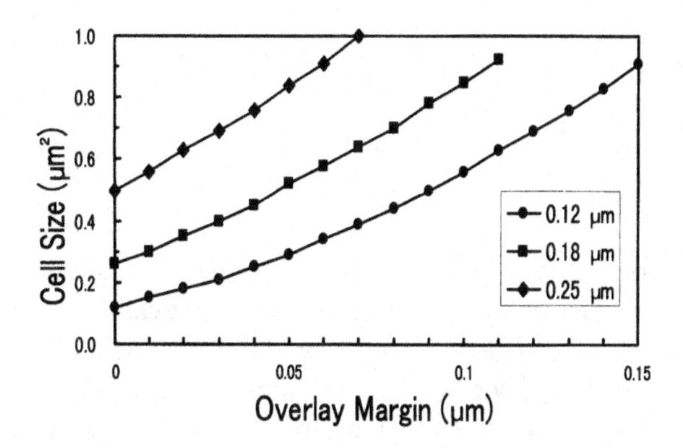

Figure 1.10 Relationship between DRAM cell size and overlay accuracy where design rule is a parameter.

device-fabrication processes, planarization is usually employed. Such a process, however, increases chip-fabrication cost.

Although many research and development activities have been devoted to lithography for device rules smaller than 0.1 μm, there is still no lithography technology suitable for devices whose design rule is less than 0.1 μm. The key factors are still resolution, CD control, overlay accuracy, field size, depth of focus, and throughput. As discussed in Section 1.1, 0.07-μm CMOS devices have been demonstrated experimentally, and simulations have predicted that it will be possible to make high-performance MOS ULSIs with a channel length of 0.01 μm. The fundamental scaling limit is thus far from the scale of present MOS devices. Practical lithography for sub-tenth-micron devices will therefore be one of the most important themes for the next 5 years.

1.3 Conclusion

Multimedia systems need high-speed, low-power and high-density ULSI devices in which sub-tenth-micron technology will be utilized. Both experimental and theoretical analyses predict that such devices could be developed within 10 years. Many problems related to fabrication processes, device structures, circuit configurations, system architectures and design methodology need to be solved, and one of the most serious is lithography. Developing practical lithography technology for Si ULSIs having design rules of 0.1 μm or less is essential if we are to establish a real multimedia society in the world of the 21st century.

1.4 References

1. M. Ono *et al.*, *Intl. Electron Device Mtg, Tech. Digest*, 119 (1993)
2. K. Takeuchi *et al.*, *Symp. on VLSI Technology, Tech. Digest*, 9 (1995)
3. M. Fukuma, *Symp. on VLSI Technology, Tech. Digest*, 7 (1988)
4. M. Izumikawa *et al.*, *IEEE J. of Solid State Circuits*, **32**, 52 (1997)
5. K. Shibahara *et al.*, *Intl. Electron Device Mtg, Tech. Digest*, 639 (1994)
6. T. Murotani *et al.*, *Intl. Solid State Circuit Conf., Tech. Digest*, 74 (1997)
7. H. Koike *et al.*, *Intl. Solid State Circuit Conf., Tech. Digest*, 368 (1996)
8. Y. Yabe *et al.*, *Intl. Solid State Circuit Conf., Tech. Digest*, 342 (1998)

Optical lithography 2

Kunihiko Kasama, Shinji Okazaki, Hisatake Sano
and Wataru Wakamiya

2.1 Introduction

Optical lithography has been widely used for fabricating semiconductor devices for more than 30 years. The principle of optical lithography is the same as that of photography: wafers are coated with a photosensitive material (resist) and patterns are exposed on the resist using optical tools.

The trends in the minimum feature size of memory devices and the major optical lithographic tools used for patterning are shown in Figure 2.1. Several types of optical tools have been used in the fabrication of semiconductor devices, and resolution has been improved by changing the exposure systems and exposure wavelengths (Figure 2.2).

In this chapter, we will describe optical lithography technology. After outlining the history and principles of optical lithography, recent representative optical lithography technologies, such as i-line- and DUV-lithography, are

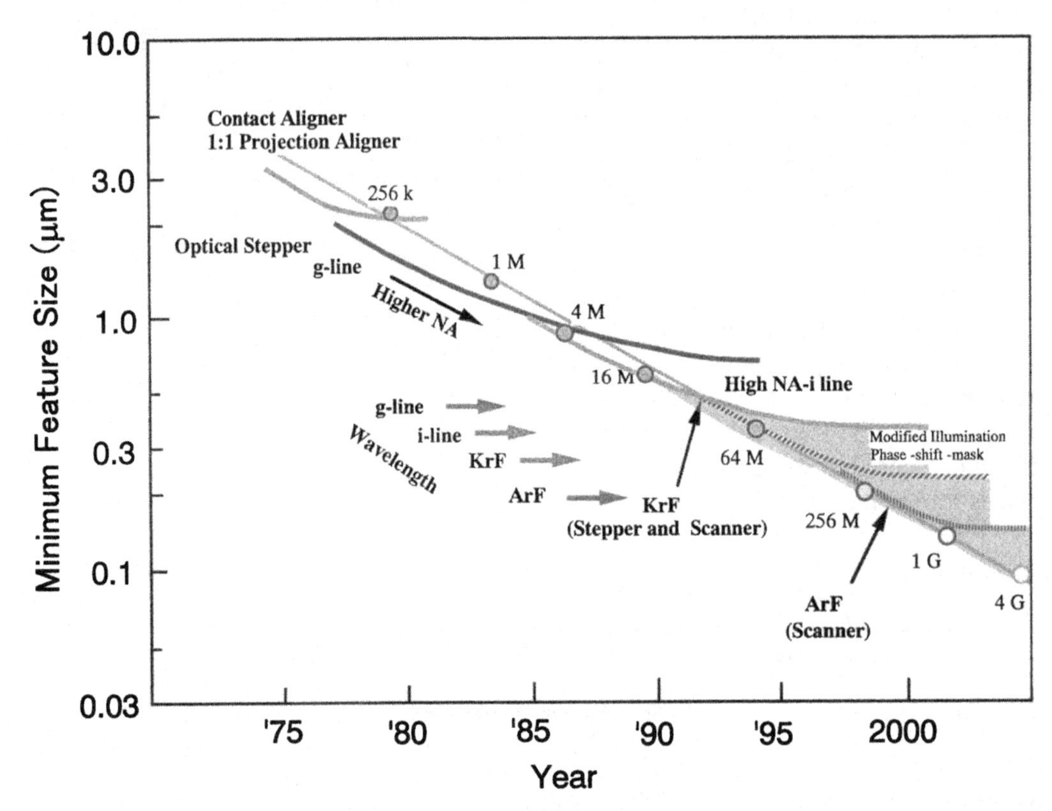

Figure 2.1 Miniaturization trends of DRAM pattern size and development of optical lithographic tools. DRAM capacity is given in bits.

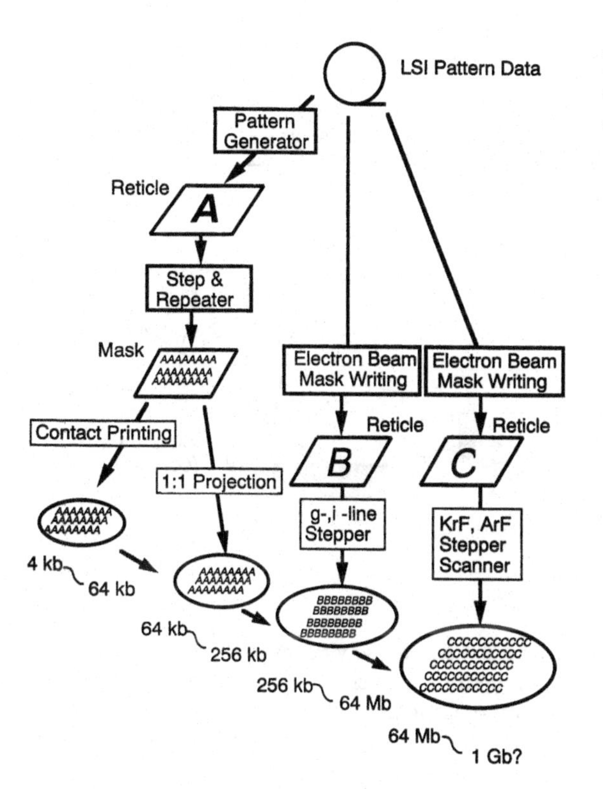

Figure 2.2 Evolution of optical lithography. DRAM capacity is given in bits.

described in detail. Resolution-enhancement technology and mask/reticle technology, which are both becoming more important, are also described.

2.2 History

2.2.1 Equipment evolution

The first optical lithography system for semiconductor-device fabrication was the contact aligner, which was used for replicating patterns with a design rule greater than 3 μm. As shown in Figure 2.3, a 1× mask is set on top of the resist layer spun on a wafer. A small proximity gap is left for aligning the patterns on the mask to those on the wafer. After alignment, the mask and wafer are placed in contact and exposed to light from a mercury lamp. Because the entire wafer is exposed in a single exposure, the throughput of this system is very high. Resolution is relatively high, but defects are frequent because the resist layer can stick to the mask.[1,2]

The next-generation optical lithography system, the one-to-one projection aligner, eliminated the sticking problems.[3] The projection aligner employed crescent-shaped illumination and reflective optics as shown in Figure 2.4. A large-mirror optical system and a scanning exposure procedure with crescent illumination enabled entire wafer exposure in a single scan. The resolution limit with this kind of system was 2–3 μm, and the system was used in the development of 64-kb and 256-kb DRAMs.

Resolution depends on the characteristics of the optical elements: a higher numerical aperture (NA) and a shorter wavelength give higher resolution. Because of the difficulty in obtaining a high NA in such a large-mirror system, research tried to increase the resolution by

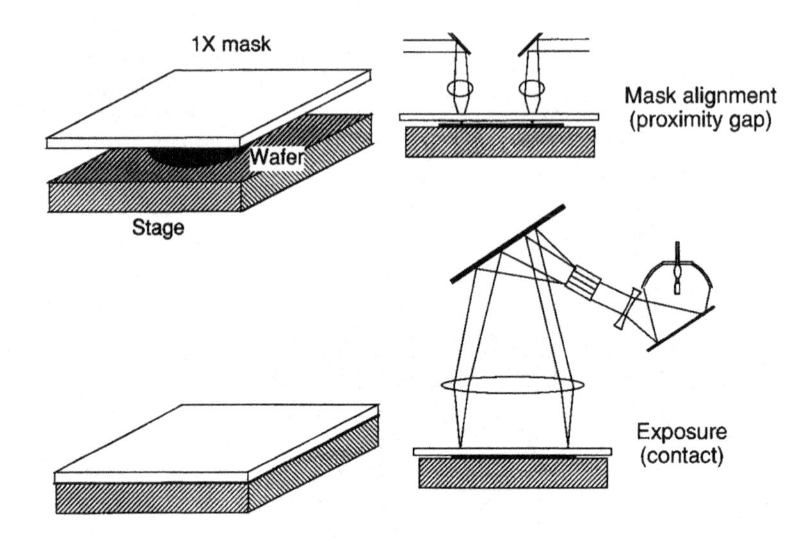

Figure 2.3 Schematic view of contact aligner.

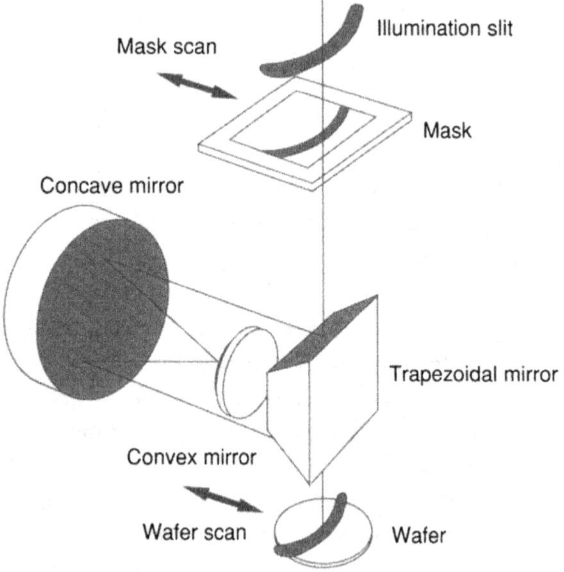

Figure 2.4 Schematic view of 1:1 projection aligner. (Courtesy Canon.)

using shorter-wavelength light (deep-UV light).[4,5] The lack of good resist materials and a bright deep-UV (DUV) light source, however, made these efforts unsuccessful.

The demand for higher resolution led to the development of the reduction projection system, primarily for the fabrication of 1× masks. The system is called a step-and-repeater, because many chip patterns are printed on a mask by repeatedly exposing and stepping. To make a 1× mask, a 10× reticle is used. The reticle is fabricated from pattern data using a pattern generator. This reticle pattern is then replicated with $\frac{1}{10}\times$ reduction on a 1× mask.[6] The step-

and-repeater was later applied in a system that directly exposes reticle patterns on a wafer. This latter system, shown schematically in Figure 2.5, is called a stepper.[7]

The resolution of the stepper, like that of the one-to-one projection aligner, depends on the characteristics of the optical system. Because the exposed area of the stepper is smaller than that of the one-to-one projection aligner, a higher-numerical-aperture optical system can be manufactured. To eliminate chromatic aberration in the all-refractive optical systems used in the optical stepper, the exposure light needed to have a narrow bandwidth. The g-line (436 nm) of a mercury lamp was originally used, resulting in a NA of less than 0.3, a reduction ratio of $\frac{1}{10}$, and a field size of around 14-mm square. With the growing requirement for higher resolution, the NA of the exposure optics became higher and the field size of the optics became larger. Over several years, the reduction ratio changed from $\frac{1}{10}$ to $\frac{1}{5}$. The NA of the latest g-line stepper is larger than 0.6, and g-line steppers with 0.8-μm resolution have been used to fabricate 256-kb, 1-Mb, and 4-Mb DRAMs.

As discussed later, however, a higher NA gives a smaller depth of focus. Obtaining higher resolution with sufficient depth of focus requires the use of a shorter-wavelength light, and the second-generation stepper used the i-line (365 nm) of the mercury lamp.[8,9] The resist materials for i-line exposure are basically the same as those for g-line exposure. The i-line steppers have resolutions as fine as 0.35 μm and are used to fabricate 4-Mb and 16-Mb DRAMs and the first generation of 64-Mb DRAM.

The requirements for even higher resolution and the progress in the development of new resist materials have led to the use of DUV (KrF excimer laser) light (248 nm), resulting in resolutions of 0.30 μm and better.[10,11,12,13] The KrF system is now being used for fabricating 64-Mb DRAMs. Because the KrF system can cover the 0.30- to 0.18-μm range, it will be the major lithographic tool even for 256-Mb-DRAM fabrication.

Technologies using light with a much shorter wavelength, such as ArF excimer laser light (193 nm), are under development.[14,15,16] It is

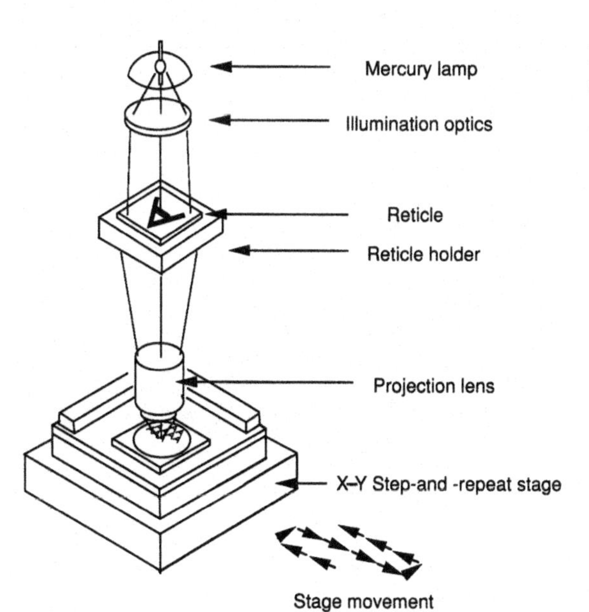

Figure 2.5 Schematic view of stepper.

expected that an ArF system will be applied to 1-Gb DRAMs and beyond, but the use of ArF excimer laser light will require the development of new optical systems, new optical materials, new light sources, new resist materials, and new process technologies.

Exposure tools need to have not only higher resolution with sufficient depth of focus, but also a larger field size and higher overlay accuracy. A larger field size has been attained by using a larger lens, but the development of larger lenses is becoming very expensive. The lens materials for the DUV system, in particular, are extremely expensive. To minimize the lens size, scan systems are again being developed. A scan system (scanner) will be used to achieve a larger field size without using a lens as large as that needed in one-to-one projection systems. To obtain higher resolution, reduction projection optics is used in the scan system. In this system, the scan speeds of the reticle and the wafer are inherently different.

There are two types of scan system: one with a catadioptric optical system,[17,18] and the other with an all-refractive optical system.[19] A schematic view of the latter system is shown in Figure 2.6. This kind of system makes it possible to obtain a larger field size with smaller optics, and the pattern-size uniformity is expected to be superior to that of a stepper. In the scan system, the optical images are created by scanning a field. Aberrations in the scan slit are averaged along the scan direction, so the pattern-size fluctuations in the scan system are suppressed.

2.2.2 Evolution of resist materials

In the early stages of optical lithography, cyclic-rubber-based negative resists were widely used.[20] Figure 2.7 shows the typical chemical structure of such a system. Because the resist is relatively soft, defects due to contact between the resist layer and mask substrate were fewer.

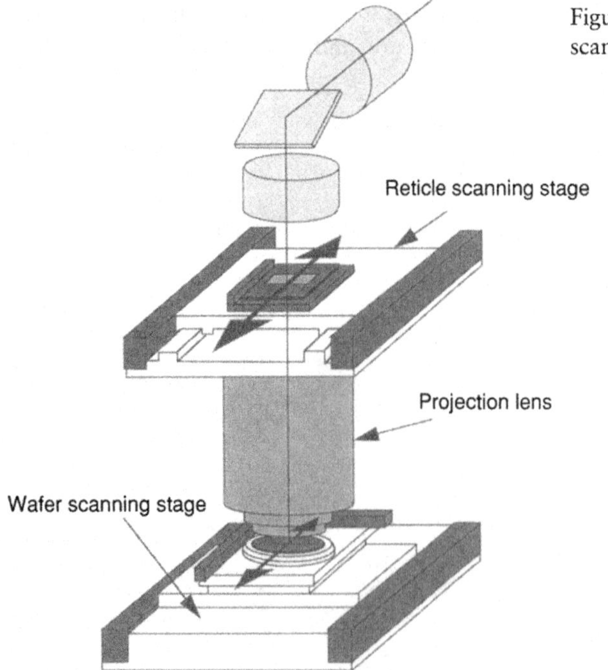

Figure 2.6 Schematic view of scanner. (Courtesy Nikon.)

The developer used for this type of resist was an organic-solvent-based solution, and the resolution attainable with this resist process was limited by the swelling that occurs during development.

To obtain higher resolutions, positive resists have been used. As shown in Figure 2.8, an alkaline aqueous soluble phenolic resin is used as the base resin.[21,22] The sensitizer naphthoquinonediazide is added to the base resin. Dissolution of the resist without UV exposure is prevented by the sensitizer. After exposure to UV light, however, the naphthoquinonediazide changes to carboxilic acid, and loses its dissolution-inhibition function. As a result, the resist becomes soluble in alkaline aqueous solutions. The advantage of this resist is the higher resolution due to swelling-free development.

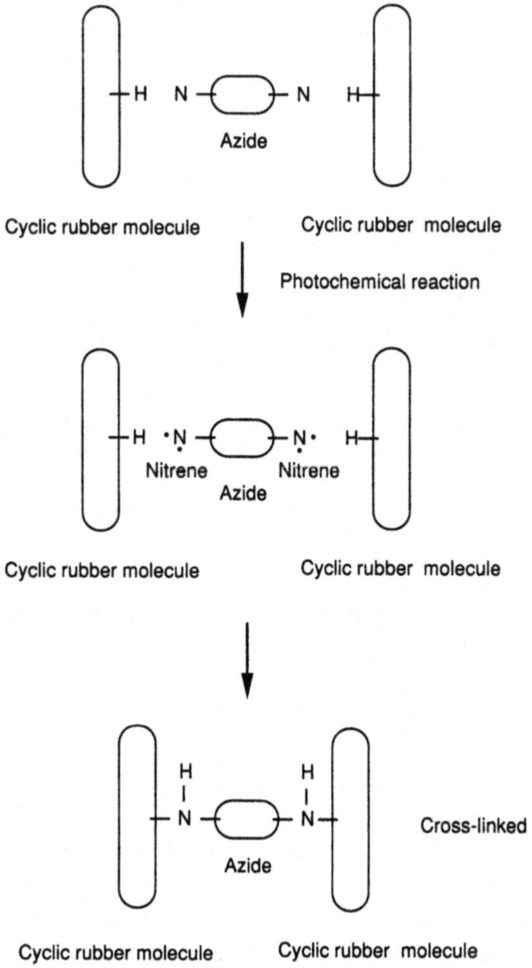

2.3 **Principles**

2.3.1 Resolution limits and depth of focus

As discussed in the previous section, there are many types of exposure tools. However, the optics of all the projection-type exposure tools can be explained by the basic optical model

Figure 2.7 Negative photoresist.

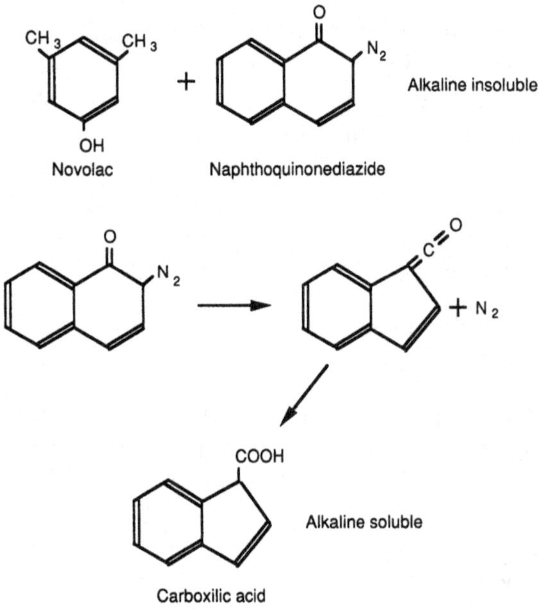

CH_3 CH_3 O N_2 + Alkaline insoluble

OH Novolac Naphthoquinonediazide

O N_2 C=O + N_2

COOH Alkaline soluble

Carboxilic acid

Figure 2.8 Reaction positive photoresist.

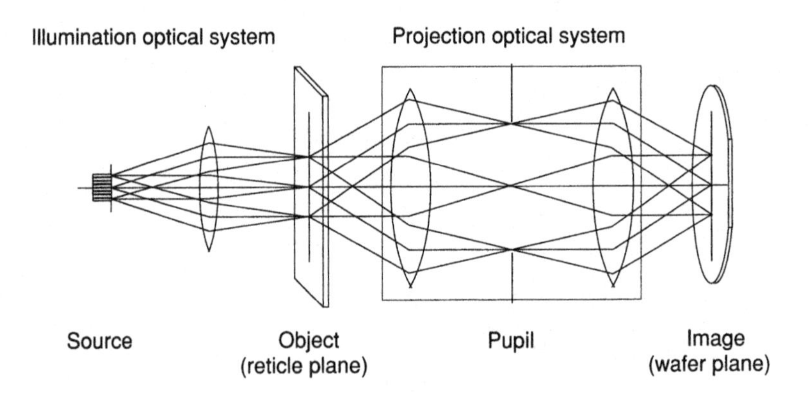

Figure 2.9 Simplified model of projection optics. The ratio of the numerical apertures of the illumination optical system and the projection system is called the coherency factor, usually represented by σ.

shown in Figure 2.9. The illumination system is assumed to be Kohler illumination and the projection system is assumed to be telecentric.

The resolution R of the projection system is given by the well-known Rayleigh equation:

$$R = k_1 \times \lambda/\mathrm{NA}, \qquad (2.1)$$

where k_1 is a constant related to the resist process, NA is the numerical aperture of the optical system, and λ is the exposure wavelength.[23]

The corresponding depth of focus (DOF) is given by:

$$\mathrm{DOF} = k_2 \times \lambda/(\mathrm{NA})^2, \qquad (2.2)$$

where k_2 is also a constant related to the resist process.

The DOF is a very important parameter in the discussion of optical lithography. As shown in Equation (2.2), DOF decreases in proportion to the square of the increase of the NA. When the NA increases to more than 0.5, the DOF decreases to less than 1.0 μm.

With progress in device integration, surface steps on the wafer have become higher. So, a larger depth of focus is required. As a result, shorter-wavelength light has been introduced in an attempt to obtain higher resolution with sufficient depth of focus.

2.3.2 Practical resolution

The step height of recent LSI devices is equal to or larger than 1.0 μm, so the DOF at the resolution limits of semiconductor-device surfaces must be larger than 1.0 μm. Several factors degrade the DOF: wafer flatness, field curvature of the projection system, focusing accuracy of the system, etc. The dependence of the DOF (defocusing level) on pattern size calculated for three kinds of projection lens systems with different NA is shown in Figure 2.10. It is clear that the resolution varies with the defocus condition: the larger the defocus, the poorer the resolution. This resolution-degradation tendency depends very much on the NA. A lower-numerical-aperture lens system gives a lower resolution on the focused plane, but resolution does not degrade significantly with increasing defocus. A higher-numerical-aperture lens system, in contrast, gives a higher resolution on the focused plane, but the resolution degrades significantly with increasing defocus.

The relationship between resolution and NA is shown in Figure 2.11.[24] If no focus error is assumed, the resolution limits are given by Rayleigh's equation. However, to achieve a sufficient DOF, the highest resolution is realized at a certain NA.

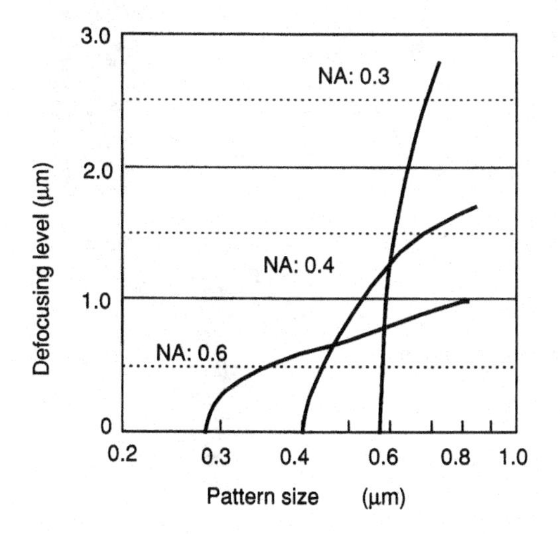

Figure 2.10 Relationship between defocusing level (DOF) and pattern size.

2.4 i-line lithography

2.4.1 Introduction

i-line light was adopted for industrial application of 0.8-μm technology in the late 1980s. It has since been widely used not only for the patterning of stacked-capacitor cells but also for trench-capacitor cells and many other integrated circuits. i-line lithography is currently the most widely used lithography system, and this section describes it and its current application.

2.4.2 Development of i-line stepper

As shown in Figure 2.12, the i-line stepper consists of various components. The stepper system

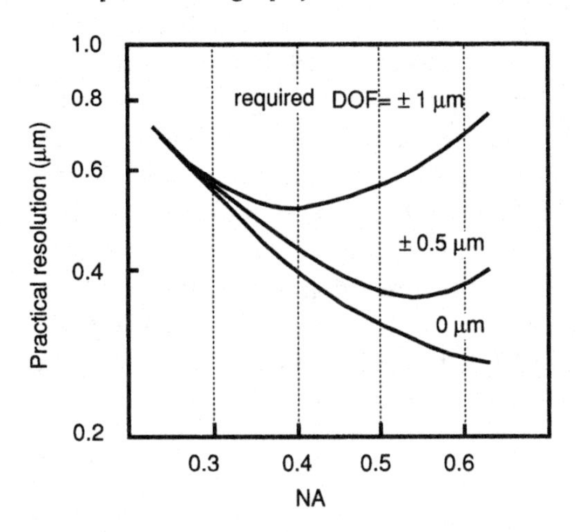

Figure 2.11 Practical resolution as a function of NA.

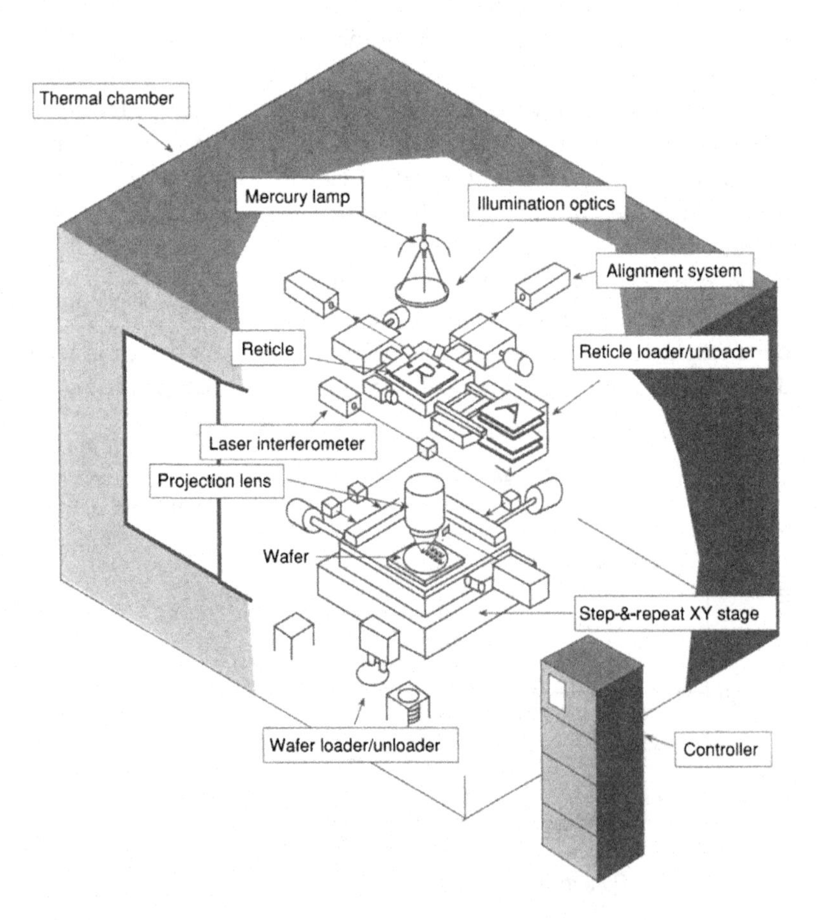

Figure 2.12 Structure and principal components of i-line stepper.

Table 2.1 *Development of the i-line stepper.*

Year	Model and supplier	NA	Field size mm × mm	Magn	Overlay ($\bar{x} + 3\sigma$) (nm)
1984	NSR-1010 i3 Nikon	0.35	10 × 10	1/10	370
1984	RA-101VL Hitachi	0.42	10 × 10	1/10	250
1989	NSR-1505 i6A Nikon	0.45	15 × 15	1/5	130
1989	PAS5000/50 ASML	0.48	15 × 15	1/5	125
1990	FPA-2000 i1 Canon	0.52	20 × 20	1/5	140
1991	NSR-2005 i8A Nikon	0.50	20 × 20	1/5	110
1992	LD 5015 iDS Hitachi	0.50	22 × 22	1/5	80
1994	NSR-2205 i11D Nikon	max. 0.63	22 × 22	1/5	70
1994	FPA-3000 i4 Canon	max. 0.63	22 × 22	1/5	60
1995	PAS5500/200 ASML	max. 0.60	22 × 22	1/5	50
1996	NSR-2205 i12D Nikon	max. 0.63	22 × 22	1/5	55

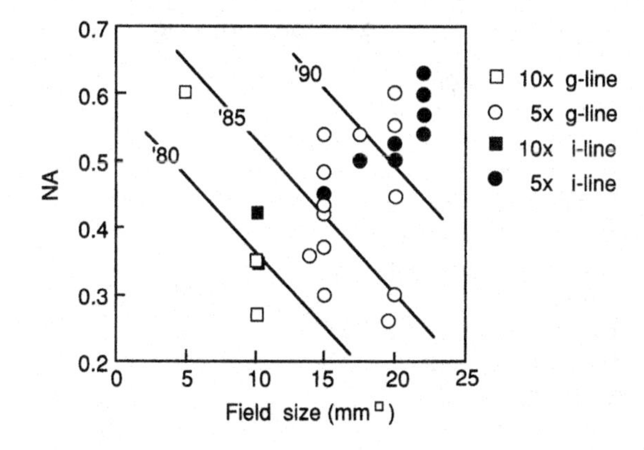

Figure 2.13 Evolution of i-line and g-line lenses.

is located in a thermal chamber in which the temperature is controlled within ±0.1 °C of room temperature, which is 20–25 °C depending on location.

The reported relationship between the field size and the NA of the lens system for i-line and g-line steppers is shown in Figure 2.13. In the first stage of i-line stepper development, the field size of the i-line lens system was 10 mm × 10 mm and the NA of the lens system was less than 0.4. The most advanced i-line steppers have lens systems with a field size of

22 mm × 22 mm and a NA larger than 0.6 (Table 2.1). The overlay accuracy has also been improved significantly.

As shown in Figure 2.14, resolution has been improved by almost 50% over that of the first i-line steppers, while overlay accuracy has been improved by almost 70%. As a result, 0.25-μm technology is now established in the industrial environment.

2.4.3 Development of i-line resist processes

At the beginning of i-line lithography development, there were no useful resists for i-line exposure. Figure 2.15 shows the patterns on different resist materials (the parameters A, B, C of these resists are listed in Table 2.2[25]). Parameter A shows the bleaching characteristics of the resist materials. A larger A means a larger bleaching

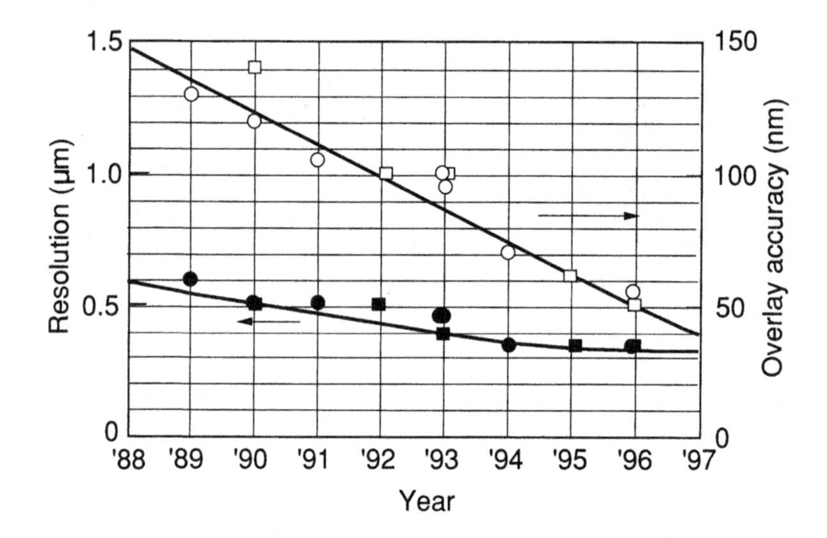

Figure 2.14 Resolution and overlay accuracy of i-line stepper.

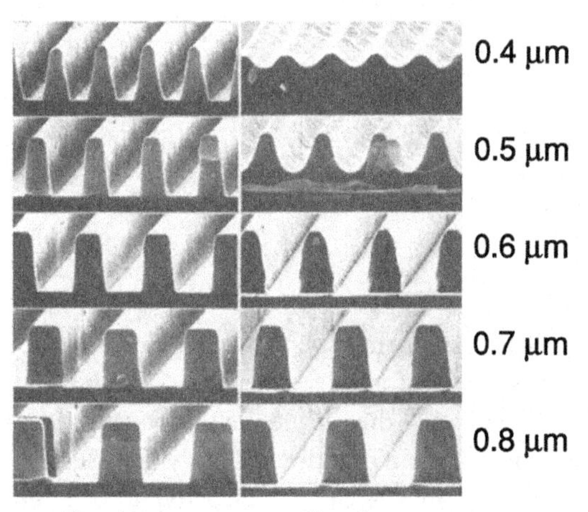

Figure 2.15 SEM photographs comparing resist pattern profiles with different A, B, C parameters (see Table 2.2) (NA = 0.42; i-line exposure).

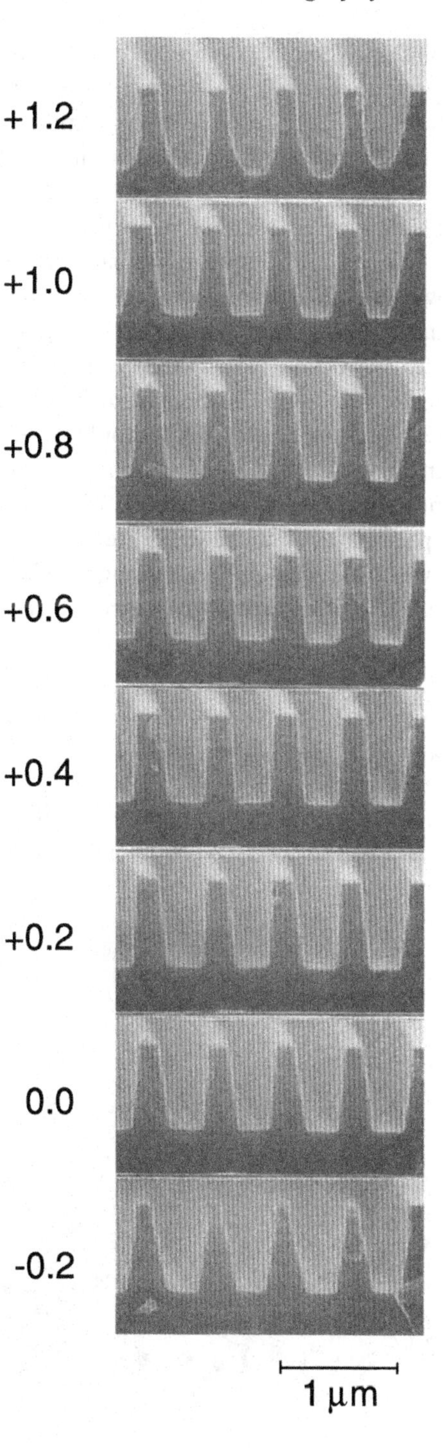

Figure 2.16 SEM photographs for 0.3-μm line-and-space patterns replicated in a recent i-line resist material by using an i-line stepper with 0.63 NA lens.

Table 2.2 *A, B, C parameters of resist materials used in experiments.*

	A (μm^{-1})	B (μm^{-1})	C (cm^2/mJ)
Resist 1	1.24	0.095	0.0282
Resist 2	1.20	0.271	0.0198

effect, which usually gives higher contrast-enhancement performance. Parameter A is heavily dependent on the characteristics of the photosensitizer. Parameter B gives the light transmission of the base materials. A larger B means a higher transmission. Generally speaking, a larger B gives a better pattern profile. Parameter B depends on the optical characteristics of the base matrix. Parameter C is related to sensitivity. Resist 1 has a smaller parameter B than does resist 2. The cross-sectional shape (profile) of resist 2 is triangular-like at smaller than 0.6-μm line-and-space pattern. This is due to strong light absorption in the resist layer. New resist materials with a smaller B have thus been developed in order to reduce the light absorption, and these resist materials have a significantly improved resist pattern profile.[26]

Not only the light absorption, but also many other characteristics of the resist material have been improved. Detail of the resist materials are discussed in Chapter 6.

2.4.4 Resolution limits of i-line lithography

The most advanced i-line stepper has a lens more than 30 mm in diameter and a NA greater than 0.6. As shown in Figure 2.16, the resolution limit of i-line exposure using the latest exposure and resist system is about 0.30 μm. Resolution can be significantly improved by applying resolution-enhancement technology.[27,28,29] Resolution-enhancement technology is described in Section 2.6.

Figure 2.17 shows resist patterns replicated with modified illumination on a halftone (attenuated) phase-shift mask: 0.26-μm patterns are clearly delineated. Alternating-phase-shifting mask technology is thought to be the ultimate resolution-enhancement technology. As shown in Figure 2.18, alternating-phase-shifting technology makes it possible for 0.175-μm line-and-space patterns to be delineated using an i-line system. This resolution is thought to be the limit of i-line lithography with current technologies.

2.5 **Deep-UV lithography**

2.5.1 Introduction

To increase resolution without sacrificing focus margin, shorter-wavelength light sources such as the KrF excimer laser (249 nm) and ArF excimer laser (193 nm) have been developed extensively.[16,30,31] Figure 2.19 shows the trend of DOF value vs design rule for DRAMs. For actual device fabrication, a DOF value of 1 μm is a minimum requirement for each design rule, and this value is governed by defocus factors, namely, device step, wafer focal-plane deviation and focus-setting error of the exposure tool. In i-line lithography, therefore, higher-NA optics and improved novolac-type resists have been used to reduce the design rule from 0.5 μm to 0.35 μm.[32] These techniques, however, are virtually saturated at around a 0.3–0.35-μm design rule, because the wavelength of i-line (365 nm) is close to the design rule. Therefore, a transfer from i-line light to KrF excimer laser light is considered to be necessary for pattern sizes below 0.3 μm. To establish this lithography, items such as KrF excimer lasers, highly transparent optical materials, and chemically amplified resists have been developed.

This section will describe the technical status of KrF excimer laser lithography for 0.25-μm

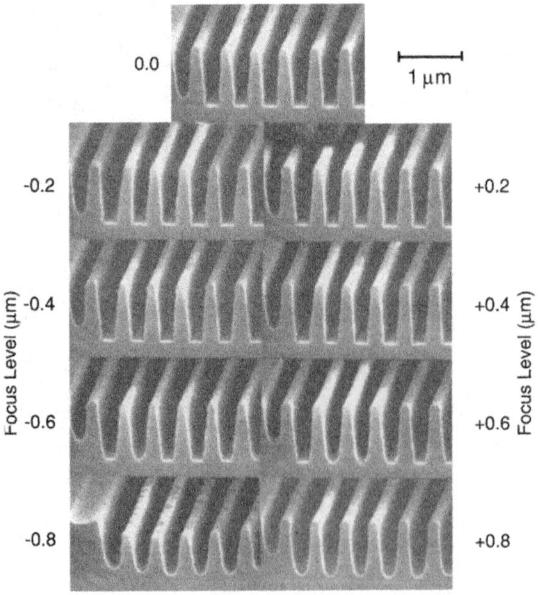

Figure 2.17 SEM photographs of 0.26-μm line-and-space resist patterns replicated using annular illumination (NA = 0.63) and an attenuated (4%) phase-shift mask.

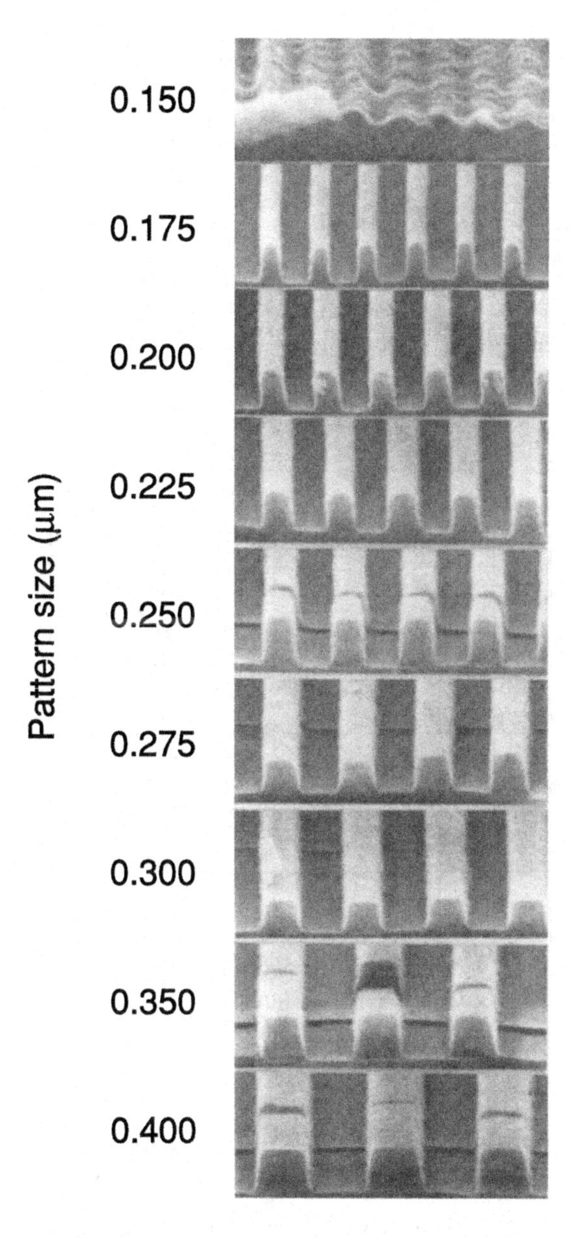

Figure 2.18 SEM photographs of line-and-space patterns replicated in i-line resist using the alternating phase-shift mask technique (NA = 0.65; σ = 0.3 (coherency factor)). (Courtesy Nikon.)

device fabrication, by considering several technical aspects, such as resolution, DOF, exposure field size, CD control, and overlay accuracy. Finally, ArF excimer laser lithography (193 nm) for 0.18-μm design rule patterning will be described briefly.

2.5.2 Requirement for 0.25-μm design rule lithography

The required performance criteria for DRAMs in various design rules are summarized in Table 2.3. For 0.25-μm device fabrication, a large DOF ($\geq$1.0 μm) is required. Moreover, the

Table 2.3 *Lithography performances required for various design rule DRAM fabrication.*

	64 Mb	256 Mb	1 Gb	4 Gb
Year of production (predicted)	1995	1998	(2001)	(2004)
Resolution (μm)	0.35 ~ 0.30	0.25 ~ 0.20	0.18 ~ 0.15	0.13
DOF (μm)	1.5	1.2	1.0	1.0
Field size (mm^2) (Exposure)	22 × 22 (2-chip)	25 × 25 (2-chip)	30 × 30 (2-chip)	36 × 18 (1-chip?)
CD control (nm)	35	25	18	13
Overlay (nm)	90	60	45	35

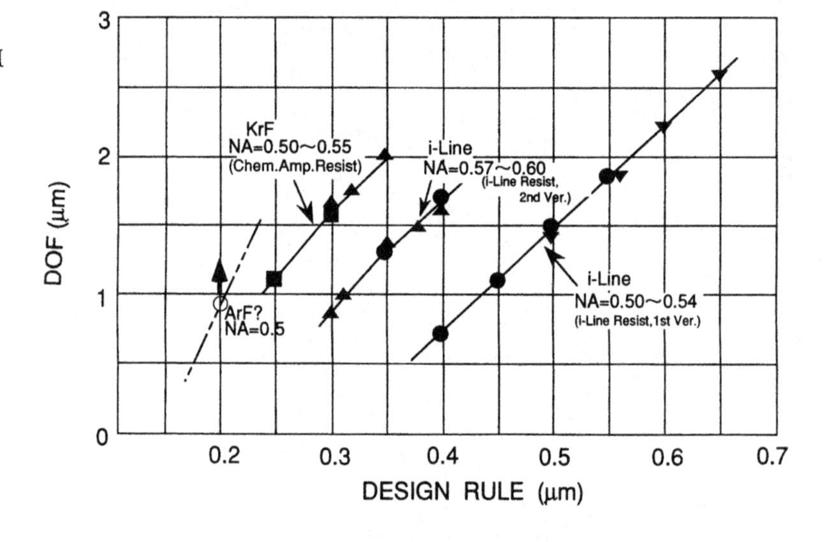

Figure 2.19 Trend of DOF value vs design rule of DRAM for various kinds of optical lithography. (▼ 4 Mb; ● 16 Mb; ▲ 64 Mb; ■ 256 Mb.)

resist pattern fidelity required, that is, pattern-size control of less than 10% of design rule and overlay accuracy of less than 25–30% of design rule, becomes very severe. A field size greater than 25 mm × 25 mm is required for 2-chip exposure of a 256-Mb DRAM.[19,33] A scan-type exposure system, instead of the conventional step-and-repeat system (stepper), has therefore been developed. The throughput of the scan-type system is the same as that of g/i-line steppers, that is, more than 50 wafers (8–12 inch (200–300 mm) diameter) per hour.

2.5.3 Basic performance of KrF excimer laser lithography

As described in Section 2.3, the resolution as well as the DOF depends on the numerical aperture (NA) and the coherence factor (σ).[34] The coherence factor is defined as NA (illumination system)/NA (projection system). Figure 2.20 shows the NA dependence of resolution and DOF. A chemically amplified resist with a thickness of 0.7 μm was used and line-and-space (L&S or L/S) patterns were evaluated. For

Table 2.4 *Coherency factor (σ) dependence of DOF for 0.25-μm L&S, isolated-line and isolated-space patterns.*

(NA = 0.5; KrF stepper; 0.7-μm-thickness chemically amplified resist.)

σ	DOF (μm)		
	0.25-μm L&S	0.25-μm line	0.25-μm space
0.3	0.6	1.0	1.0
0.5	0.8	1.0	1.0
0.7	1.0	1.0	1.0

coarse patterns (>0.25 μm), optics with a lower NA gives a greater DOF, whereas for fine patterns a higher NA gives a greater DOF. Therefore, it is considered that high-NA optics is effective for the formation of fine patterns.

Figures 2.21 and 2.22 show the σ-dependence of the CD vs defocus characteristics for L&S and isolated-line patterns. In these figures, the exposure dose was fixed to the optimum dose for 0.25-μm L&S patterns. Table 2.4 summarizes the σ-dependence of the DOF value for 0.25-μm L&S, isolated-line and isolated-space patterns. The DOF values for the L&S pattern increase with increasing coherency factor. On the other hand, the DOF values for both isolated patterns are almost constant, but the width of an isolated line was different from that of the lines in the L&S patterns because of the optical proximity effect. Figure 2.23 shows the proximity effects expressed as the pattern-size difference between L&S lines and isolated lines. Simulation and experimental results both show that higher coherency or oblique illumination tends to reduce the size of the optical proximity effect. These results are similar to those observed in i-line lithography.[32]

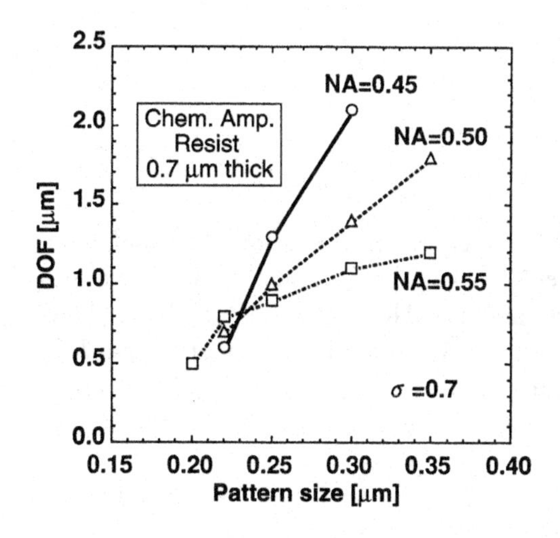

Figure 2.20 Numerical aperture (NA) dependence of resolution and DOF for i-line stepper.

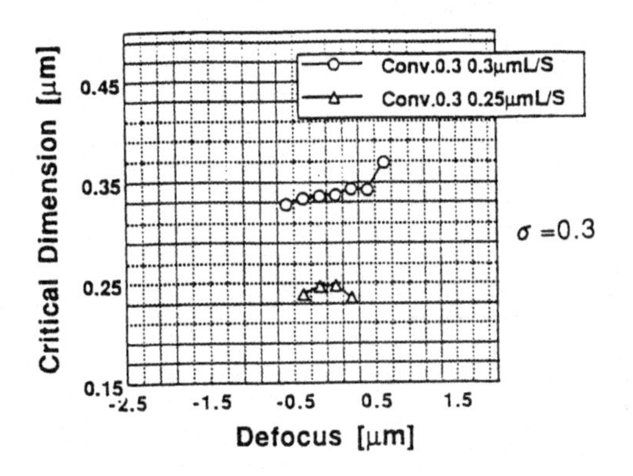

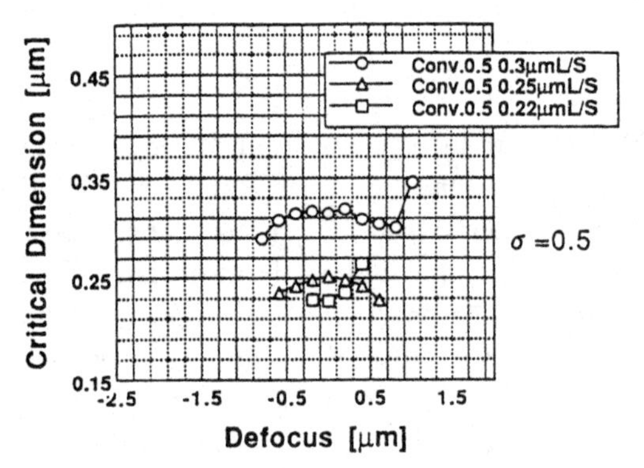

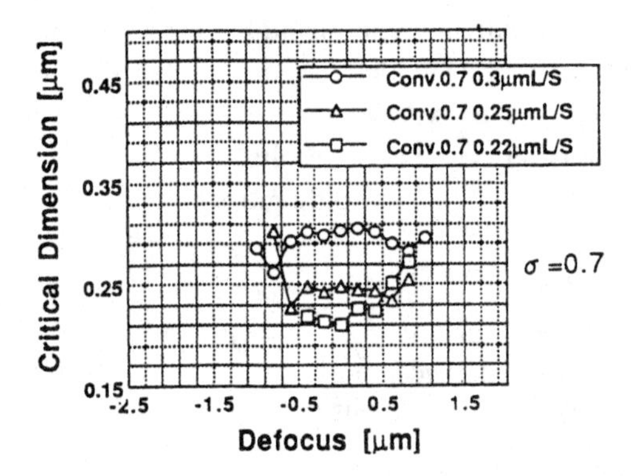

Figure 2.21 Coherency factor (σ) dependence of CD vs defocus characteristics for L&S patterns. Conventional illumination.

To use KrF excimer laser lithography in actual device production, the resolution of chemically amplified resists needed to be improved. Figure 2.24 shows an example of dissolution-rate characteristics obtained experimentally.[35,36] Generally, these log(dissolution rate)–log(exposure dose) plots are S-shaped: dissolution rate is almost constant at low doses, increases abruptly at the threshold exposure-dose position E_{th} and then saturates to a constant rate. The dissolution rate can, therefore, be characterized with parameters: R_{min} (minimum dissolution rate), R_{max} (maximum dissolution rate), E_{th}, and a selectivity parameter n (steepness of slope), using the Mack dissolution-rate model and the PROLITH/2

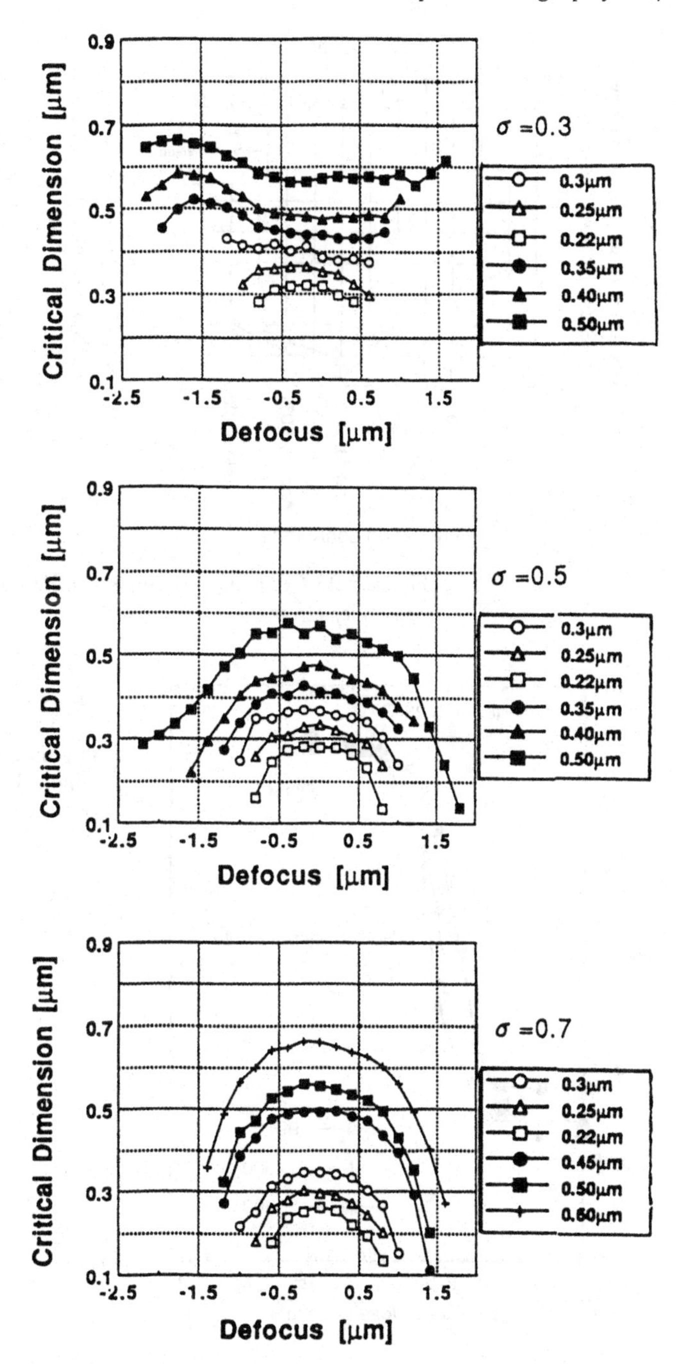

Figure 2.22 Coherency factor (σ) dependence of CD vs defocus characteristics for isolated-line patterns. Conventional illumination.

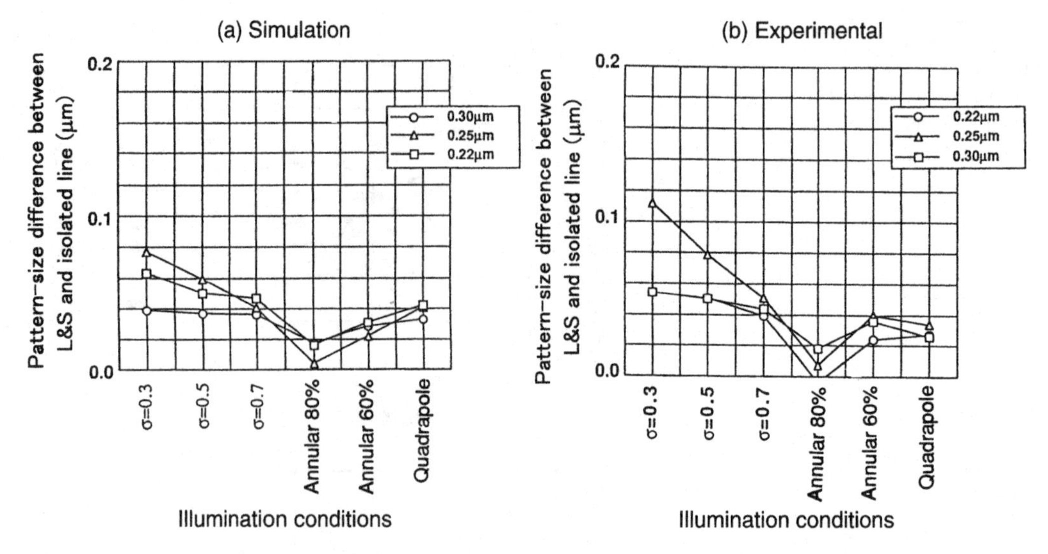

Figure 2.23 Coherency factor (σ) and illumination dependence of optical proximity effect.

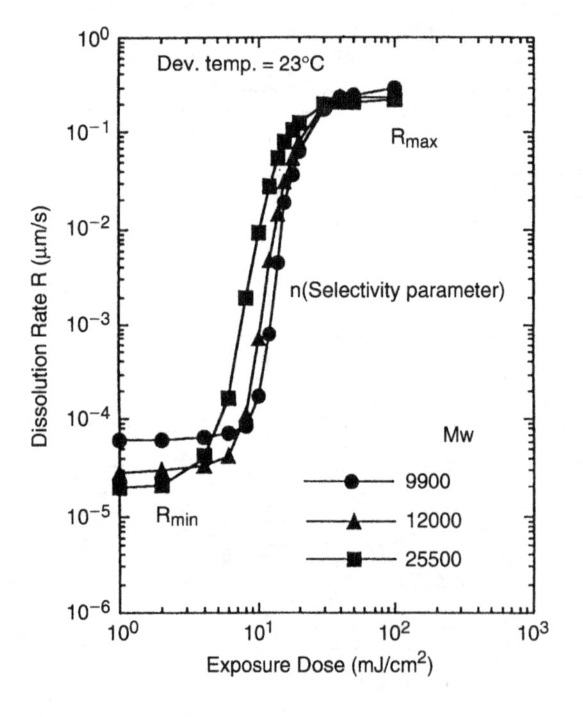

Figure 2.24 Dissolution-rate characteristics of chemically amplified resist (molecular weight dependence).

resist profile simulator.[37] The selectivity parameter n was found to be the most important parameter determining resolution. Figure 2.25 shows L&S pattern profiles calculated for various selectivity values. Both pattern profiles and resolution initially improve with increasing selectivity n, but when selectivity is very high, the pattern profile deteriorates because of side-

Table 2.5 *Improvements of chemically amplified resist materials.*

PVP-type resin
Molecular weight and dispersion
Copolymerization
Protection group (Ratio; bulkiness; hydrophobicity; acid catalysis efficiency)
PAG
Acid generation yield
Acid strength
Diffusion length
Deprotection rate
Blend of different PAGs
Additive
Dissolution inhibitor
Acid quencher
Dye
Others
Solvent
Viscosity
Solubility
Safety

wall roughness. Figure 2.26 shows resolution and focus margin calculated assuming ideal dissolution-rate characteristics: 0.20-μm resolution and a focus margin of 1.0–1.25 μm for a 0.25-μm L&S pattern are expected with a good profile. These values are sufficient for 0.25-μm level resist patterning.

Table 2.5 indicates the items that have been investigated for each component of the KrF positive-type chemically amplified resist system. (The chemistry and resist process for KrF resists will be explained in detail in Chapter 6.) Figure 2.27 shows SEM photographs indicating resolution and focus margin for 0.25-μm L&S patterns. The current resist system is considered to provide sufficient resolution for 0.25-μm device fabrication.

2.5.4 Oblique illumination and attenuating phase-shift mask

As described above, KrF excimer laser lithography can provide the resolution needed for 0.25-μm patterning, even with conventional illumination, but its process margin is still small. Various resolution-enhancement techniques, such as oblique illumination,[38,39,40,41,42] phase-shift mask,[43,44] pupil filtering[29] and their combination[45,46] have therefore been proposed. These techniques will be described in detail in Section 2.6. Oblique illumination and an attenuating phase-shift mask for window pattern formation[47] are explained here. Introduction of these technologies into actual device fabrication is relatively easy and very effective.

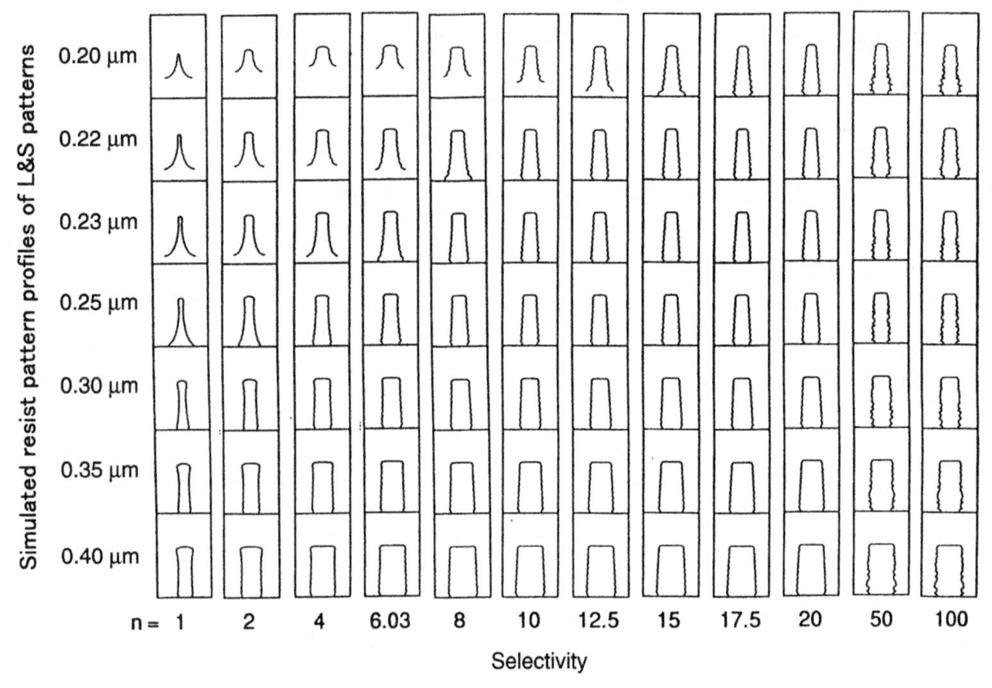

Figure 2.25 Selectivity value dependence of L&S resist pattern profile.

Figure 2.26 Resolution and focus margin in the case of ideal dissolution-rate characteristics ($R_{max}/R_{min} = 10\,000$; $n = 17.5$; $DL = 35$).
n: selectivity parameter
DL: diffusion length

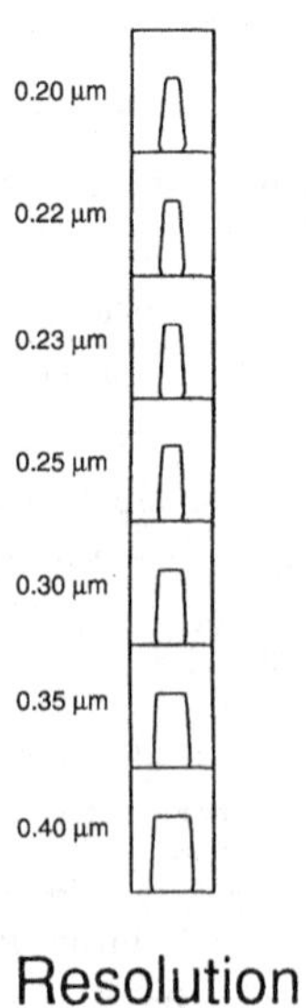

Defocus (μm)	0.25 μm	0.30 μm
−1.25		
−1.00		
−0.75		
−0.50		
−0.25		
±0.00		
+0.25		
+0.50		
+0.75		
+1.00		
+1.25		

Focus Margin

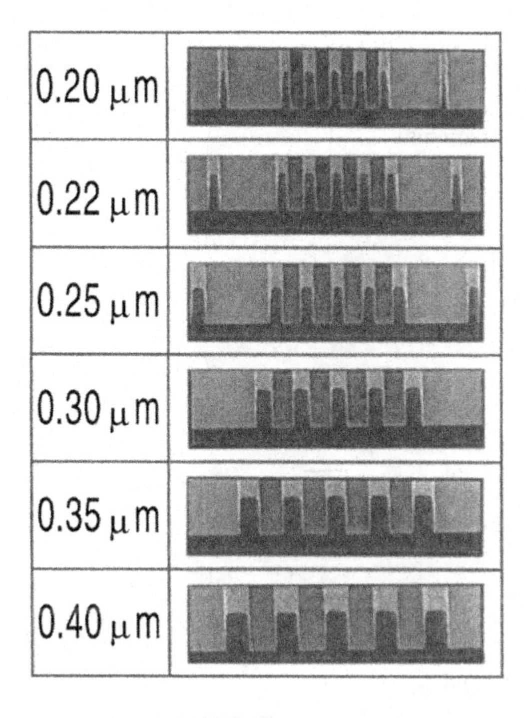

Figure 2.27 SEM photographs showing resolution and focus margin for 0.25-μm L&S patterns in chemically amplified resist.

Figure 2.28 shows the illumination dependence of the CD vs defocus characteristics of L&S patterns. The DOF is obviously improved by using annular and quadruple illuminations. Table 2.6 lists the DOF values for 0.25-μm L&S, isolated-line, and isolated-space patterns. Oblique illumination is effective for L&S patterns, but not for the isolated patterns. The DOF difference between conventional and oblique illuminations is due to the optical proximity effect (line/space width difference).

An attenuating phase-shift mask has been proposed for improving the DOF value of window pattern formation. Figure 2.29(b) shows this mask structure, having a partially transparent (5–20%) phase shifter. The light amplitude at the mask and the intensity of the wafer are modified with this mask. Figure 2.29 also shows the defocus dependence of the light-intensity distribution. Intensity contrast degradation is much suppressed by using the attenuating mask, rather than a conventional mask. Single-layer materials for the attenuating layer have been developed recently.[48,49] The mask fabrication process is, except for the blind portion, the same as that of the conventional mask. In order to avoid resist thickness loss at the four corners near the blind area, due to multiple-exposure by penetrating light through the blind portion, fine dotted pattern formation or an additional Cr-layer on the blind portion have been investigated. Figure 2.30 shows the DOF comparison of 0.25-μm window patterns between a conventional and an attenuating

Figure 2.28 Illumination dependence of CD vs defocus characteristics of L&S patterns.

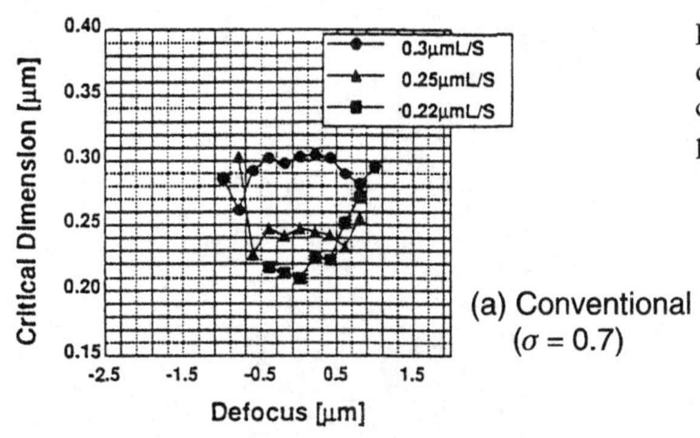

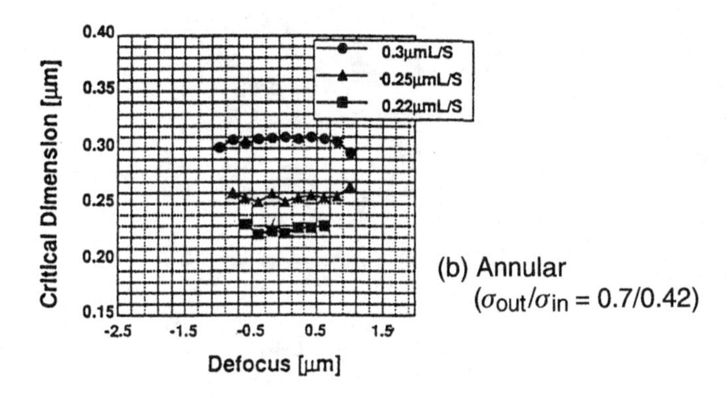

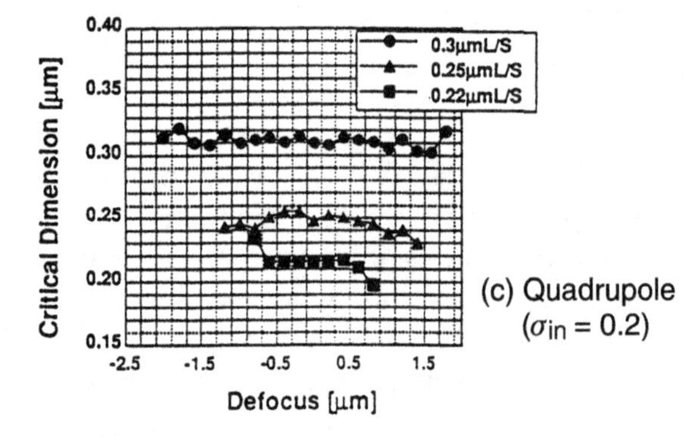

phase-shift mask. A DOF of about 1.5 μm was obtained by the latter technique. Figure 2.31 shows the σ-dependence of the DOF value. Smaller σ brings about a larger DOF value. Since the demand for fine window patterning is increasing rapidly, this phase-shift mask will be very important for window pattern formation.

2.5.5 CD control

Linewidth fluctuation (CD-error) must be controlled to less than 10% of the device design rule. There are three major factors causing CD-error: environmental instability, the optical proximity effect and the pattern topography effect (Table 2.7).

Table 2.6 *Illumination dependence of DOF for 0.25-μm L&S, isolated-line and isolated-space patterns.*

(NA = 0.5; $\sigma_{max} = 0.7$; 0.7-μm-thick chemically amplified resist.)

	DOF (μm)		
	0.25-μm L&S	0.25-μm line	0.25-μm space
Conventional ($\sigma = 0.7$)	1.0	1.0	1.0
Annular 60%	1.4	0.8	0.6
Annular 80%	1.4	0.8	0.6
Quadrupole ($\sigma = 0.2$)	1.6	0.8	0.6

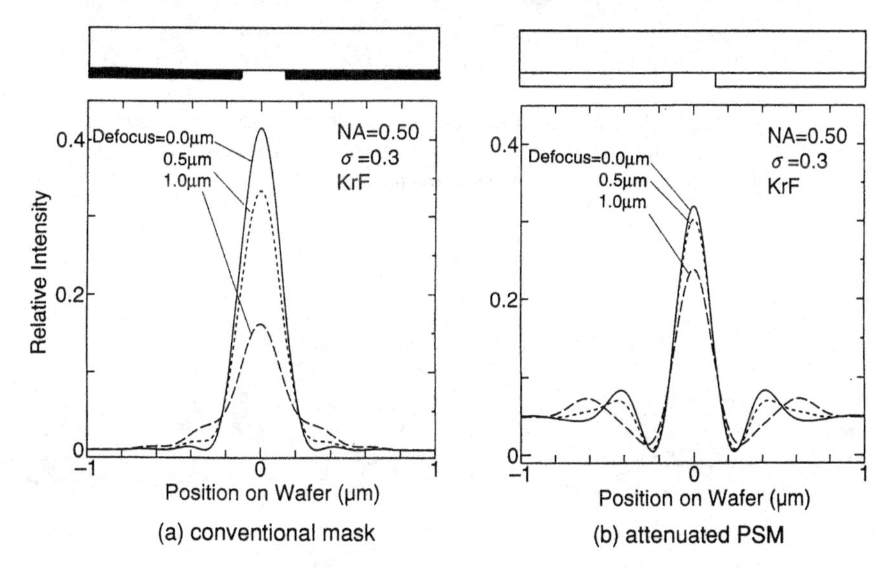

Figure 2.29 Illumination-intensity distribution in the case of (a) conventional binary mask and (b) attenuated phase-shift mask (PSM).

Environmental instability such as the post-exposure delay (PED) effect[50] and substrate dependence,[51,52] characteristic of chemically amplified resists, is due to the neutralization of photo-generated acid by bases in the atmosphere or on the substrate. Overcoats and chemical filters have been used to prevent the formation of a T-top shape profile.[53,54,55] Figure 2.32 shows the effectiveness of a chemical filter capable of absorbing basic species (NH_3, NAP, etc.). By applying a chemical filter, the concentration of basic species is reduced down to 1 ppb. As for the substrate dependence, surface treatments such as an O_2 plasma and high-temperature baking are being considered together with the improvement of the resist material.[56,57]

The proximity effect is caused by light-intensity differences and depends on pattern density.[58] To reduce this effect, high-NA lenses and shorter-wavelength exposures are effective. And, mask modification such as mask bias and

Conventional Mask

-1.5μm | -1.25μm | -1.00μm | -0.75μm | -0.5μm | -0.25μm | 0.00μm | 0.25μm | 0.50μm | 0.75μm | 1.0μm | 1.25μm | 1.50μm

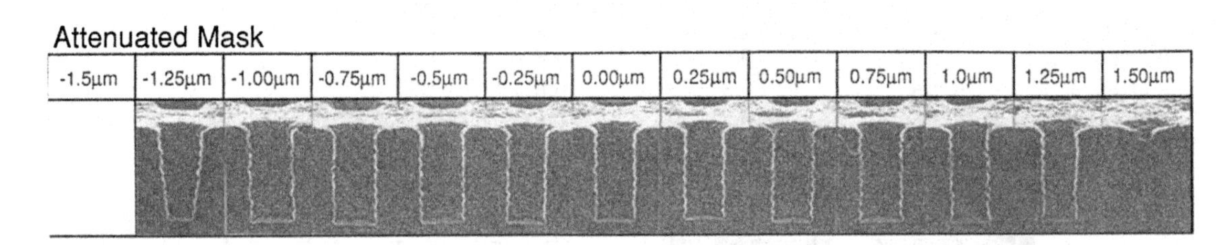

Figure 2.30 SEM photographs showing practical DOF comparison between attenuated phase-shift mask and conventional binary mask (NA = 0.5, $\sigma = 0.3$; resist thickness = 0.7 μm; mask transmittance = 5%).

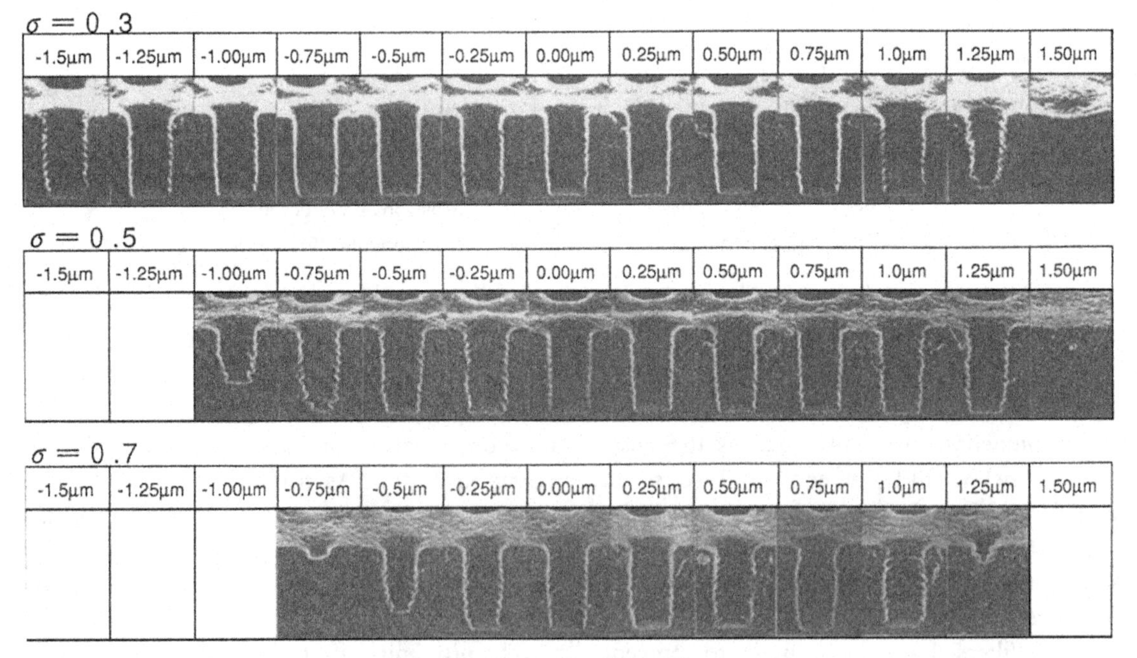

Figure 2.31 SEM photographs showing coherency factor (σ) dependence of DOF for KrF excimer laser lithography with an attenuating phase-shift mask (NA = 0.5; λ = 248 nm; 0.7-μm-thick resist).

Table 2.7 *Principal linewidth fluctuation factors.*

Linewidth control

Goal:
  $\Delta L < 10\%$ of design rule

Main linewidth fluctuation factors:

  Environmental instability
    PED effect (ammonia gas-induced T-top profile)
    Substrate dependence
    → Overcoat/chemical filter/Improvement of resist material
      Surface treatment of substrate

  Proximity effect
    → High NA
      Mask modification by simulation analysis (OPC, Serif etc.)

  Pattern topography effect
    Interference/Bulk effect
    Reflection from device slant (Halation)
    → Dyed resist/TAR/ARC

PED: Post-exposure delay (time between exposure and development processes).
OPC: Optical proximity correction.
TAR: Top anti-reflection coating (a technique overcoating a thin film on top of the resist film to reduce the exposure-light reflection at the resist surface).
ARC: Anti-reflection coating (a technique coating a thin film between the substrate and resist to reduce the exposure-light reflection from the substrate).

serif-pattern† addition using simulation analysis becomes very useful. Figure 2.33 shows one example of optical proximity effect correction (OPC).[59,60,61] The use of corner mask bias greatly alleviates pattern shortening. Near the resolution limit, resist dissolution characteristics also influence the proximity effect. Generally, a high-resolution resist brings about a small proximity effect, because of high dissolution-rate-contrast characteristics (see Subsection 6.1.4).

The topography effect is now the dominant cause of CD fluctuation in practical device fabrication. Problems include the interference effect and resist side-wall notching due to the reflection from a slanted reflected substrate. To reduce the severity of these reflection-induced effects, various techniques such as those using a dyed resist or an ARC (anti-reflective coating under the resist material) or a TAR (top anti-reflector) are being developed.

The light intensity at the bottom of the photoresist 'swings', depending on the phase of the standing wave, and its swing ratio S[62] is a fundamental measure of the interference effect:

$$S = 4(R_1 R_2)^{1/2} \cdot \exp(-\alpha d), \qquad (2.3)$$

where R_1 and R_2 are the reflectivity at the air/resist interface and the resist/substrate interface respectively, α is the absorption coefficient, and d is the thickness of the resist. The following improvements in the resist process can reduce this ratio: reduce R_1 through the use of a TAR, reduce R_2 through the use of an ARC, and increase absorption in the resist by using a

†Small pattern added to the corner of the original pattern.

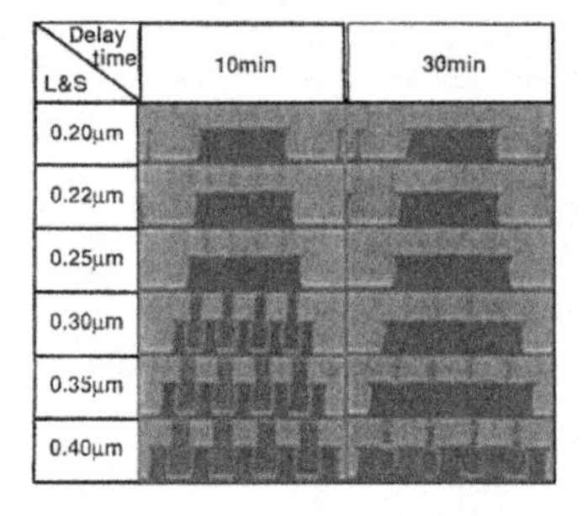

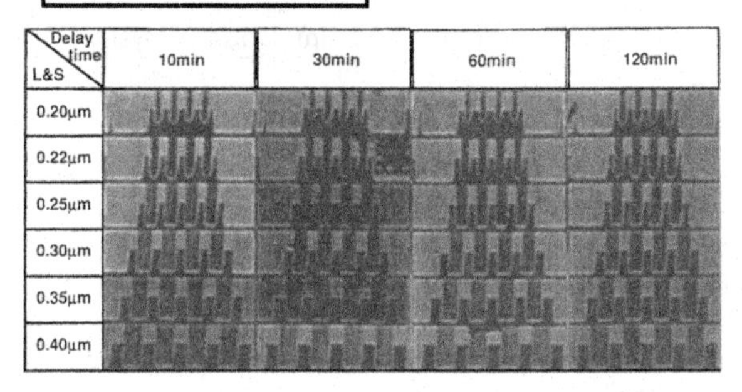

Figure 2.32 SEM photographs showing that chemical filter improves post-exposure-bake (PEB) dependency of chemically amplified resist.

dye. The dyed-resist process is simple and effective, since the swing ratio S decreases exponentially with increasing αd. Such a high absorption, however, tends to cause the formation of a tapered resist profile. Resist absorption and profile have therefore to be optimized.[63]

An ARC is the anti-reflective layer coated under the resist material, and various ARC materials, such as organic type a-C:H, inorganic type SiO_xN_y, and so on, have been investigated.[64,65,66] Figure 2.34 shows that an ARC is effective in suppressing the interference effect. Figure 2.35 shows SEM photographs of the line pattern on a stepped WSi substrate. With the ARC film, the resist pattern can be resolved without linewidth fluctuation and side-wall notching phenomena.

A TAR, top anti-reflector, is coated onto the resist surface.[67] To reduce reflection from the top of the resist, this film induces destructive interference between light reflected from two interfaces (air/TAR and TAR/resist). The optimum condition for the TAR layer is given by the following equations:

$$n = (n_R)^{1/2}, \tag{2.4}$$

$$d = \lambda / 4n, \tag{2.5}$$

where n and n_R are the real parts of the refractive index (refractivity) of the TAR and resist, respectively, d is the thickness of the resist, and λ is the wavelength of the exposure light. In general, the ideal refractivity value of a TAR is about 1.3 because most resist materials (novolac type and polyvinyl-phenol type) have a refractivity of around 1.65–1.70.

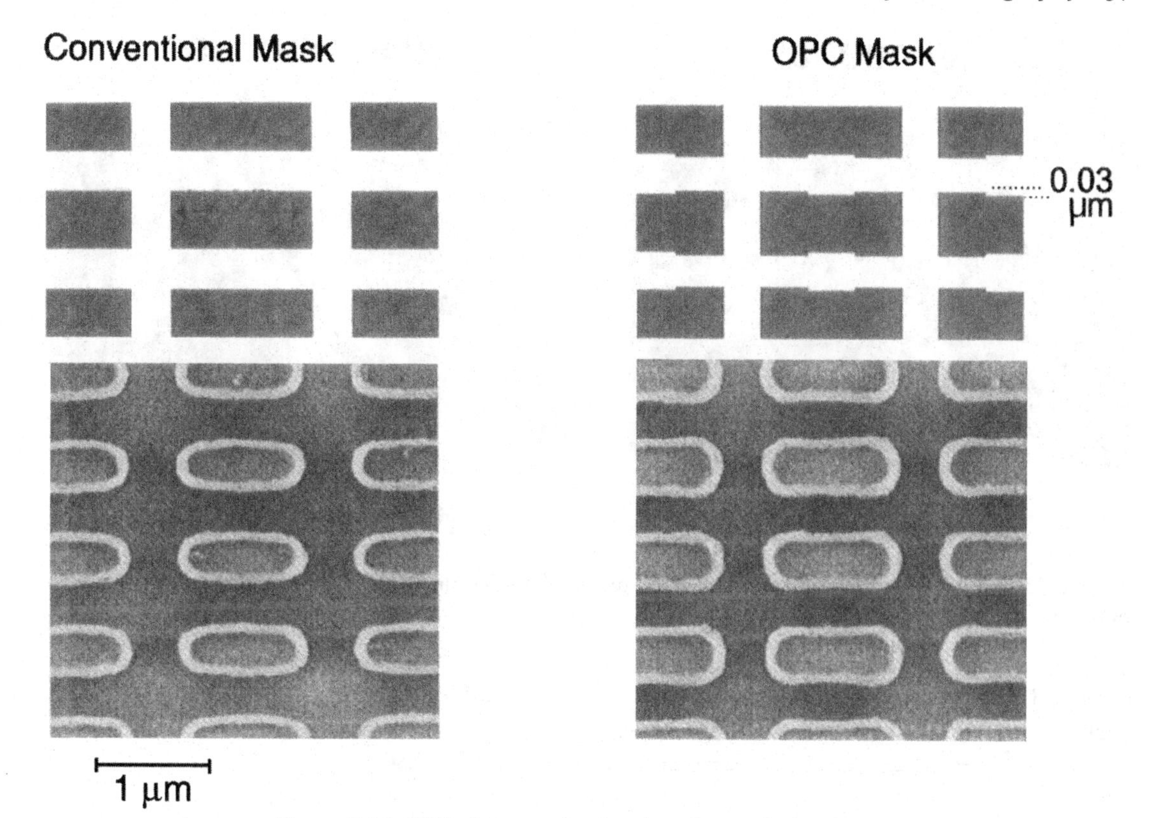

Figure 2.33 SEM photographs showing that optical proximity correction (OPC) improves pattern-replication fidelity (NA = 0.5; $\sigma_{out}/\sigma_{in} = 0.7/0.42$; 0.7-µm-thick chemically amplified resist).

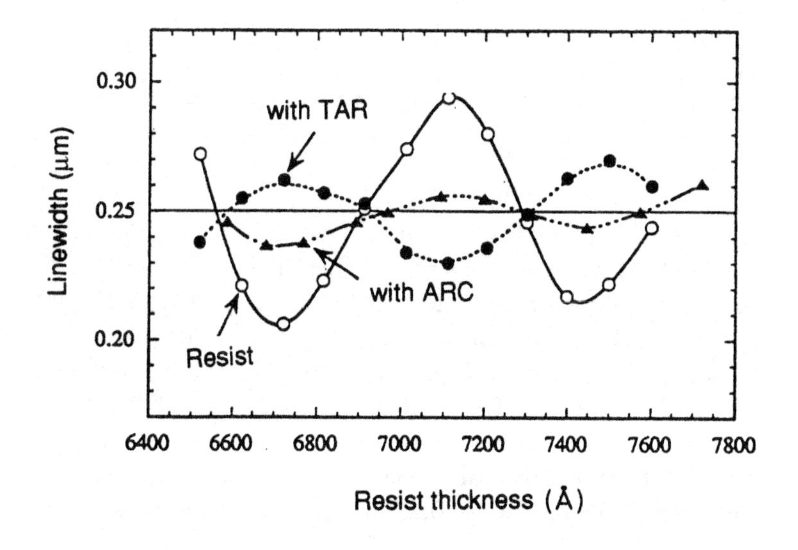

Figure 2.34 Linewidth fluctuation as a function of resist thickness. Both TAR and ARC processes reduce interference effect. TAR: top anti-reflector; ARC: anti-reflection coating.

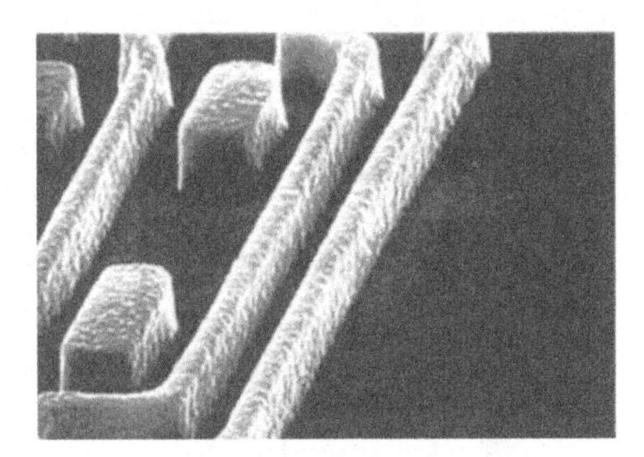

Figure 2.35 SEM photographs showing that ARC process improves linewidth narrowing at a step on a topographic substrate. Linewidth is 0.25 μm.

Table 2.8 *Analysis of linewidth fluctuation for 0.25-μm pattern. Each value tabulated indicates the range of the fluctuation.*

Item	CD error (nm)
Exposure tool:	
Dose control	5
Lens aberration	20
Pattern density: Proximity effect	10
Topography: Interference & Halation	30
Resist process: Coater & Developer Defocus	20
Reticle	10
TOTAL ERROR	44

Unfortunately, it is difficult to find a TAR material with such a low refractivity value. The refractivity values of current TAR materials are about 1.4–1.5. Although a TAR has no effect on halation, the TAR process is simpler than an ARC process. The TAR film is easily removed using a water treatment.

Table 2.8 summarizes various factors related to linewidth (CD) control for 0.25-μm patterns.

2.5.6 Overlay accuracy

The overlay accuracy required is 25–30% of the design rule, so the accuracy necessary for 0.25-μm devices is ±60–75 nm.

Table 2.9 classifies alignment methods from the standpoints of process, system, and detection methods. In global alignment, alignment marks are detected at several shot-positions on

Table 2.9 *Classification of alignment methods.*

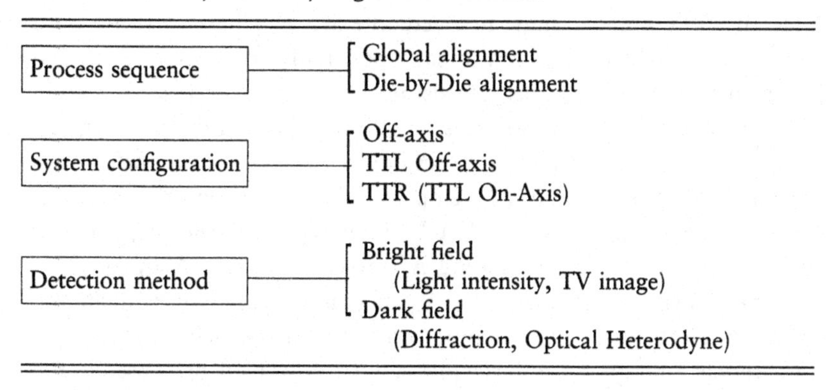

Table 2.10 *Summary of overlay deterioration factors (mean $+3\sigma$).*

Item	Overlay error (nm)
Exposure tool	
Alignment accuracy:	
Mark detection	10
Stepping	25
Baseline	15
Dynamic distortion	10
Lens-distortion stability	3
Reticle rotation	10
Reticle-pattern placement	15
Wafer processes	25
TOTAL ERROR (one machine)	45
Machine to machine:	
Static lens distortion	15
Dynamic-distortion difference	5
Stepping-accuracy difference	20
TOTAL ERROR (mix-and-match)	$49 + 15 = 64$

a wafer and all shot-positions are set on the basis of averaging these measured positions. Then, exposure is carried out by stepping the wafer stage. In die-by-die alignment, on the other hand, an alignment procedure is carried out for each exposure cycle. There are three types of system configuration: off-axis; through-the-lens (TTL); and through-the-reticle (TTR). TTR is basically the most reliable because it offers the possibility of stage-free alignment, but the correction of color aberration is difficult in both the TTR and TTL configurations, especially in the DUV region. Most of the current KrF exposure systems thus use off-axis alignment, so the baseline must be controlled carefully.

Total overlay accuracy is degraded by many factors,[68,69] as shown in Table 2.10. The value of each factor was estimated by supposing a scan-type exposure system, in which there are specific factors related to the simultaneous movement of the wafer and reticle stages: dynamic distortion and the dynamic-distortion difference from one machine to another. Alignment accuracy of the exposure tool, which includes (i) the sensitivity of the mark detector, (ii) stepping accuracy and (iii) baseline stability, is the most fundamental factor. At present, static lens-distortion changes, between different tools or between different optical conditions. Reticle rotation and reticle-pattern-placement error are also important factors. To improve total overlay accuracy, all these errors must be reduced.

2.5.7 Target specifications for 0.25-μm patterning

Tables 2.11 and 2.12 show the specifications of recent KrF exposure tools (stepper and scanner) and KrF excimer lasers. These specifications are considered to be almost enough for 0.25-μm design rule device fabrication. However, pattern formation accuracy (linewidth control and overlay) depends not only on the exposure tool but also on resist-process robustness.

2.5.8 ArF excimer laser lithography

ArF excimer laser lithography is expected to be the pattern formation method used at the 0.15-μm design rule. But the technical barriers that must be overcome are very high, because both conventional optical materials and resist systems, consisting of aromatic resins, are highly absorptive at 193 nm. Currently, highly transparent optics having wider exposure fields have been developed. Table 2.13 shows the specifications of ArF small-exposure-field steppers for developing ArF resist materials. Alicyclic polymers, that is, methacrylate polymers including dry-etching resistant groups such as adamantyl and tricylodecanyl groups have been used as the base resins for an ArF chemically amplified resist, and 0.17–0.2-μm resolution with this type of resist has been reported.[70,71,72] A top-surface imaging (TSI) process is also being developed.[73]

2.6 **Resolution-enhancement technologies**

2.6.1 Introduction

Since LSI pattern size is now approaching the wavelength of the exposure light, there is a need for better resolution-enhancement technologies. Among them, oblique-illumination technologies,[39,41,42] phase-shifting mask technologies,[43,74,75] and pupil-filter technology[29] appear to be very important.

According to conventional imaging theory, an optical image is given by the formula in Figure 2.36. This formula is easily understood by comparing it with the optical system also shown in that figure. Ultraviolet light from the light source impinges upon the ‘fly’s-eye’ lens and divides into small elements. Many light-source images are formed at the exit of the fly’s-eye lens. This results in the formation of a secondary source plane. Every source element illuminates the whole of the photomask. Because of the averaging effect, the photomask is uniformly illuminated; this corresponds to the

Table 2.11 *Specifications of newly developed KrF excimer laser exposure tools.*

Manufacturer	Nikon	Canon	ASML	Nikon	SVGL
Model	NSR-2205EX14C	FPA-3000EX5	PAS5500/300(4×)	NSR-S203B	Micrascan III
Lens NA	0.40–0.60 (software-controlled)	0.45–0.63 (software-controlled)	0.40–0.57 (software-controlled)	0.68–0.60 (software-controlled)	0.40–0.60 (software-controlled)
Illumination NA	0.45	0.50	0.20–0.46	0.51	0.48
Partial coherence (@ max NA)	0.75	0.8	0.8	0.75	0.8 (0.3)
Modified illumination	Software-controlled 6 revolvers	Software-controlled 6 revolvers	Software-controlled zoom lens	Software-controlled 6 revolvers	Software-controlled 4 array units
Field size (mm × mm)	22 × 22	22 × 22	22 × 22	25 × 33	26 × 32.5
Distortion (nm)	30	20/22	35	20/25	35
Overlay accuracy (nm)	40	30	45	35	55
Illumination power (mW/cm^2)	250	380		625	
Illumination uniformity (nm)	±1.2%	±1.0%, RET ±1.5%	±1.0%, Ann ±1.4%	±1.2%	±1.0%
Stage accuracy (nm)	30	15			$3\sigma < 20$
Light source	Cymer ELS-5400	Cymer ELS-5300	Cymer ELS-5600	Cymer ELS-5410	Cymer ELS-5000 (1kHz, 15W)
Water scan speed (mm/s)	–	–	–	175	125

Table 2.12 *Specification of KrF excimer laser systems.*

Laser manufacturer	Cymer	Cymer	Lambda Phisik	Komatsu
Model	ELS-4000F	ELS-5000	LAMBDA LITHO	KLES-G7
Wavelength (nm)	248.385	248.385	248.327	248.385
Wavelength calibration accuracy (pm)	$< \pm 0.1$	–	$< \pm 0.15$	$< \pm 0.1$
Repetition rate (Hz)	600	1000	600	600
Pulse energy (mJ)	12	10	13.5	12.5
Average output power (W)	7.2	10	8	7.5
Spectral bandwidth (FWHM) (pm)	< 1.3	< 0.8	< 0.9	< 0.8
Integrated energy stability (%)	$< \pm 7$	$< \pm 1$	$< \pm 5$	$< \pm 7$

Table 2.13 *Specification of small-field ArF excimer laser exposure tools for resist system.*

Manufacturer	Nikon	Integrated Solutions, Inc.	Exitech
Model		ArF MicroStep	Series 8000 193 nm Microstepper
Lens NA	0.55	0.4–0.6	0.5
Projection lens	Refractive	Catadioptric	Refractive
Illumination NA	0.44	0.6	0.5
Partial coherence (@ max NA)	0.3–0.8	0.3–1.0	0.3–1.0
Modified illumination	Manual	Manual	Manual
Reduction ratio	1/20	1/10	1/10
Field size (mm × mm)	3 × 3	1.5 × 1.5	2 × 2
Alignment technique	No	Global	Global
Illumination power (mW/cm^2)	100	80	10
Illumination uniformity (%)	±5	±2.5	±5
Light source	400 Hz, 1.4 W	400 Hz, 8W	Lambda Phisik LPX210i 100 Hz, 0.5W

summation of the formula in Figure 2.36. At the secondary source plane, intensity modulation is possible; oblique-illumination technology is achieved by selecting specific illumination sources. When the light passes through the photomask, diffraction from the mask pattern occurs; this corresponds to F in the formula in Figure 2.36. Since F has a phase term, a phase-shifting effect can be introduced by utilizing a photomask whose clear regions have a phase shift of 180 degrees. The image is formed by the interference of diffraction. Depth of focus can be significantly improved by putting an absorber or phase filter into the projection lens pupil. This is called pupil filtering.

2.6.2 Oblique-illumination technology

By using only obliquely incident beams to illuminate a photomask, we can enhance resolution and increase the depth of focus. This technology is called 'oblique-illumination technology' or 'modified-illumination technology.' The expression 'off-axis illumination technology' (OAI) has been widely used but is physically incorrect.

Depth of focus enlargement by this technology can be explained by reference to Figure 2.37(a), where imaging of a large pattern ($R = \lambda/2NA$) is assumed. A vertically incident beam A (broken line) is diffracted by the photomask and several diffraction beams are generated. In the following, for convenience, only the one 0th-order and the two 1st-order diffraction beams are mentioned. These three diffraction

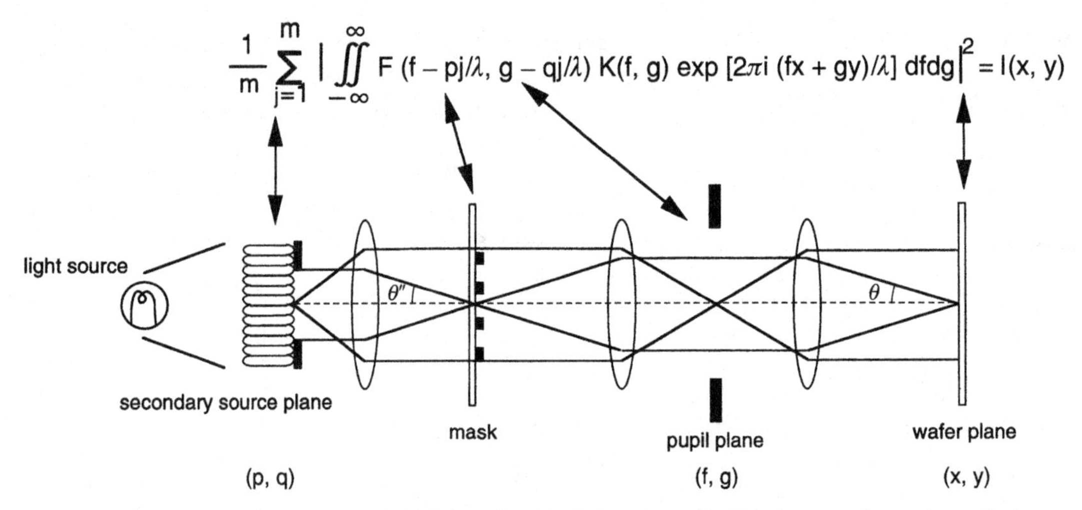

Figure 2.36 Principle of optical imaging. F: Fourier-transformed mask image; K : pupil plane function; (p, q), (f, g), (x, y): coordinates of secondary source plane, pupil plane and wafer plane, respectively; m: reduction ratio of projection lens (normally $\sim \frac{1}{5}$ to $\frac{1}{4}$). The coherence factor σ is given by $(\sin\theta''/\sin\theta)m$.

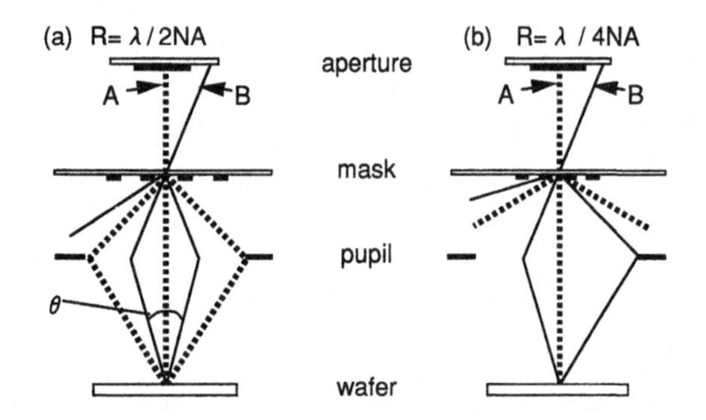

Figure 2.37 Principle of oblique-illumination technology. A: Vertically incident beam from aperture center. B: Obliquely incident beam from aperture edge.

beams can go through the projection-lens pupil. Aerial images are formed by three-beam interference on the wafer plane. Though the obliquely incident beam B is also diffracted in the same way, one of the 1st-order diffraction beams is cut off by the NA of the projection lens, and aerial images are formed by two-beam interference on the wafer plane. The phase difference between the diffraction beams at the wafer plane in the defocused condition becomes smaller with oblique illumination since the incident angles of the diffraction beams to the wafer plane are smaller. Oblique-illumination technology thus results in a larger depth of focus.

Resolution enhancement can be explained by reference to Figure 2.37(b), where the imaging of a small pattern ($R = \lambda/4\mathrm{NA}$) is assumed. The vertically incident beam A (broken line) no longer contributes to imaging because both 1st-order diffraction beams are cut off by the NA of the projection lens. The 0th-order diffraction beam from incident beam A cannot take part in the imaging and merely lowers aerial-image contrast and so diminishes resolution. From the incident beam B, in contrast, even for a small pattern, two diffraction beams (i.e., the 0th-order diffraction and one of the ±1st-order diffractions) can go through the projection-lens pupil and form aerial images on the wafer plane. Thus, resolution is enhanced by oblique-illumination technology.

The configuration in which the photomask is illuminated only by obliquely incident beams is set up by putting a shade at the center of the secondary illumination source of the projection system. To optimize the secondary illumination source shape, the diffraction distribution at the projection-lens pupil plane should be considered. Distributions of 0th-order and 1st-order diffractions, from periodic patterns, obtained by several illumination technologies are shown in Figure 2.38(a)–(d).

In the case of normal illumination (Figure 2.38 (a)), a reduction of the pattern size leads to an increase of background light, which worsens the resolution because the background does not contribute to the imaging and just lowers the image contrast. Furthermore, even for large patterns, the three-beam interference that occurs in the center portion of the 0th-order diffraction vitiates against the depth of focus. Therefore, the shade should be set to cut off the background and the three-beam-interference portion.

Figure 2.38(b) shows annular illumination, for which a circular shade is set at the center of the secondary illumination source. Figure 2.38 (c) and (d) show dipole and quadrupole types of modified illumination, for which a spindle-type shade having the curvature of the projection lens NA is set at the center of the secondary illumination source. The dipole type is suitable for a one-dimensional pattern configuration and the quadrupole type is suitable for an orthogonal pattern configuration. The background and the three-beam-interference portion are more effectively eliminated by modified illumination than by annular illumination, so we expect resolution and depth of focus to be improved more with modified illumination.

This has been confirmed experimentally. Figure 2.39 shows the dependence of the CD-focus characteristics of a 0.25-μm L/S pattern on the illumination condition (when $\lambda = 248$ nm, NA = 0.55, and $\sigma = 0.8$). Filled triangles and squares respectively represent the results obtained by annular illumination ($\sigma_{in}/\sigma_{out} = 0.4/0.8$) and quadrupole-type modified illumination (Nikon Shrinc). Assuming that the targeted CD is set up as ±10% of designed value, the depths of focus for a 0.25-μm L/S pattern,

Pattern \ Illumination	(a) normal	(b) annular	(c) modified	(d) modified
		shade	shade	shade
Large				
Medium				
	background light			
Small				
	-1st 0 +1st	-1st 0 +1st	-1st 0 +1st	-1st 0 +1st

Figure 2.38 Distribution of diffraction in the projection-lens pupil plane.

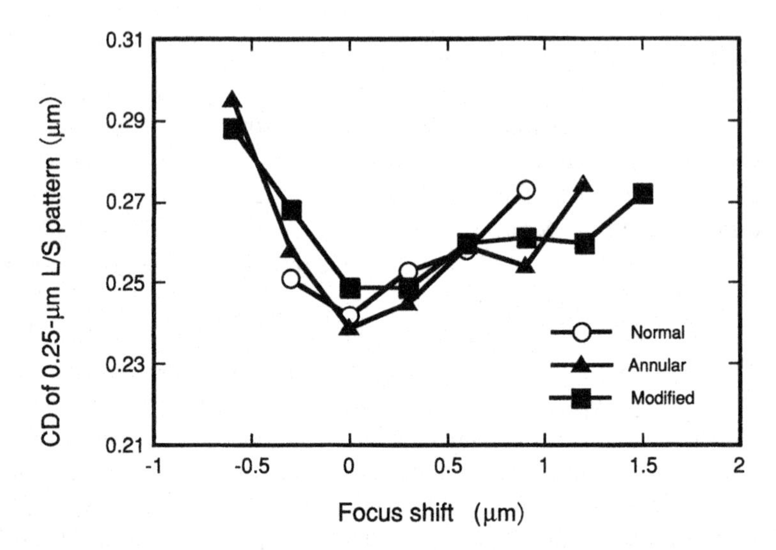

Figure 2.39 Improvement of depth of focus for 0.25-µm L/S patterns by oblique-illumination technologies ($\lambda = 248$ nm; NA $= 0.55$; $\sigma = 0.8$).

obtained by normal, annular, and quadrupole-type illuminations are respectively 1.2, 1.6, and 2.0 µm. Quadrupole-type modified illumination almost doubles the depth of focus for 0.25-µm L/S patterning.

But oblique-illumination technology has some disadvantages. Compared to normal illumination, oblique illumination decreases illumination power and uniformity. This means that oblique illumination lowers productivity and CD controllability. Modified illumination also has a special problem in that when patterns are placed obliquely to aimed orthogonal coordinates the imaging performance is scarcely improved and may instead be degraded.

So, with oblique-illumination technology there is a trade-off between an improvement in resolution and depth of focus of some specific patterns and the problems mentioned above. Its application to actual device fabrication will require optimization of the secondary illumination source shape, considering the objective pattern configuration as well as the required imaging performance and productivity.

Furthermore, it should be noticed that an improvement of imaging performance by obilque-illumination technology is limited to periodic patterns; isolated and outermost patterns are scarcely improved.

2.6.3 Phase-shifting mask technology

2.6.3.1 *Spatial-frequency modulation type*

The Levenson-type phase-shifting mask is representative of the spatial-frequency modulation type. The principle is shown in Figure 2.40. By placing a 180-degree phase shifter in every other opening of periodic patterns, the spatial frequency of a mask pattern near the resolution limit achievable by normal illumination technology is halved. As a result, these patterns can be resolved easily. This situation is shown more clearly in Figure 2.41, in which the generation of diffraction beams is represented schematically. The use of a Levenson-type phase-shifting mask eliminates 0th-order diffraction and halves the diffraction angles. Elimination of 0th-order diffraction is equivalent to depth of focus enlargement because two-beam interference becomes dominant. Halving the diffraction angles means that reso-

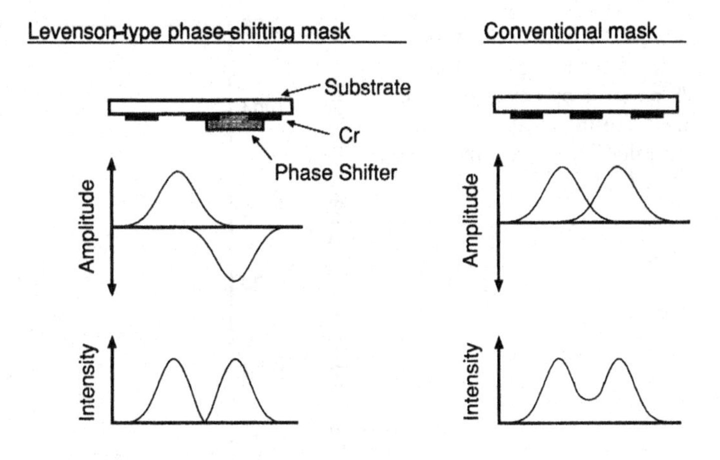

Figure 2.40 Principle of Levenson-type phase-shifting mask.

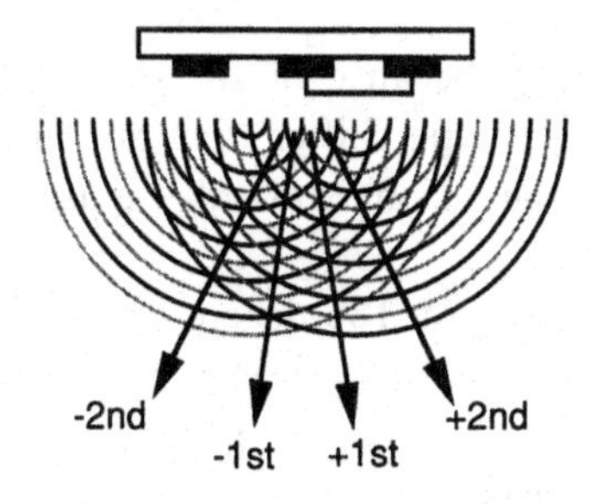

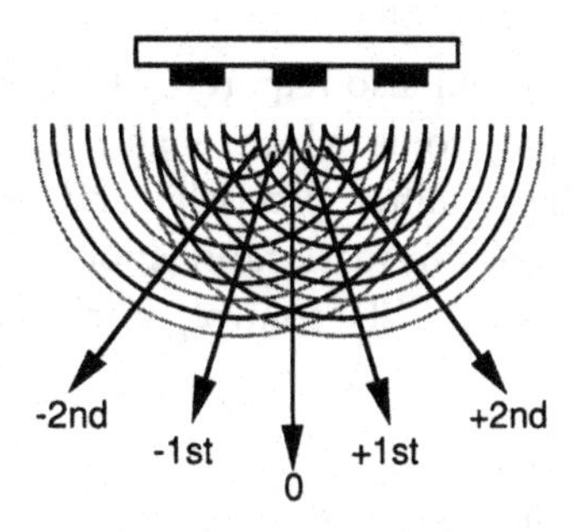

Figure 2.41 Schematic representation of the generation of diffraction. (a) Levenson-type phase-shifting mask. (b) Binary-intensity mask.

lution is enhanced, theoretically to a level twice as fine as that obtained by normal illumination technology.

Enlargement of the depth of focus has been confirmed experimentally. Figure 2.42 shows the dependence of CD-focus characteristics for a binary-intensity mask ($\sigma = 0.8$) and for a Levenson-type phase-shifting mask ($\sigma = 0.2$). Assuming that the targeted CD is set as $\pm 10\%$ of the designed value, the depth of focus for 0.20-μm L/S patterns obtained by binary-intensity mask exposure is 0.6 μm and that obtained by Levenson-type phase-shifting mask exposure is 2.4 μm. This result shows that Levenson-type phase-shifting mask technology is one of the most attractive candidates for 0.20-μm lithography.

But there are some disadvantages to this technology. As with oblique-illumination technology, imaging improvement is limited to periodic patterns while the imaging of isolated and outermost patterns is degraded. Furthermore, the dependence of depth of focus on pattern duty (linewidth-to-pitch ratio in L/S patterns) cannot be disregarded. Therefore, some protective measures should be taken in order to use Levenson-type phase-shifting mask technology for actual device fabrication. Implementation

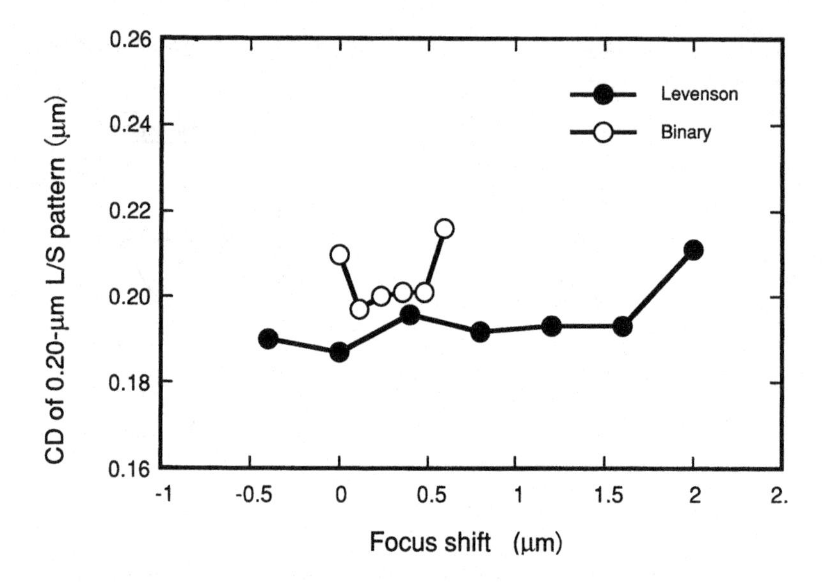

Figure 2.42 Improvement of depth of focus for 0.20-μm L/S patterns by Levenson-type phase-shifting mask technology ($\lambda = 248$ nm; NA = 0.55).

of this technology will also require CAD (computer-aided design) tools which enable us to position the phase shifter precisely, and will also require progress in mask-fabrication technology such as mask CD control, inspection and repair.

2.6.3.2 *Edge-enhancement type*

Improvement of the aerial image contrast around the mask-edge positions is the basic principle of edge-enhancement-type phase-shifting mask technology. This technology can be roughly divided into three types: (1) auxiliary pattern, (2) rim, and (3) attenuated. The improvement of imaging performance derived from this technology is inferior to that derived from spatial-frequency-modulation-type phase-shifting mask technology, but edge-enhancement phase-shifting mask technology is more likely to be practical because conventional positive-resist processing can be used. Furthermore, there is less restriction on shifter layout. Among these technologies, the attenuated type of phase-shifting mask technology is the most attractive and has already been used practically in i-line lithography because of the development of a single-layer-type shifter that can be fabricated using the conventional mask-fabrication techniques.

By placing a 180-degree phase shifter with less-than-10% transparency in the same configuration as the opaque patterns in a conventional mask, improvement of the aerial image contrast around the mask-edge position can be improved, as shown in Figure 2.43. The light phase that goes through the absorptive phase-shifter portion is shifted 180 degrees and has a negative amplitude relative to the 0th-order diffraction generated at mask openings. Therefore, light intensity, which is given by the square of light amplitude, becomes zero around the mask-edge position. As a result, aerial image contrast is improved. Improved aerial image contrast prevents the quality deterioration due to deviation from the optimum focus and exposure conditions. Thus, exposure and focus latitudes are enlarged. According to the change in the aerial image intensity profile, mask pattern size, which gives maximum depth of focus, varies from the designed value. This deviation is called mask biasing, the value of which is one of the most

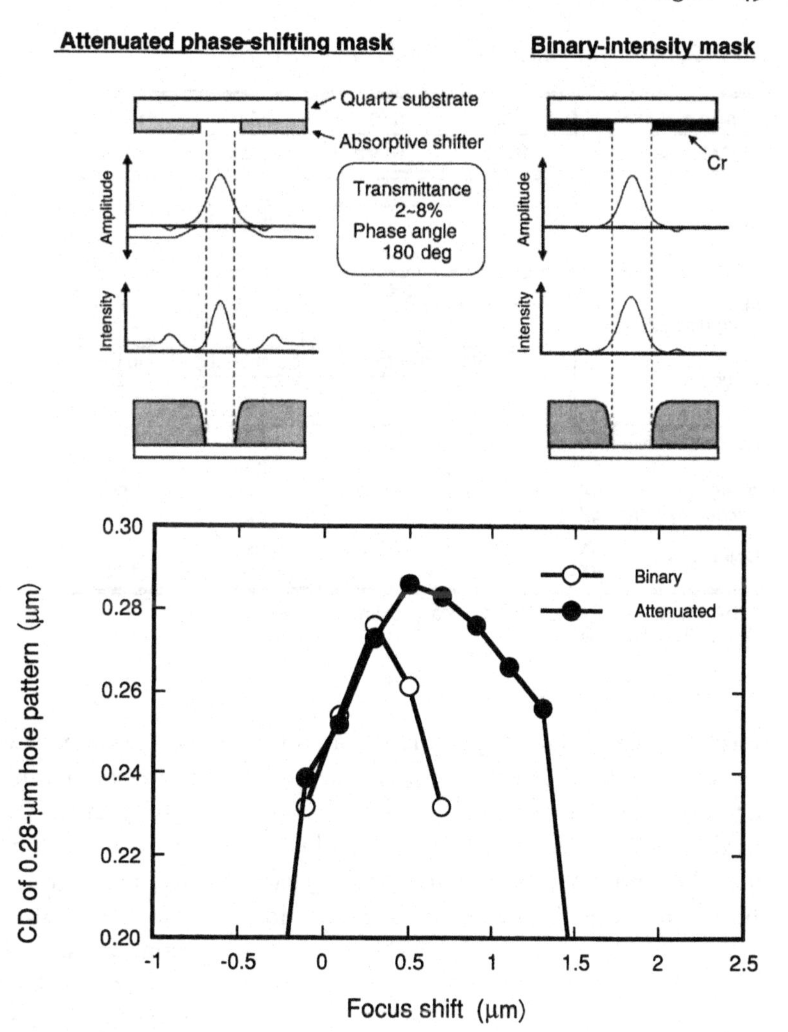

Figure 2.43 Principle of attenuated phase-shifting mask technology.

Figure 2.44 Improvement of the depth of focus for 0.28-µm hole pattern by attenuated phase-shifting mask technology ($\lambda = 248$ nm; NA $= 0.55$; $\sigma = 0.4$).

important parameters of the attenuated phase-shifting mask.

Enlargement of depth of focus has been confirmed experimentally. Figure 2.44 shows the effect of an attenuated phase-shifting mask on the CD-focus characteristics of a 0.28-µm hole pattern ($\lambda = 248$ nm, NA $= 0.55$, $\sigma = 0.4$). Open and filled circles respectively represent the results obtained by a binary-intensity mask and an attenuated phase-shifting mask consisting of MoSiON single-layer shifters with a transparency of 6% ($\lambda = 248$ nm). Assuming that a CD-error of down to -10% of the designed value is acceptable, the depth of focus for a 0.28-µm hole pattern obtained by a binary-intensity mask is 0.5 µm and that obtained by an attenuated phase-shifting mask is 1.2 µm.

With an attenuated phase-shifting mask, the higher the transparency and the smaller the biasing, the larger the depth of focus that can be obtained. But at the same time, the higher transparency and smaller bias increase the side-peak intensity, leading to a decrease in

Table 2.14 *Requirements for masks and their trend.*

Design rule (nm)		500	350	250	180
DRAM generation		16Mb	64Mb	256Mb	1Gb
Source		i-line	i-line	i-line	
				DUV (KrF)	DUV (KrF)
					DUV (ArF)
Mask magnification		5X	5X	5X	
				4X	4X
Structure		Halftone PSM	Halftone PSM	Alternate-shifter-type PSM	Alternate-shifter-type PSM
				OPC	OPC
CD control (nm)	5X	±70	±50	±35	±25
	4X			±30	±20
Position accuracy (nm)	5X	±120	±100	±70	±50
	4X			±50	±40
Defect size (nm)	5X	<500	<350	<250	<180
	4X			<200	<150

PSM: Phase-shift mask; OPC: Optical proximity correction;
CD: Critical dimension

resist thickness and to pattern deformation. Therefore, considering the required depth of focus and the tolerance level of pattern deformation, both mask transparency and biasing must be optimized before attenuated phase-shifting masks can be used in device fabrication.

2.7 Mask/reticle technology

2.7.1 Introduction[76,77]

The importance of photomasks and reticles (both referred to hereafter as masks) in the semiconductor industry is growing. Table 2.14 lists requirements for masks and shows their trend over four design-rule or DRAM 'generations'. Exposure wavelength has been getting shorter: from the i-line (365 nm) to the wavelengths of deep-UV sources such as the KrF-excimer laser (248 nm) and the ArF-excimer laser (193 nm). Mask magnification will get smaller from 5× to 4× to cope with an increasing chip size. New structures or features such as phase-shift masks (PSMs) and optical proximity correction (OPC) are being introduced. Typical specifications on critical dimension (CD) control, position accuracy, and minimum intolerable defect size are listed. Note that even within the same DRAM 'generation' the design rule gets smaller, leading to tighter mask specifications. A similar list of mask specifications has been published by the Semiconductor Industry Association in the US.[78,79]

2.7.2 Characteristics of blanks and masks

Six-inch (152-mm)-square, 0.25-inch (6.35-mm)-thick quartz plates are commonly used in steppers for both i-line and KrF exposure. Unless otherwise stated, the term 'plates' used here refers to this plate size. The use of 9-inch (230-mm)-square plates is also being considered for future scanner application.

A mask blank consists of an opaque film made up of one to three layers on a quartz substrate. Quartz is used as a substrate material because of its high transmittance at wavelengths below the i-line, its very low thermal expansion coefficient (0.52×10^{-6}/K), and its chemical stability. The substrate is finely polished so that it has a flatness of 0.5 µm over the whole area except within 5 mm of the edges. Its edges are cut round and polished so that the film will not easily peel off from the edges.

The opaque film has to have an optical density > 3: this means that the transmittance of the light must be less than 0.1%. It should also have the following properties: (1) high chemical stability (i.e., resistance especially to sulfuric acid and ammonia solutions), (2) high durability against irradiation, (3) strong adhesion to the substrate, (4) moderate electrical conductivity (so that charging-up during electron-beam (EB) writing can be avoided), and (5) easiness of forming and of patterning. Of the materials with those properties, chromium and its compounds are those most widely used.

The Cr monolayer has high reflectance on both the film side and the glass side. The high reflectance of the film side causes trouble because of stray light on the wafer, whereas the high reflectance on the glass side may cause low contrast between the alignment mark on the a wafer and that on the mask. Therefore, mask structures that result in low reflectance at the exposure wavelength (or alignment wavelength) have been developed and such film-side low-reflectance masks have become standards. Reflectance is lowered by adding an anti-reflection layer (such as CrO_xN_y) using the thin-film interference effect as shown in Figure 2.45.

2.7.3 Manufacturing process

Figure 2.46 shows a typical process flow of mask manufacturing. The details of each process step and key points in the manufacturing are as follows.

2.7.3.1 *Resist patterning*[80]

EB writers have been mainly used in production, but recently laser writers have also taken up an important role. Table 2.15 lists the specifications of three types of advanced mask writers. Note that the acceleration voltage in vector EB writers is being increased from 20 kV to 50 kV in order to increase position accuracy and reduce the beam size. As is evident from Table 2.15, the writers listed meet the needs for masks for 0.35-µm design rule devices, and almost meet the need for those for 0.25-µm design rule devices. Resists with target thicknesses of 300–500 nm are spun on the blanks with an accuracy of 3–5 nm except within 5 mm of the edges. Both positive and negative tone resists are used depending on the pattern types. Spin-spray development, rinsing, and descumming follow the exposure.

2.7.3.2 *Etching*

Wet-etching is used most, but dry-etching is gradually taking the place of wet-etching. The most commonly used wet-etchant of chromium

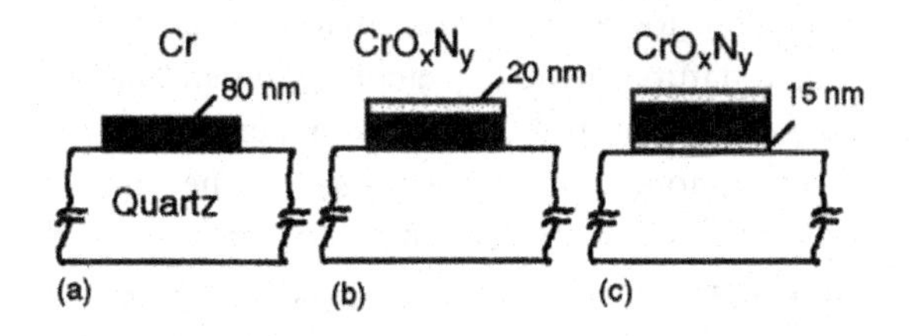

Figure 2.45 Structure of binary-intensity mask blanks: (a) mono-layer, (b) bi-layer, (c) tri-layer.

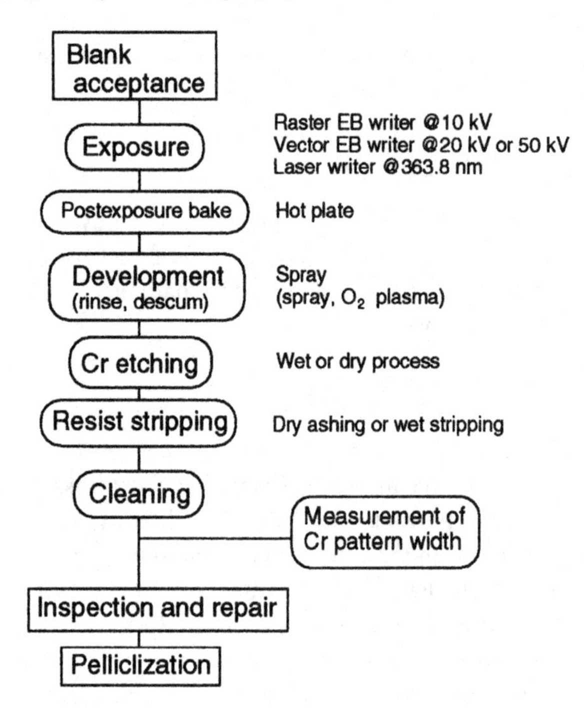

Figure 2.46 Process flow in mask manufacturing.

is an acid aqueous solution of ceric ammonium nitrate, $(NH_4)_2Ce(NO_3)_6$, and perchloric acid, HClO. A surfactant is often added to increase the wettability of the blank surface. The reaction scheme is as follows:

$$3Ce^{4+} + 3e^- \rightarrow 3Ce^{3+}$$

$$\underline{Cr \rightarrow Cr^{3+} + 3e^-}$$

$$Cr + 3Ce^{4+} \rightarrow Cr^{3+} + 3Ce^{3+}$$

Spin-spray etching is used most. Termination of etching is controlled by adjusting the etching time. Over-etching is common in order to assure the complete removal of the Cr layer over the whole plate. Over-etching time is 20–50% of just-etching time. Since wet-etching proceeds isotropically, the amount of side-etching can be as much as 120–180 nm per side.

The most commonly used dry-etching technique is reactive ion etching, which uses a glow discharge plasma of a gas excited by a radio-frequency electric field between two parallel plates.[81] The reaction of Cr or CrO with O^* and Cl^* yields a volatile compound, chromyl chloride CrO_2Cl_2:

$$Cr + 2O^* + 2Cl^* \rightarrow CrO_2Cl_2$$

$$CrO + O^* + 2Cl^* \rightarrow CrO_2Cl_2$$

Therefore, a mixture consisting of O_2 and a Cl source such as Cl_2, CCl_4, $CHCl_3$, CH_2Cl_2, or CH_3Cl should be used for etching Cr or CrO. Ions produced in the plasma also assist the above reaction by sputtering and activating the sample surface. Termination of etching is again controlled by adjusting the etching time. Operators often determine the over-etch time by observing the procedure and finding the just-etch time for each mask. Owing to the anisotropic nature of the dry-etching, the amount of side-etching is only 20–40 nm per side. This small amount of side-etching is one of the advantages of dry-etching. Dry-etching will play a major role in the tighter CD control

Table 2.15 *Specifications of three types of advanced mask writers.*

Type: *Beam*	Electron	Electron	Laser
Type: *Scanning*	Raster	Vector	Enhanced raster
Type: *Beam shape*	Gaussian point beam	Shaped beam	32 Gaussian point beams
Beam source	Thermal field emission gun	LaB_6 gun	Ar ion laser
Acceleration voltage/wavelength	10 kV	50 kV	363.8 nm
Maximum plate size (in × in) [mm × mm]	7 × 7 [180 × 180]	7 × 7 [180 × 180]	9 × 9 [230 × 230]
Writing strategy	4-off set pass writing		4-offset pass writing
Resolution (minimum feature size (nm))	400	300	500
Address range (nm)	15–275	10–2500	1000
minimum increment (nm)	25	10	5
Position accuracy (nm)	30 (3σ)	40 (3σ)	35 (max)
Overlay accuracy (nm)	30 (3σ)	30 (3σ)	25 (max)
Stripe butting (nm)	25 ($+3\sigma$)	30 ($+3\sigma$)	15 (+range/2)
Linewidth (CD) control ($\bar{x}$-target) (nm)	35 (max)	30 (3σ)	25 (max)
Linewidth (CD) uniformity (max-$\bar{x}$)	35	30	25

$\bar{x}$: mean

necessary for masks used to make 0.25-µm-rule or smaller devices.

2.7.3.3 *CD control*

CD control is evaluated by measuring the sizes of several monitor patterns and/or main patterns and calculating the deviation from the target size. A typical target size is 1.5–2.0 µm. Pattern sizes are measured with transmitted light or reflected light.

Three optional control methods have been used to improve CD control (Figure 2.47). In all of them the width of a resist pattern or a Cr pattern is measured and used for adjustment of the related process. When control method 3 is used, the width of the Cr pattern should be measured under a resist pattern. However, this is impossible from the pattern side. This problem has recently been solved by measuring the Cr edges from the glass side using a newly developed confocal optical microscope.[82] Figure 2.48 shows a result obtained when control method 2 was used. This result is sufficient for the masks needed to make 0.25-µm design rule devices.

2.7.3.4 *Position accuracy*

Position accuracy represents correctness of the pattern placement in a single mask and among a set of masks and is also called placement or registration accuracy. It is expressed in terms of the (maximum) deviation of the coordinates of monitor patterns for their ideal (or absolute) coordinates. The coordinates of the monitor patterns are measured optically with an *x*–*y* coordinate-measurement system equipped with a laser interferometer. The accuracy of measurement of advanced systems has reached 4–6 nm (3σ). Position accuracy is often degraded in pattern writing owing to deformation of the plate and inaccurate beam positioning. The example of position-accuracy results in Figure 2.49 shows that the result is sufficient for 0.35-µm-rule devices (see Table 2.15).

2.7.3.5 *Inspection*[83]

Mask defects are classified as 'soft' or 'hard'. Soft defects are any that can be removed by a cleaning process, such as particles and contamination, whereas hard defects cannot be removed.

Various types of hard defects that occur in contact-hole patterns are shown in Figure 2.50, and they can be grouped according to geometry: (1) extra geometry, such as pinspots (or black spots), extrusions (or protrusions), and bridges; (2) missing geometry, such as pinholes and intrusions; and (3) the deviation of pattern width or geometry from the design.

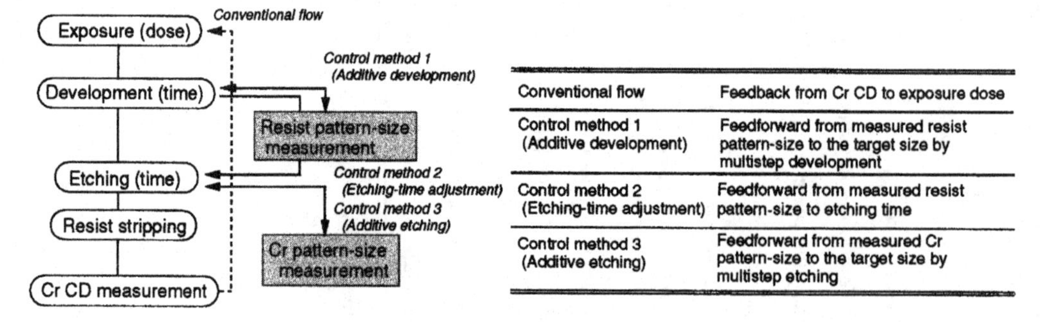

Conventional flow	Feedback from Cr CD to exposure dose
Control method 1 (Additive development)	Feedforward from measured resist pattern-size to the target size by multistep development
Control method 2 (Etching-time adjustment)	Feedforward from measured resist pattern-size to etching time
Control method 3 (Additive etching)	Feedforward from measured Cr pattern-size to the target size by multistep etching

Figure 2.47 Process flow based on three control method options.

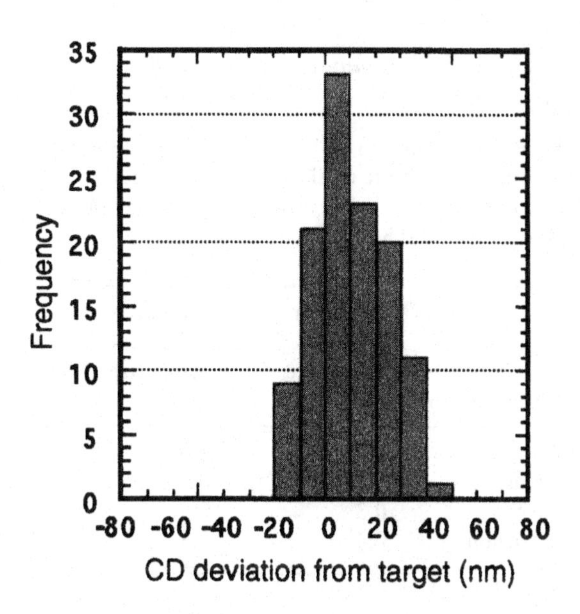

Figure 2.48 Improved CD control by etching-time adjustment; control method 2. (Number of measured plates: 17; Number of measurement points for each plate: 7.)

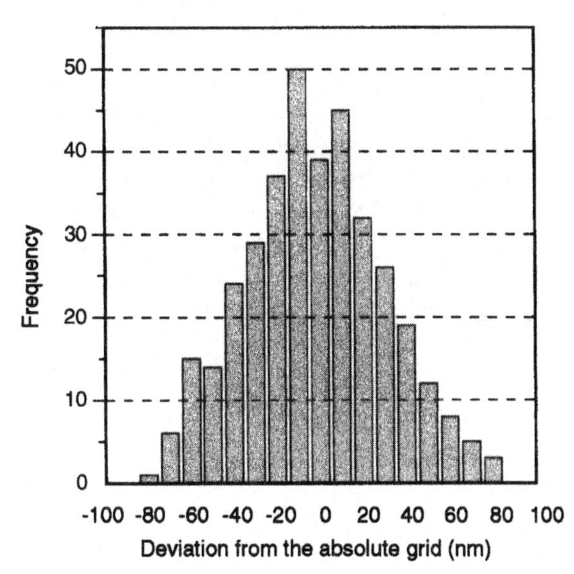

Figure 2.49 Pattern-position accuracy for a reticle.

The effect of a defect on the pattern size depends on mask type, pattern type and printing conditions. For each type of defect, a printability test can determine the minimum intolerable defect size, which is defined as the size at which the deviation of the pattern size caused by the defect is 10% of the normal pattern size. One example of the minimum intolerable defect size for a contact-hole layer of halftone phase-shift masks printed at the i-line is listed in Table 2.16. The printing conditions are: 0.4-μm hole on wafer (i.e., 2-μm hole on mask), and NA = 0.57. Recently, a microscope that can simulate the characteristics of a particular optical stepper[84] has been developed. It is useful to evaluate the printability of the pattern of interest.

The inspection can be done in a die-to-die mode or die-to-database mode (see Section 7.3) and the speed and sensitivity are greater in the former than in the latter. Both modes can be used for one mask. For a typical DRAM mask with a multi-die layout, main memory areas are inspected in a die-to-die mode and the peripheral area is inspected in a die-to-database mode. Operators have to classify the defects detected by the inspection tool by observing them in a review mode.

The sensitivity of an inspection tool can be determined using a SEMI (Semiconductor Equipment and Materials International) standard test mask with various types of defects on it. Table 2.16 lists the minimum defect sensitivity of a tool with a nominal sensitivity of 0.35 μm. Note that both the minimum detectable defect size and the minimum intolerable defect size are larger for a pinhole than those for the other categories. As long as the former is smaller than the latter, however, this will not cause trouble. The sensitivity of advanced inspection tools is 0.2 μm in a die-to-die mode, and this is sufficient for inspection of the masks for 0.25-μm-rule devices.

Table 2.16 *Minimum defect size to be inspected.*

Hole size: 0.4 μm on wafer, 2.0 μm on mask; Mask: halftone phase-shift mask; Printing conditions: $\lambda =$ 365 nm, NA = 0.57; Sensitivity of the inspection tool (specification): 0.35 μm.

Defect category	Minimum detectable defect size (μm)	Minimum intolerable defect size (μm)
Pinhole	0.8	1.0
Intrusion	0.4	0.5
Pinspot	0.3	0.4
Extrusion	0.4	0.5

Soft defects consist of foreign materials such as metallic particles and organic materials, and the impact of a soft defect on printing varies according to its position on a mask as shown in Figure 2.51.[85] Soft defect A is the most effectively printed; soft defects B and C may work as A in rare cases when they move onto the glass on the film side; and soft defects D and E are the

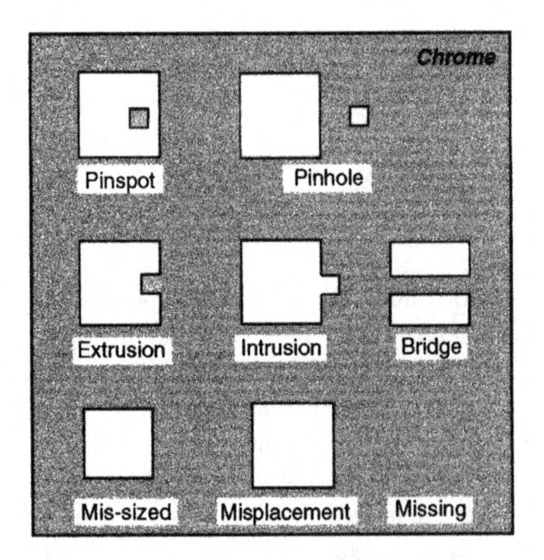

Figure 2.50 Various types of hard defects on contact-hole patterns.

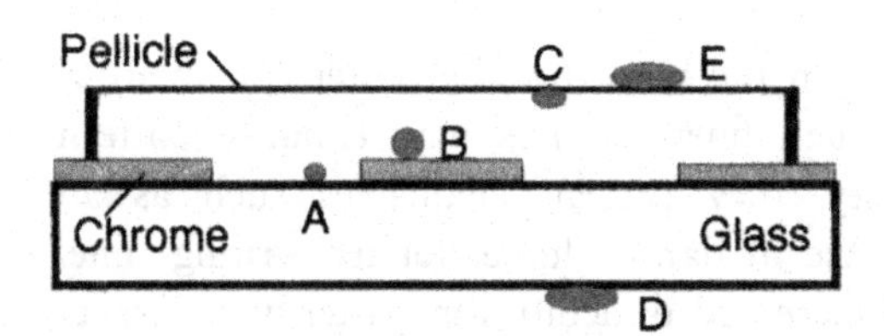

Soft defect	Location	Minimum intolerable defect size (μm)
A	Glass on film side	0.5
B	Chrome	2
C	Under pellicle	2
D	Glass side	20
E	Top of pellicle	20

Figure 2.51 Soft-defect specifications of pelliclized masks for 0.35-μm design rule devices.

least effectively printed because they are out of focus. Soft defects are inspected optically, that is, by illuminating the mask with a light and detecting the intensity of the light scattered by them.

2.7.3.6 *Repair*[84]

Hard defects are, according to their types, repaired by either a laser repairer (or laser 'zapper') or a focused-ion-beam (FIB) repairer (see Section 7.4). Extra geometry is removed by a laser repairer which irradiates the defect with a shaped laser beam at a wavelength of 530 nm. The removal of the opaque Cr layer is based on both thermal evaporation and ablation. Since the quartz substrate is transparent to the repairing light, it is not damaged. This type of tool is widely used. As the features that are to be repaired are getting smaller, however, further improvements in optical imaging, beam shape, and beam placement accuracy are necessary.

Missing geometry is repaired by a FIB repairer that uses Ga^+ ions. The missing area is covered with an opaque patch of carbon film that is deposited from an organic molecule like pyrene by irradiation. Since the image of a Cr pattern is clearly observed by a scanning ion microscope (SIM) and the position of the ion beam can be controlled easily, the position accuracy of the beam is quite high. Therefore, in order to take advantage of high-precision FIB technology, its application to removing extra geometry is being investigated.

2.7.3.7 *Cleaning*[86]

The cleaning process is getting more important with the decreasing size of intolerable soft defects. Soft defects can be removed by a cleaning process that includes various operations such as immersion in heated sulfuric acid, a brush scrub, and a scrub with high-pressure water. These operations are performed in an automated processor.

2.7.3.8 *Pelliclization*[86]

After the mask is inspected, repaired, and cleaned, a pellicle is mounted on it. A pellicle is composed of a thin, transparent, self-supporting film and a 6.3-mm-high frame that holds the film at its edges (see Figure 2.51). Its role is to prevent foreign materials (mainly particles) from attaching to the film side of a mask. Since the pellicle stays 6.3 mm above the mask surface, the foreign materials (if any) on it are not printed because they are out of focus – unless they are as large as 20 µm. The film is made such that its transmittance at the exposure wavelength is almost 100% and, also, such that it is quite resistant to a high dose of irradiation.

2.7.4 Optical proximity correction[77,86]

When a feature to be printed on a wafer is smaller than one-quarter of the resolution limit, or when two adjacent features are that close to each other, the pattern fidelity is greatly degraded because of the strong interference of the light diffracted by the features. This phenomenon is called an optical proximity effect. Original patterns have to be corrected to obtain the designed printed image. Sizing of patterns has often been employed as a correction. For example, a line end is elongated since a printed line gets shorter than its design. Recently, however, more complex features such as serifs, jogs, and sub-resolution features like those shown in Figure 2.52 (in both positive and negative tones) have been employed as correction elements. Such a correction is called an optical proximity correction (OPC), and this technique is thought to be indispensable for masks for 0.25-µm- and smaller design rule devices. Figure 2.53 shows an example of OPC to a line end. This kind of correction is optimized with the help of experiment or simulation with respect to the aerial image or the printed image in the resist for the pattern of interest. Although finer features may enhance pattern fidelity, they present challenges such as an increase in data volume, longer writing time, and increased difficulties in patterning, inspection, and repair. Therefore, optimization is quite important.

2.7.5 Phase-shift mask

Phase-shift technology is one of the resolution-enhancement technologies that have been introduced recently. It takes advantage of the destructive interference of two light beams 180 degrees out of phase and provides higher resolution and a larger depth-of-focus (DOF). Figure 2.54 shows the structures of five typical types of phase-shift masks (PSMs). The thickness d of the shifter is so designed that the phase difference between the light beam passing through the clear areas and the light beam passing through the adjacent shifter areas is to be 180 degrees. This thickness d should satisfy the relation $d = \lambda/[2(n-1)]$, where λ is the wavelength of the exposure light and n is the refractive index of the shifter at this wavelength.

2.7.5.1 *Alternate-shifter PSMs*

An alternate-shifter (Levenson[27]) phase-shift mask provides the full potential of PSMs, such as higher resolution and a larger DOF latitude as described in Subsection 2.6.3.1. Figure 2.55

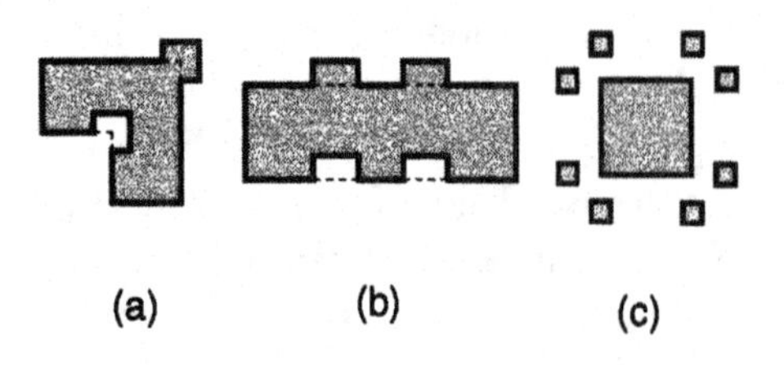

Figure 2.52 Three types of features used as elements in optical proximity correction: (a) serif, (b) jog, (c) sub-resolution.

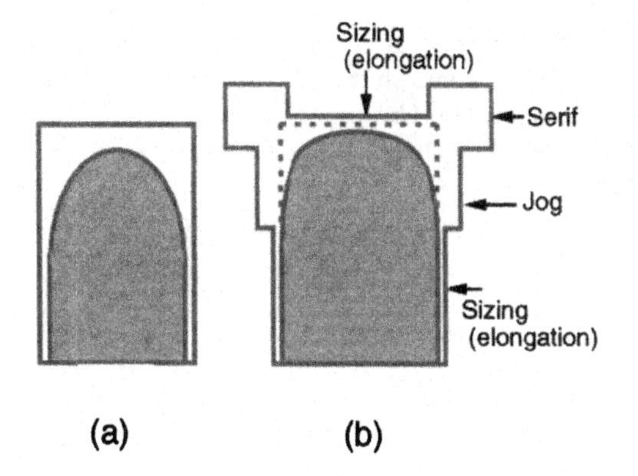

Figure 2.53 Optical proximity correction to a line end: (a) without optical proximity correction, (b) with optical proximity correction (Dotted line: target pattern; Solid line: designed pattern; Shaded area: resist pattern).

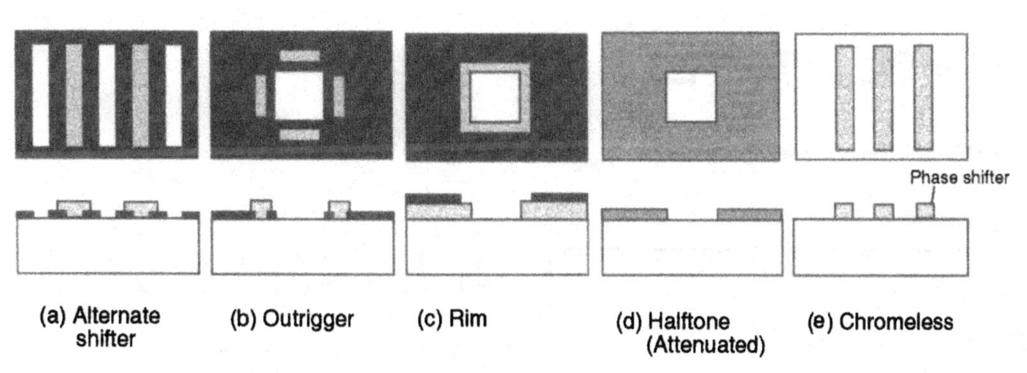

Figure 2.54 Five types of phase-shift masks. Plan view and cross-section showing each type of structure.

shows three types of alternate-shifter PSM structures. The etch-stop layer may or may not be used. In the structures shown in (a) and (b), spin-on glass was used as a shifter material. In DUV lithography, however, the use of spin-on glass makes defect control difficult and results in masks that are easily damaged by DUV irradiation. Note that in (b) and (c) the Cr patterns have undercuts so that the effect of the side wall of the shifter is reduced. The etched-quartz type of shifter is thought to be a promising candidate mask for DUV lithography and the object of intense development efforts.

Although an alternate-shifter PSM, type (a), has been applied to i-line lithography for 0.30-μm-rule memory devices,[87] its application is not so easy owing to several difficulties. One is in designing because of the limitation that any two adjacent spaces should have 180-degree phase difference. Another is in constructing a three-dimensional structure. Other difficulties are in inspection and in the repair of phase defects. Since this type of mask is expected to be used widely in DUV lithography, efforts to overcome these difficulties are being made.

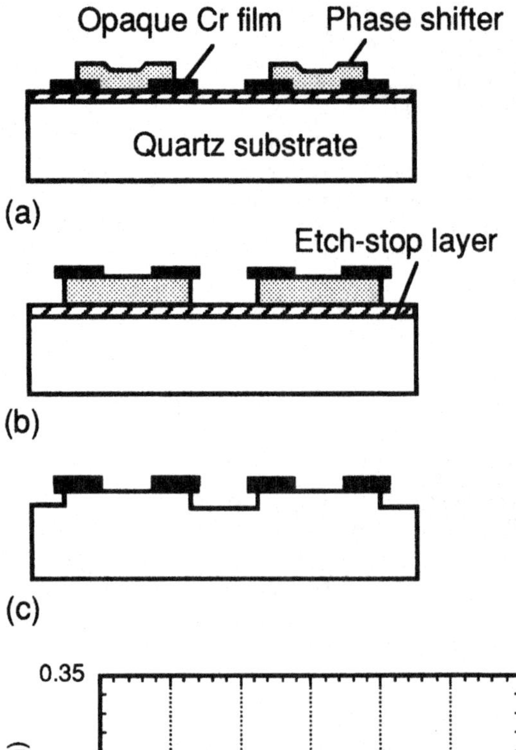

Figure 2.55 Structure of alternate-shifter phase-shift masks: (a) shifter-on-Cr, (b) Cr-on-shifter, (c) etched-quartz shifter.

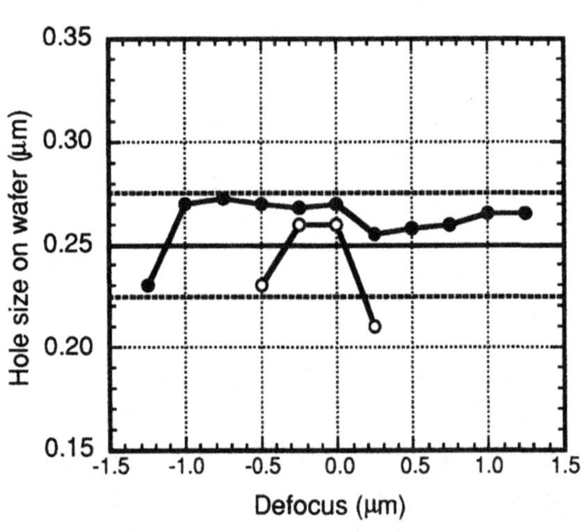

Figure 2.56 Comparison in depth-of-focus between a binary-intensity mask (open circles) and a halftone phase-shift mask (filled circles).

2.7.5.2 *Halftone (or attenuated) PSMs*

A halftone (or attenuated) PSM is one in which the shifter is semitransparent; the transmittance at the wavelength of the exposure light is 4–8% (air or quartz reference). The principle of the phase-shift effect is illustrated in Figure 2.43. Exposure dose has to be set properly to avoid the influence of the sidelobes of the intensity. Note that the space printed on the resist is smaller than the space on the mask (scaled on wafer). This mask is effective for printing isolated features, especially contact holes. As shown in Figure 2.56, the DOF for a halftone PSM is about twice that for a binary-intensity mask.

The greatest advantage of halftone PSM in mask making is that its structure, processes, and quality-assurance tools are almost the same as those for a binary-intensity mask. Therefore, quality-assured halftone PSMs for i-line lithography are routinely manufactured. The special feature of the phase shifter of a halftone mask is that it works as both a shifter and an attenuator. This means that the shifter material has to have a suitable refractive index and extinction coefficient at the exposure wavelength. The material should also have the film properties necessary for it to be well-patterned on a mask (see Subsection 2.7.2). A dozen candidate materials have been reported so far, and SiN,[88] MoSiON,[75,89] and CrON[75,89,90,91,92,93] are used in masks for i-line lithography. Recently, SiN,[94] MoSiON,[95] and CrF[96] have been reported to be promising shifter materials for KrF lithography. Research on a new shifter material for ArF lithography is under way. Figure 2.57 shows the structure of a CrF-based tri-layer shifter for KrF lithography and also shows its transmittance spectrum. One reason for using the multilayer structure is to lower the transmittance at an inspection wavelength (488 nm) below 20% so that defect inspection tools can work well. For SiN and MoSiON, on the other hand, a mono-layer structure is used.

For halftone PSMs, a dry-etching process is necessary because of the poor results obtained when wet-etching the shifters mentioned above. CrON and CrF shifters and MoSiON shifters are dry-etched based on Cl-plasma chemistry (see Subsection 2.7.3.2) and F-plasma chemistry, respectively. A cross-sectional SEM photograph of the hole pattern of a CrF-based shifter

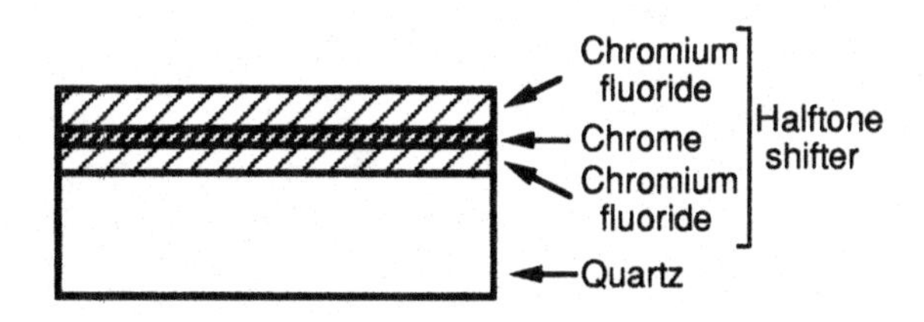

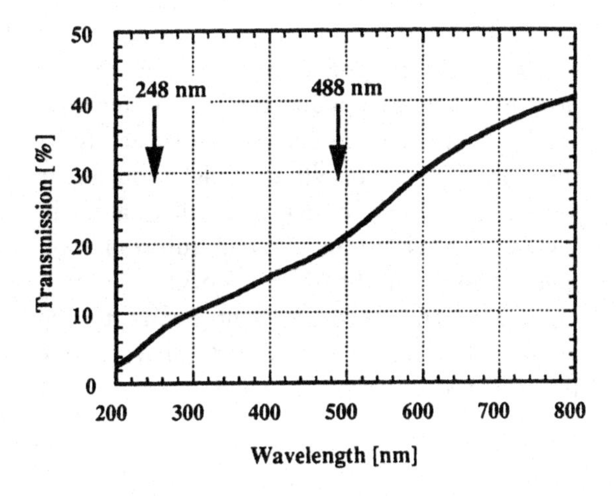

Figure 2.57 Structure and transmittance spectrum of a halftone shifter (Shifter: CrF; Application: KrF lithography; Inspection wavelength: 488 nm).

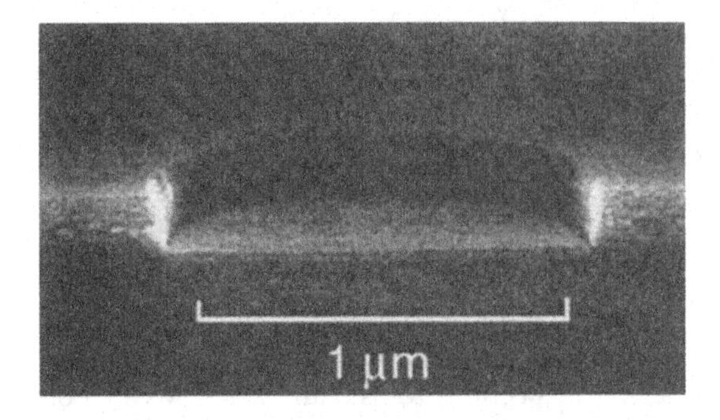

Figure 2.58 Cross-sectional SEM photograph of a CrF-based shifter pattern (Pattern: contact hole; Application: KrF lithography).

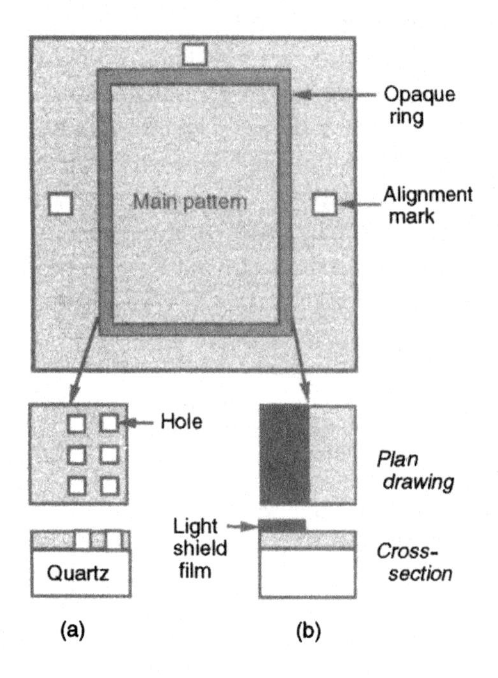

Figure 2.59 Two types of opaque rings: (a) shifter type, (b) additional-film type.

is shown in Figure 2.58, where it is evident that the side wall of the shifter is nearly vertical.

The two new items of quality assurance for a halftone mask are transmittance and phase difference. Typical specifications for them are 8 (or 4) $\pm$ 0.5% (air or quartz reference) and 180 $\pm$ 5 degrees for both i-line and KrF lithography. Transmittance can be measured with a spectrometer, whereas phase difference can be measured directly with newly developed tools.[97,98]

One problem specific to a halftone PSM is how to make an opaque ring (or a ring that works as an opaque ring) in order to avoid the multi-exposure effect. This problem is solved by the two methods illustrated in Figure 2.59: either an array of sub-resolution holes formed in the shifter works as a ring opaque to the exposure light owing to the phase-shift effect,[88] or an additional opaque film is formed on the shift layer.[99]

2.7.5.3 *Others*[77]

A chromeless PSM has a special application in printing a very narrow single line (or space) for a gate-pattern. A line 100–125-nm wide can easily be formed by the use of the shifter edge. An outrigger PSM is suitable for printing contact holes. A rim PSM is useful for isolated lines although its application is being delayed because of the complexity of mask making.

Because the phase-shift effect is rather complex, the importance of both aerial-image simulation and automatic generation of mask layout is growing.

2.8 References

1. H. I. Smith *et al.*, *J. Electrochem. Soc.*, **121**, 1503 (1974)
2. T. Matsuzawa *et al.*, *J. Electrochem. Soc.*, **128**, 184 (1981)
3. B. J. Lin, *Microelectronic Eng.*, **6**, 31 (1987)
4. J. H. Bruning, *J. Vac. Sci. Technol.*, **16**, 1925 (1980)
5. B. J. Lin, *J. Vac. Sci. Technol.*, **12**, 1317 (1975)
6. H. R. Rottman, *Solid State Technology*, **18**, (No. 6), 29 (1975)
7. J. Roussel, *Solid State Technology*, **21**, 67 (1978)
8. H. L. Stover *et al.*, *Proc. SPIE*, **470**, 22 (1984)
9. N. Hasegawa *et al.*, *Symp. on VLSI Technology, Tech. Digest*, 78 (1985)
10. G. M. Dubroecq *et al.*, *Abst. Microcircuits Eng.*, **82**, 73 (1982)
11. V. Pol *et al.*, *Proc. SPIE*, **663**, 6 (1986)
12. M. Nakase *et al.*, *Proc. of SPIE*, **773**, 226 (1987)
13. M. Sasago, M. Endo, Y. Tani, K. Ogawa and N. Nomura, *IEDM Tech. Digest*, 316–19 (1986)
14. H. Nakagawa *et al.*, *Symp. on VLSI Technology, Tech. Digest*, 9 (1989)
15. R. R. Kunz *et al.*, *Proc. SPIE*, **1466**, 218 (1991)
16. R. Schenker *et al.*, *Proc. SPIE*, **2726**, 698 (1996)
17. J. D. Buckley, D. N. Galburt and C. Karatzas, *J. Vac. Sci. Technol. B*, **7**, (No. 6), 1607–12 (1989)
18. D. M. Williamson *et al.*, *Proc. SPIE*, **1088**, 424 (1989)
19. K. Suzuki, S. Wakamoto and K. Nishi, *Proc. SPIE*, **2726**, 767 (1996)
20. A. Reiser, *Photoreactive Polymers*, John Wiley & Sons, p. 22 (1989)
21. M. Hanabata *et al.*, *J. Vac. Sci. Technol. B*, **7**, (No. 4), 640 (1988)
22. J. Pakansky and J. R Lyerla, *IBM J. Res. Develop.*, **23**, 42 (1979)
23. H. Hopkins, *Proc. Roy. Soc.* (London), **A217**, 408 (1953).
24. H. Fukuda *et al.*, *IEEE Trans. Electron Devices*, **ED-38**, 67 (1991)
25. F. H. Dill *et al.*, *IEEE Trans. Electron Devices*, **EDD-26**, 445 (1975)
26. T. Tanaka *et al.*, *J. Vac. Sci. Technol. B*, **7**, (No. 2), 188 (1989)
27. M. D. Levenson *et al.*, *IEEE Trans. Electron Devices*, **ED-38**, 67 (1991)
28. S. Matsuo *et al.*, *IEDM Tech. Digest*, 970 (1991)
29. H. Fukuda, T. Terasawa and S. Okazaki, *J. Vac. Sci. Technol. B*, **9**, 3113–16 (1991)
30. V. Pol *et al.*, *Proc. SPIE*, **633**, 6–16 (1986)
31. M. Hibbs and R. Kunz, *Proc. SPIE*, **2440**, 40–8 (1995)
32. K. Yamanaka, H. Iwasaki, H. Nozue and K. Kasama, *Proc. SPIE*, **1927**, 310–19 (1993)
33. D. Williamson *et al.*, *Proc. SPIE*, **2726** (1996)
34. W. N. Partlo, S. G. Olson, C. Sparkers and J. E. Connors, *Proc. SPIE*, **1927**, 320–32 (1993)
35. T. Itani, H. Iwasaki, M. Fujimoto and K. Kasama, *Proc. SPIE*, **2195**, 126–36 (1994)
36. T. Itani *et al.*, *Proc. SPIE*, **2438**, 191–201 (1995)
37. C. A. Mack, *Proc. SPIE*, **538**, 207–20 (1985)
38. D. L. Fehrs, H. B. Lovering and R. T. Scruton, *KTI Micro-electronics Seminar*, 217–30 (1989)
39. K. Kamon *et al.*, *Jpn. J. Appl. Phys.*, **30**, 3021–9 (1991)
40. K. Tounai, H. Tanabe, H. Nozue and K. Kasama, *Proc. SPIE*, **1674**, 753–64 (1992)
41. N. Shiraishi, H. Hirukawa, Y. Takeuchi and N. Magome, *Proc. SPIE*, **1674**, 741–52 (1992)

42. M. Noguchi, M. Muraki, Y. Iwasaki and A. Suzuki, *Proc. SPIE,* **1674**, 92–104 (1992)
43. M. D. Levenson, N. S. Viswanathan and R. A. Simpson, *IEEE Trans. Electron Devices,* **ED-29**, 1828–36 (1982)
44. T. Terasawa, N. Hasegawa, T. Kurosaki and T. Tanaka, *Proc. SPIE,* **1088**, 25–8 (1989)
45. T. Horiuchi, Y. Takeuchi, S. Matsuo and K. Harada, *IEDM Tech. Digest*, 657–60 (1993)
46. D. M. Newmark, J. Garofalo and S. Vaidya, *Proc. SPIE,* **1927**, 63 (1993)
47. D. M. Terasawa, N. Hasegawa, H. Fukuda and S. Katagiri, *Jpn. J. Appl. Phys.*, **30**, 2991–7 (1991)
48. N. Yoshioka *et al.*, *IEDM Tech. Digest*, 653–6 (1993)
49. H. Miyashita *et al.*, *Proc. SPIE*, 248–60 (1994)
50. S. A. MacDonald *et al.*, *Proc. SPIE*, **1466**, 1–12 (1991)
51. O. Suga, H. Yamaguchi and S. Okazaki, *Microelectronic Eng.,* **13**, 65–8 (1991)
52. J. S. Petersen *et al.*, *J. Photopolym. Sci. Technol.*, **8**, 571–97 (1995)
53. T. Kumada, *Proc. SPIE*, **1925**, 31–42 (1993)
54. A. Oikawa *et al.*, *Proc. SPIE*, **2538**, 599–608 (1995)
55. J. C. Vigil, M. W. Barrik and T. H. Grafe, *Proc. SPIE,* **2438**, 626–43 (1995)
56. K. J. Przybilla *et al.*, *Proc. SPIE,* **1925**, 76–91 (1993)
57. H. Ito *et al.*, *Proc. SPIE*, **1925**, 65–75 (1993)
58. K. Tounai, S. Hashimoto, S. Shiraki and K. Kasama, *Proc. SPIE*, **2197**, 31–41 (1994)
59. S. Shioiri and H. Tanabe, *Proc. SPIE,* **2440**, 261–9 (1995)
60. J. G. Garofalo *et al.*, *Proc. SPIE*, **2440**, 302–12 (1995)
61. N. Cobb and A. Zaklor, *Proc. SPIE,* **2440**, 313–27 (1995)
62. T. A. Brunner, *Proc. SPIE,* **1466**, 297–308 (1991)
63. H. Yoshino *et al.*, *Proc. SPIE,* **2724**, 216–26 (1996)
64. S. Sethi *et al.*, *Proc. SPIE,* **1463**, 30–40 (1991)
65. Y. Suda, T. Motoyama, H. Harada and M. Kanazawa, *Proc. SPIE,* **1674**, 350–61 (1992)
66. T. Ogawa, H. Nakano, T. Gocho and T. Tsumori, *Proc. SPIE*, **2197**, 722–32 (1994)
67. T. Tanaka, N. Hasegawa, H. Shiraishi and S. Okazaki, *J. Electrochem. Soc.*, **137**, 3900–5 (1990)
68. D. J. Cronin and G. M. Gallatin, *Proc. SPIE,* **2197**, 932–42 (1994)
69. N. Magome and H. Kawai, *Proc. SPIE,* **2440**, 902–12 (1995)
70. Y. Kaimoto, K. Nozaki, S. Takechi and N. Abe, *Proc. SPIE,* **1672**, 66–73 (1992)
71. K. Nakano *et al.*, *Proc. SPIE,* **2195**, 194–204 (1994)
72. R. D. Allen, G. M. Wallraff, R. A. DiPietro and D. C. Hofer, *Proc. SPIE,* **2438**, 474–85 (1995)
73. M. A. Hartney, D. W. Johnson and A. C. Spencer, *Proc. SPIE,* **1446**, 238 (1991)
74. M. Shibuya, Japan Patent 62052 (1982)
75. M. Nakajima *et al.*, *Proc. SPIE,* **2197**, 111–21 (1994)
76. *64- to 256-Megabit Reticle Generation: Technology Requirements and Approaches*, ed. G. K. Hearn, **CR51** (1953)
77. F. C. Lo *et al.*, *Proc. SPIE*, **2254**, 2–13 (1994)
78. *Lithography*, Semiconductor Industry Association – Workshop; Working Group Reports, 35–45 (1993)
79. *The 1997 National Technology Roadmap For Semiconductors*, ed. Semiconductor Industry Association, 87 (1997)
80. C. N. Berglund, E-beam and optical reticle generation, in *64- to 256-Megabit Reticle Generation: Technology Requirements and Approaches*, ed. G. K. Hearn, **CR51**, pp. 142–155 (1993)
81. T. Coleman and P. Buck, *Proc. SPIE,* **2621**, 62–72 (1995)
82. M. Morita, S. Tachikawa and M. Iida, *Proc. SPIE,* **2793**, 513–19 (1996)
83. R. R. Singh, Reticle defect inspection and repair, in *64- to 256-Megabit Reticle Genera- tion: Technology Requirements and Approaches*, ed. G. K. Hearn, **CR51**, 201–12 (1993)
84. R. A. Budd *et al.*, Proc. SPIE, **2197**, 530–40 (1994)
85. S. V. Daugherty, Cleaning and pelliclization, in *64- to 256-Megabit Reticle Generation: Technology Requirements and Approaches*, ed. G. K. Hearn, **CR51**, 213–24 (1993)
86. H. Kawahira *et al.*, *Proc. SPIE,* **2793**, 22–33 (1996)
87. M. Hoga, Y. Koizumi, F. Mizuno and H. Nakaune, *Proc. SPIE,* **2254**, 14–25 (1994)
88. S. Ito *et al.*, *Proc. SPIE,* **2197**, 99–110 (1994)

89. Y. Saito *et al.*, *Proc. SPIE*, **2254**, 60–3 (1994)
90. F. D. Kalk, R. H. French, H. U. Alpay and G. Hughes, *Proc. SPIE*, **2254**, 64–70 (1994)
91. H. Mohri *et al.*, *Proc. SPIE*, **2254**, 238–47 (1994)
92. H. Miyashita *et al.*, *Proc. SPIE*, **2254**, 248–60
93. T. Yokoyama *et al.*, *Proc. SPIE*, **2254**, 261–74 (1994)
94. K. Kawano *et al.*, *Proc. SPIE*, **2512**, 349–55 (1995)
95. G. Dao *et al.*, *Proc. SPIE*, **2512**, 333–42 (1995)
96. K. Mikami *et al.*, *Proc. SPIE*, **2512**, 333–42 (1995)
97. H. Kusunose *et al.*, *Proc. SPIE*, **2254**, 294–301 (1994)
98. H. Kusunose *et al.*, *Proc. SPIE*, **2793**, 251–60 (1996)
99. Y. Yamada *et al.*, *Proc. SPIE*, **2621**, 266–72 (1995)

X-ray lithography 3

Kimiyoshi Deguchi, Teruo Hosokawa,
Sunao Ishihara and Katsumi Suzuki

3.1 Principles

X-ray lithography (XRL) is a simple proximity printing technology for replicating 1× mask patterns directly onto a wafer with 0.5–1-nm-wavelength soft X-rays. When XRL was developed by D. L. Spears and H. I. Smith in 1972, an electron bombardment target was used as an X-ray source.[1] In 1976, a synchrotron light source was first used for XRL and excellent resolution as small as 70 nm was demonstrated.[2] The basic set up of a typical XRL system using a synchrotron as an X-ray source is shown schematically in Figure 3.1.[3] Extremely fine patterning with a proximity gap of 15–40 μm has been achieved using X-rays with wavelengths more than two orders of magnitude shorter than deep-UV wavelengths. Synchrotron-based XRL provides a large depth of focus and a wide exposure-dose window, both of which are very important in ULSI fabrication. Insensitivity to dust is another advantage and will become increasingly important because pattern defects due to extremely small particles will be a more serious problem as feature size decreases.

The principal mechanism by which soft X-rays interact with materials is the photoelectric effect (Figure 3.2). Other possible interactions such as the Compton effect and Thomson scattering are negligibly small at a wavelength near 1 nm. When an X-ray photon interacts with an orbital (inner-shell) electron, the electron gains enough energy to be ejected as a free electron. The ejected free electrons will also collide with other atomic electrons and, in turn, eject them. This reaction will be repeated until the energy of the electrons falls below the ionizing energy.

An inner-shell vacancy created by the photoelectric effect, is filled with an outer-shell (higher-energy) electron through two possible processes. One is a radiative transition which accompanies the energy-release by radiating fluorescent X-rays. The other is a non-radiative transition in which the energy is transferred to another atomic electron by the Coulomb interaction. The electron which receives the energy is ejected as a free electron. This reaction is called the Auger effect and the ejected electron is called an Auger electron. Auger electrons will also react with other electrons in a way similar to that in which the photoelectrons do.

Both the photoelectric effect and the Auger effect result in a number of atoms or molecules being excited or ionized. As an example, a cross-section for hydrogen ionization is shown in Figure 3.3[4] as a function of the energy of the incident electron. The maximum cross-section is near 70 eV. So secondary electrons, rather than the X-ray photons or primary electrons ejected by the photoelectric effect, will play the principal role in the patterning of the resist. The mechanism of resist patterning in XRL is therefore considered to be almost the same as that in electron-beam lithography (EBL).

The resolution of XRL is determined mainly by Fresnel diffraction and scattering of secondary electrons in a resist film. The calculated resolution limit in XRL is shown in Figure 3.4[5] as a function of wavelength. Here, W_D

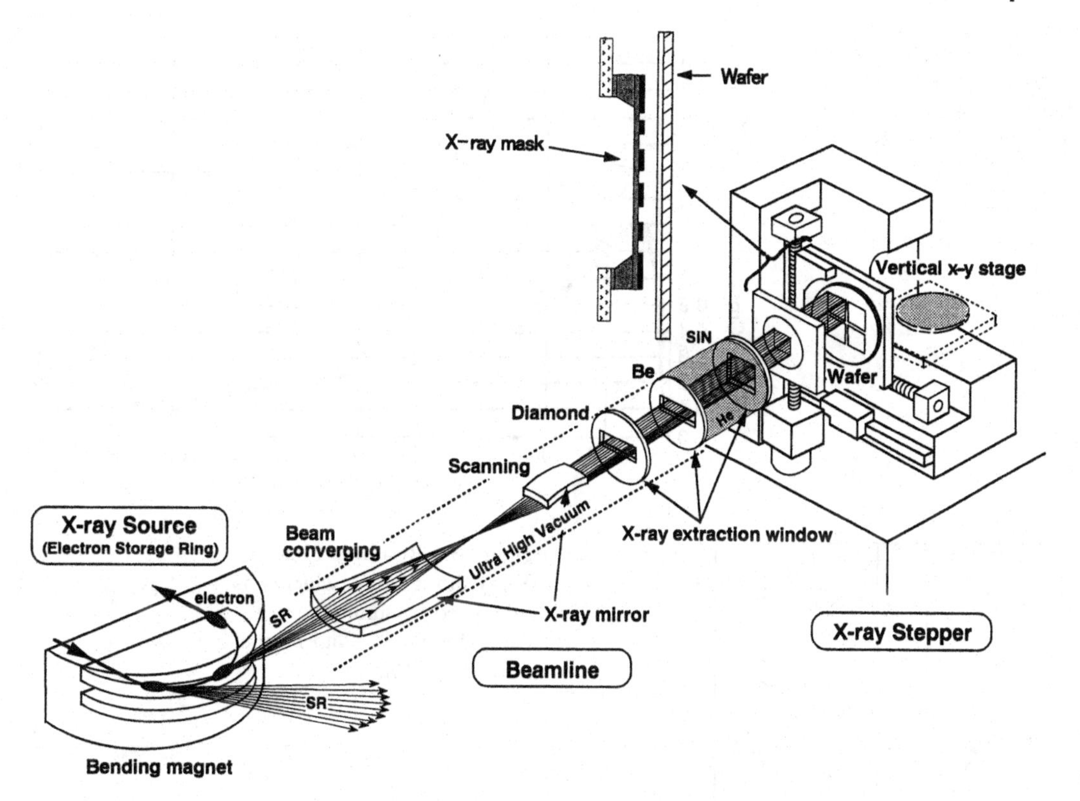

Figure 3.1 Diagram for a synchrotron-based X-ray lithography system.

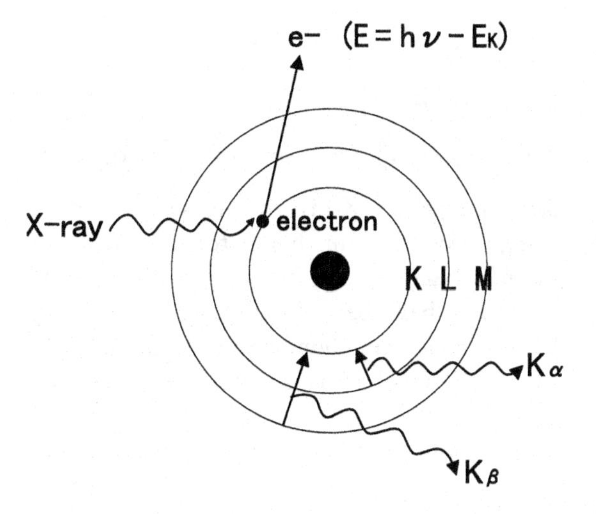

Figure 3.2 Diagram illustrating photoelectric effect. An X-ray photon interacts with an inner-shell electron and a free electron is ejected.

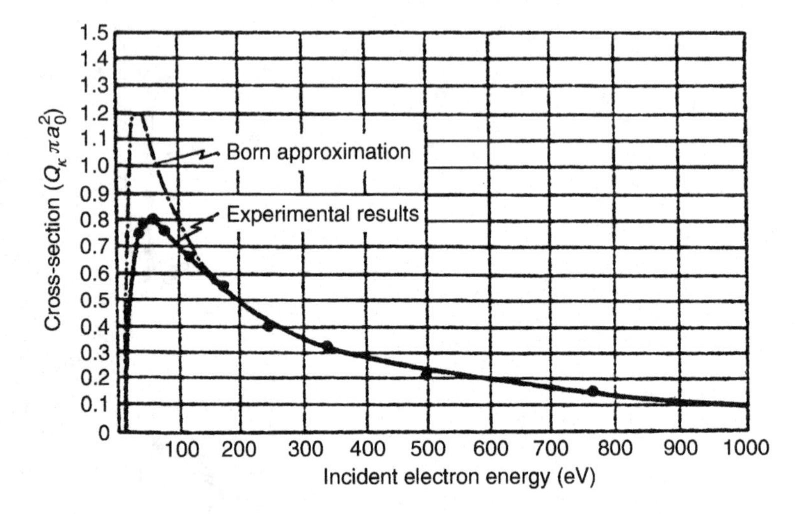

Figure 3.3 A cross-section for hydrogen ionization with an electron as a function of incident electron energy.

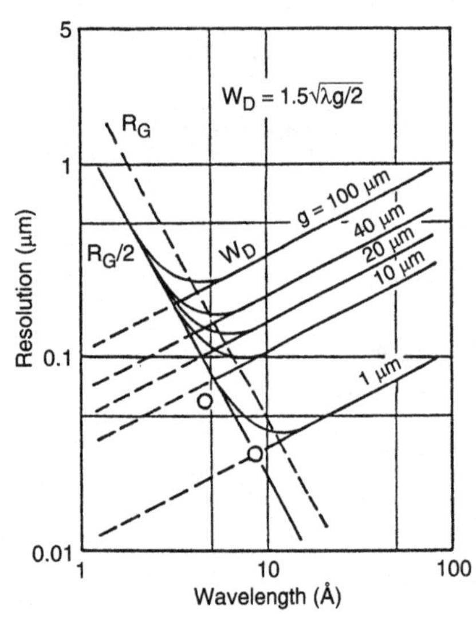

Figure 3.4 Resolution limit in X-ray lithography, which was calculated by summation of Grün Range (R_G) and Fresnel diffraction W_D.

represents Fresnel diffraction given by the equation $W_D = 1.5(\lambda g/2)^{1/2}$ (where λ is wavelength and g is the gap between the X-ray mask and a wafer), and R_G represents the electron-scattering range, which is determined by the Grün range for wavelengths shorter than 0.5 nm (energies higher than 2.5 keV) or by measuring experimental data for wavelengths longer than 0.5 nm. The open circles in the figure show the experimental results reported by Feder *et al.*[6] These data fit the summation of $R_G/2$ and W_D.

3.2 X-ray source

3.2.1 History of X-ray sources for lithography

Several types of X-ray source, such as those using electron-beam bombardment, gas-puff plasma, laser plasma, and synchrotron radiation (SR), have been developed for XRL use. In electron-beam bombardment, 10–20-keV electron beams bombard a rotating target. The energy of the electron beam is converted to characteristic soft X-rays and to low-intensity white X-rays. The characteristic X-rays are used for XRL. This method, however, is not suitable for use in lithography because of its low X-ray intensity, large divergence of X-ray flux, a large spot-size which causes penumbra blur, and contamination of extraction windows or X-ray masks with debris.

When a gas-puff plasma source is used, an X-ray intensity of about ten times that produced by electron-beam bombardment is obtained. In addition, the penumbra blur is decreased to less than one-third of that in EB-bombardment. On the other hand, the amount of exposure is hard to control because the X-ray intensity fluctuates pulse-by-pulse. Moreover, not only is the X-ray intensity still too low for practical use, but this source also contaminates the Be window or X-ray mask with plasma debris. This contamination reduces X-ray power, degrades exposure uniformity, and makes it difficult to align the mask.

In the laser-plasma X-ray source, micro-plasma is formed by focusing a high-power laser beam on the target. The penumbra blur is therefore remarkably decreased. This X-ray source, however, suffers from the effects of plasma debris. In addition, the large divergence of the X-ray flux is an obstacle to extremely fine patterning.

Synchrotron radiation (SR) is an electromagnetic wave generated when high-energy electrons are accelerated. Its wavelength is determined by the electron energy and acceleration (or radius of the electron orbit) ranging from those of γ-rays to infrared and submillimeter radiation. X-ray intensity obtained from a synchrotron is approximately one order of magnitude higher than that from a plasma X-ray source, and the divergence of the X-rays is only about 1 mrad, which makes geometrical run-out negligibly small. The only problems with this method are the initial cost and the size of the required floor space. SR is, however, the most promising technology today. Only with SR has it been possible to fabricate ULSIs having a design rule less than 0.2 μm with a reasonable throughput of 10–20 wafers (6″ (152-mm) ϕ] per hour. For these reasons, the rest of this section will focus on SR systems.

3.2.2 Characteristics of synchrotron radiation

Figure 3.5 shows a schematic view of a typical SR system, which consists of two main parts: a storage ring section and an injection accelerator. Electrons with energies of more than 10 MeV are generated in the injection accelerator and injected into the storage ring. The injection scheme is divided into three classes: low-energy, medium-energy and full-energy injections. After electron-beam injection, the former two need acceleration in the ring to a final energy typically between 600 MeV and 1 GeV. Either a linac, a microtron or a synchrotron are used as the injectors.

When electrons emit SR at the bending magnet, they lose kinetic energy. This lost energy is compensated for by an RF cavity installed in the ring; thus, electrons continuously travel in the orbit with almost constant energy. Beamlines (BL) connected to the ring at the bending magnet transport the SR to the exposure stations (see Figure 3.14).

As an X-ray source for lithography, SR has the following excellent features:

- High X-ray intensity,
- Theoretically calculable X-ray intensity (see Subsection 3.3.2),
- Small divergence of X-ray flux; radiation angle is in the order of 1 mrad (see Figure 3.6),
- Wide spectrum and continuous wavelength distribution (see Figures 3.7, 3.8),

- Cleanness (no debris or other contaminants),
- Small source size (less than 1 mm, depending on the ring lattice design),
- Stable source position (normally less than 50-μm positional variation),
- Stable X-ray intensity.

3.2.3 Synchrotron radiation distribution

Synchrotron radiation characteristics can be derived from Maxwell's equation. The energy W radiated from one electron in a circular orbit of radius ρ for angular direction Ψ (Figure 3.5) is expressed as:

$$W = \frac{27}{8\pi^2 \mu_0 c^2}\left(\frac{e}{4\pi\varepsilon_0}\right)^2 \frac{\gamma^8}{\rho^2}\left(\frac{\lambda_c}{\lambda}\right)^4 \left[1+(\gamma\Psi)^2\right]^2 \times \left\{K_{2/3}^2(\xi) + \frac{(\gamma\Psi)^2}{1+(\gamma\Psi)^2}K_{1/3}^2(\xi)\right\} \times \Delta\theta\Delta\Psi\Delta\lambda \text{ [J]},$$

and

$$\gamma = \frac{E}{mc^2} \simeq 1.957 \times 10^3 E[\text{GeV}],$$
$$\rho = \frac{\sqrt{E(E+2mc^2)}}{ceB} \simeq \frac{3.335E[\text{GeV}]}{B[\text{T}]} \text{ [m]},$$
$$\lambda_c = \frac{4\pi\rho}{3\gamma^3} \simeq \frac{1.864}{B[\text{T}]E^2[\text{GeV}]} \text{ [nm]},$$
$$\xi = \frac{\lambda_c}{2\lambda}\left[1+(\gamma\Psi)^2\right]^{3/2}, \quad (3.1)$$

where γ is the normalized electron energy, E is electron energy, m is electron mass, $-e$ is electron charge, c is light speed, B is bending magnetic flux density, ε_0 and μ_0 are respectively the dielectric constant and magnetic permeability of a vacuum, and K_n is a modified Bessel function. The wavelength dependency of the radiation characteristics is a function of normalized wavelength λ/λ_c where λ_c is the critical wavelength. Regarding the terms contained within the braces, { }, the first term corresponds to the polarized part that is parallel with the orbital plane, and the second term to the polarized part that is normal to the orbital plane.

Figure 3.6 shows the intensity distribution for angular direction Ψ. Note that for each wavelength, intensity is normalized at $\Psi = 0$.

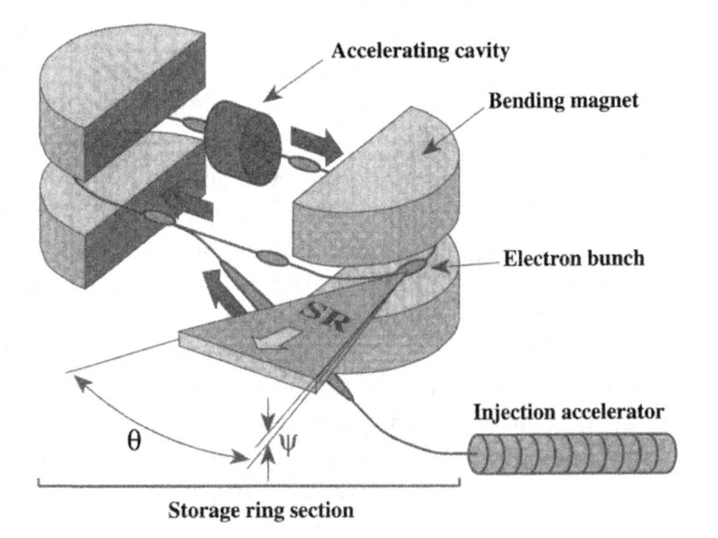

Figure 3.5 Schematic diagram of a compact storage ring.

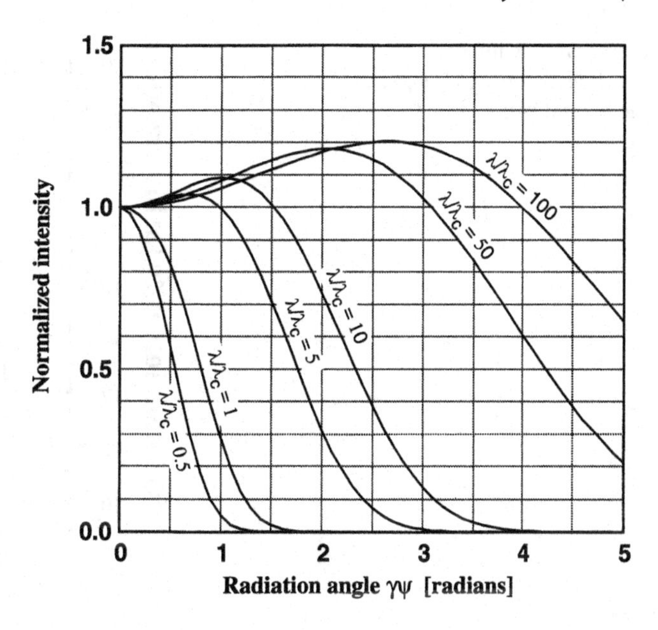

Figure 3.6 SR angular distribution. For each wavelength, SR intensity is normalized at $\Psi = 0$.

It is shown that the radiation angular range for wavelengths near λ_c is about $1/\gamma$. The higher the photon energy (or the shorter the wavelength), the smaller the divergence. The total intensity takes the maximum in the orbital plane ($\Psi = 0$) and is expressed as:

$$P_{\Psi=0} = 3.9185 \times 10^{-11} \frac{\gamma^8}{\rho^2} \left\{ \left(\frac{\lambda_c}{\lambda}\right)^4 K_{2/3}^2 \left(\frac{\lambda_c}{2\lambda}\right) \right\} \times \Delta\Psi[\text{rad}]\Delta\theta[\text{rad}]\Delta\lambda[\text{m}]/I[\text{A}]\ [\text{J/s}], \quad (3.2)$$

where I is stored beam current.

The intensity distribution in the orbital plane is shown in Figure 3.7 as a function of wavelength normalized by λ_c. The wavelength for the peak intensity is at about $0.34\lambda_c$. Because the radiation's vertical angular range is very small, the irradiation area has to be vertically expanded for lithography. One of the most common ways to do this is by using a wobbling X-ray mirror to scan the X-ray flux vertically. Another way is to wobble the electron trajectory vertically. In a third method, instead of expanding SR flux, a mask and a wafer are moved vertically across the SR irradiation area. The power per second per unit electron-beam current, per unit horizontal angle, and per unit wavelength for the entire range of the vertical direction is obtained by integrating Equation (3.2):

$$P_{\Psi\text{all}} = \frac{9\sqrt{3}}{64} \frac{e}{\pi^3 c^2 \mu_0 \varepsilon_0^2} \frac{\gamma^7}{\rho^2} \left(\frac{\lambda_c}{\lambda}\right)^3 \times \int_{\lambda_c/\lambda}^{\infty} K_{5/3}(x)\,\mathrm{d}x = 1.4121 \times 10^{-10} \frac{\gamma^7}{\rho^2} \left(\frac{\lambda_c}{\lambda}\right)^3 \times \int_{\lambda_c/\lambda}^{\infty} K_{5.3}(x)\,\mathrm{d}x\ [\text{J/s} \cdot \text{m} \cdot \text{rad} \cdot \text{A}]. \quad (3.3)$$

Figure 3.8 shows the dependency of $P_{\Psi\text{all}}$ on the normalized wavelength.

3.2.4 Transmission in a beamline

X-ray exposure is carried out either in an air environment or in a He environment in order

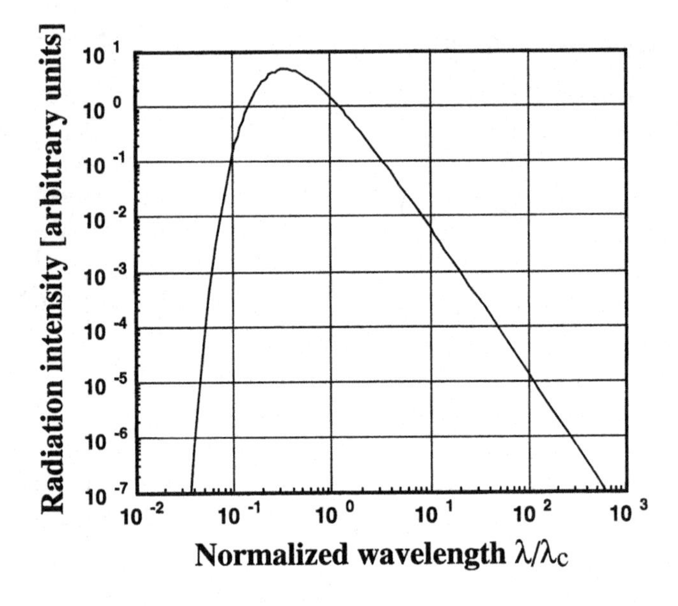

Figure 3.7 SR intensity dependency on wavelength in the orbital plane, $\Psi = 0$.

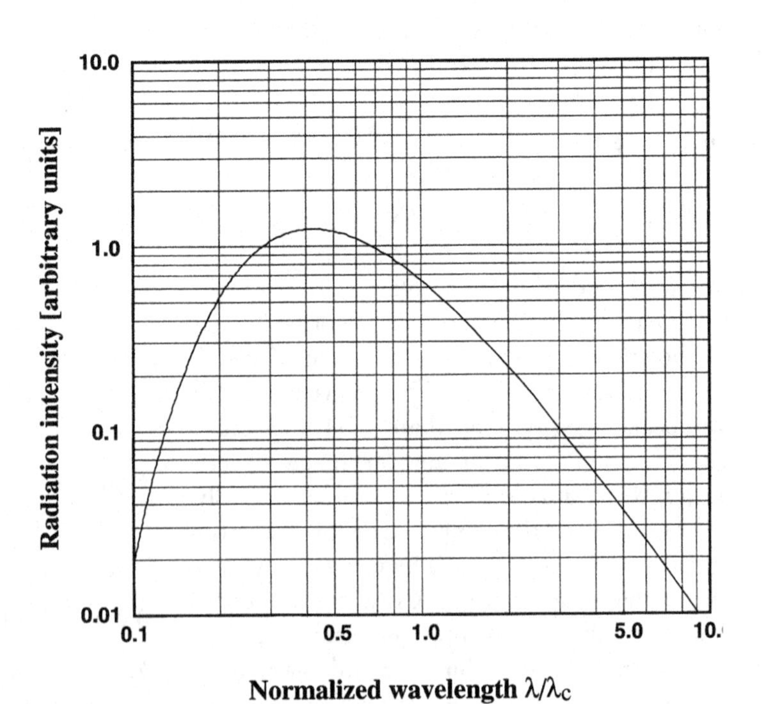

Figure 3.8 Dependency on wavelength of SR intensity integrated for all Ψ.

to avoid heating the mask or wafer excessively. The wavelength range used is determined to obtain sufficient mask contrast and, at the same time, to suppress the Fresnel diffraction as much as possible. One optimum wavelength range is 0.7–1.0 nm, and SR in this range is extracted from the ultra-high-vacuum ring to an exposure area through a beamline (see Figure 3.1). The beamline includes X-ray mirrors, windows (Be, SiN_x, diamond, and SiC), and a He chamber for installing alignment optics (see Figure 3.1). All these components absorb X-rays and this absorption leads to a great loss of X-ray power. The absorption coefficient is a function of wavelength. X-rays with wavelengths longer than about 1 nm are absorbed by vacuum windows, and those with wavelengths shorter than about 0.6 nm are absorbed by the mirrors. A 0.7-nm to 1.0-nm range of SR is thus extracted through the beamline.

To increase the area of the exposure field and the density of exposure power, X-ray mirrors are used for collecting SR beams spread over a large range of horizontal angles and for scanning the SR beam vertically. Theoretically calculated and experimentally measured reflectivities of X-ray mirrors are shown in Figure 3.9.[7] The reflectivity falls rapidly with increasing beam grazing angle θ. There is an optimum collecting angle which maximizes power intensity because the grazing angle is enlarged as the collecting angle increases. The optimum grazing angle for a one-mirror system has been found to be about 1.8 degrees.[8]

When the X-rays are incident normal to the film surface, reflection at the interface is negligible. In this case, transmission T, the ratio of outgoing power to incident power, is obtained by using the following equation:[9]

$$T = e^{-\mu_m \rho z_t}, \tag{3.4}$$

where $\mu_m[\mathrm{cm^2/g}]$, $\rho[\mathrm{g/cm^3}]$, and $z_t[\mathrm{cm}]$ are mass absorption coefficient, density of the film material and film thickness, respectively. For compound materials the coefficient is given by:

$$\mu_m = \frac{\sum_i n_i A_i \mu_{m,i}}{\sum_i n_i A_i}, \tag{3.5}$$

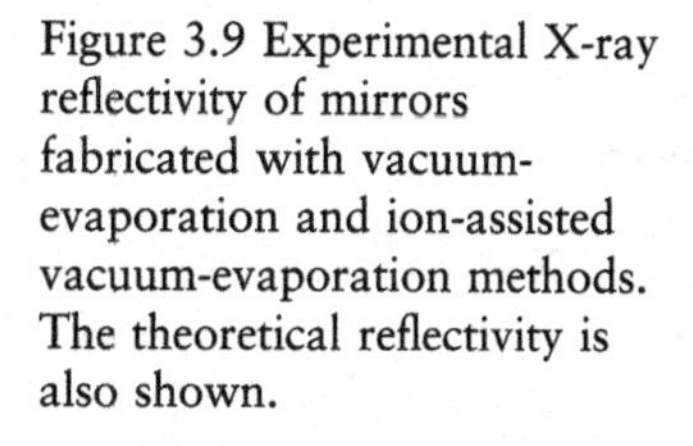

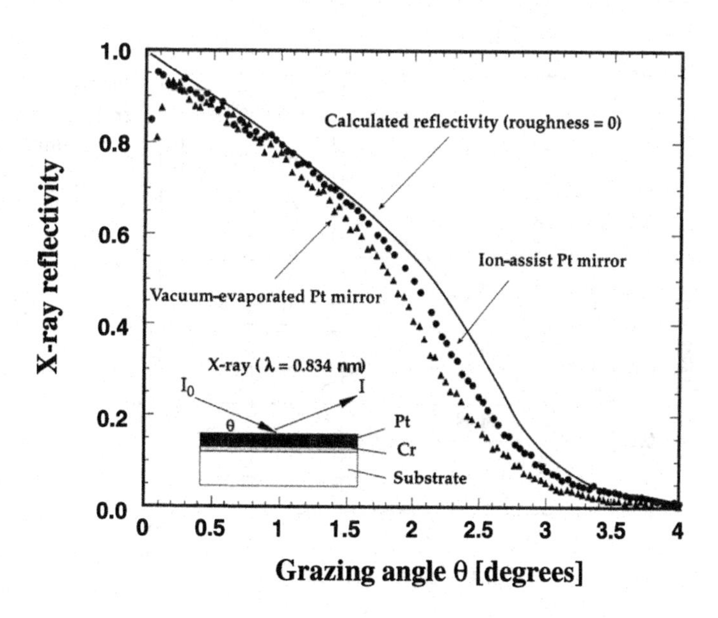

Figure 3.9 Experimental X-ray reflectivity of mirrors fabricated with vacuum-evaporation and ion-assisted vacuum-evaporation methods. The theoretical reflectivity is also shown.

where A_i, $\mu_{m,i}$, and n_i are atomic weight, mass absorption coefficient of each atom, and number of atoms per molecule. The coefficient μ_m depends on the wavelength and is given in Reference 9 for various materials. Values of μ_m can also be obtained through the 'Internet'.[10] Figure 3.10 shows the relation between transmission and thickness for the principal materials commonly used in SR-lithography beamlines. The total transmission depends on the design of the beamline. As an example, the total transmission of a beamline at Super-ALIS (Super-advanced Lithography Imaging System)[11] is 3.4%.

3.2.5 Ring design

In the first stage of designing a ring, it is important to select electron-beam energy E and bending magnetic flux B. Figures helpful for designing a ring are obtained using Equations (3.1)–(3.3). Figure 3.11 gives the critical wavelength λ_c and the bending radius ρ. Figure 3.12 shows the radiated power P (0.7–1.0 nm), from 0.7 to 1.0 nm per unit beam current, which is effective for lithography. Figure 3.13 shows the ratio of P (0.7–1.0 nm) to total radiated power P_{total} and this ratio corresponds to the power conversion efficiency. The maximum conversion efficiency is 15% in this wavelength range.

These figures are helpful for determining E and B. For example, if we assume a beam current of 300 mA, a beamline transmission efficiency of 0.034, a resist sensitivity of 100 mJ/cm^2, an exposure time of 4 s for one shot, and an exposure field 2-cm square (corresponding to a throughput of about fifteen 6″ (152-mm) ϕ wafers per hour), we conclude that the necessary P (0.7–1.0 nm) is 2000 W/A. A maximum trajectory radius is determined from the viewpoint of space effectiveness. For the upper limit of the bending radius of 1 m, for example, one choice of E and B is 0.65 GeV and 2.2 T, respectively. In this case, magnetic flux-density becomes a minimum.

3.2.6 Compact SR instruments

An SR instrument for industrial use must be compact. There are two ways of reducing the

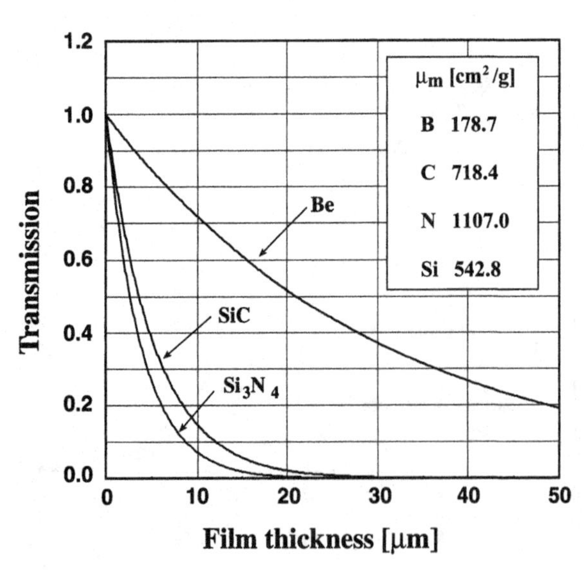

Figure 3.10 X-ray transmission of the materials commonly used in the lithography beamline. The X-ray wavelength is 0.834 nm.

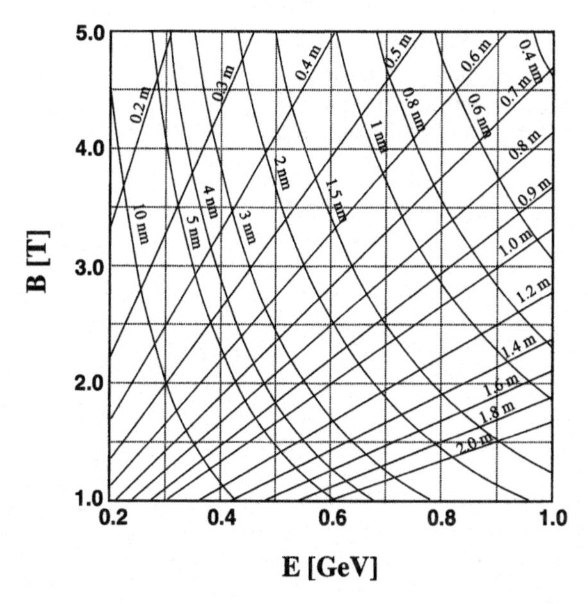

Figure 3.11 Critical wavelength (nm) and bending radius (m) vs electron-beam energy *E* (GeV) and magnetic flux-density *B* (T) of the bending magnet.

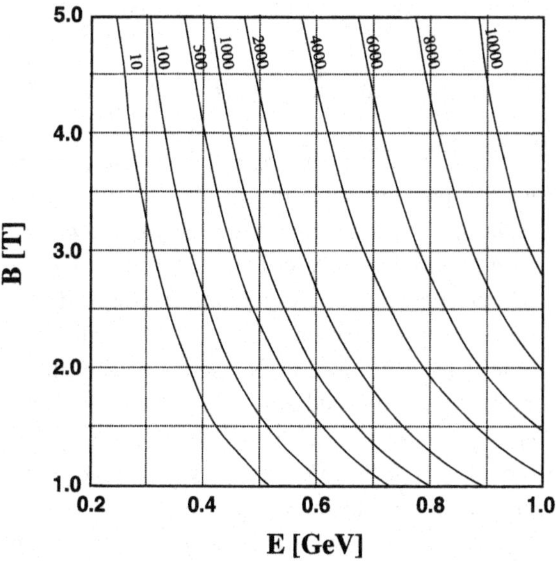

Figure 3.12 Radiated power *P* (W/A) in the 0.7–1.0-nm wavelength range for 2π horizontal angle per unit beam current.

size of the instrument: (1) using superconducting bending magnets; and (2) using low-energy injection. Superconducting magnets decrease the radius of the beam trajectory and reduce power consumption. Low-energy injection requires a smaller injection accelerator, which saves space and is cost effective.

The technological problem of employing a superconducting magnet is the difficulty in achieving a uniform distribution of bending magnetic flux for large-bending angle magnets. Flux uniformity is necessary for maintaining stable electron circulation in a ring. There are two types of superconducting magnet for compact rings: a magnet with a yoke; and one without a yoke. Recent studies reveal that sufficient uniformity in the order of 10^{-4} has been achieved.

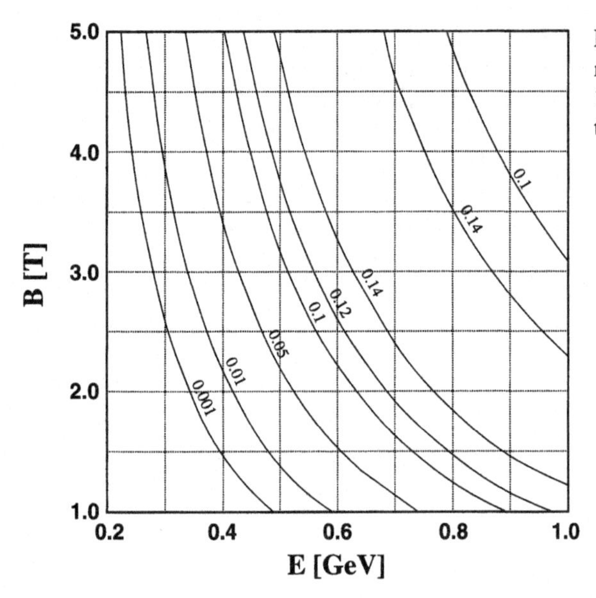

Figure 3.13 The ratio of radiated power P in the 0.7–1.0-nm wavelength range to the total radiated power P_{total}.

The injection energies so far used in compact rings can be divided into two classes: 100–200 MeV (medium-energy); and about 15 MeV (low-energy). In the medium-energy range, beam stacking can be used, that is, the stored current can be increased with an increasing number of pulse beam injections. Low-energy injection, however, needs to be a one-pulse injection. A low-energy injection accelerator is small and inexpensive. High-energy radiation, such as gamma radiation, is generated in the injection process because a large number of high-energy electrons spill out of the ring during the process. The one-pulse injection scheme is expected to generate much less radiation than medium-energy injection. So, for the one-pulse injection scheme, radiation shielding is expected to be much simpler. The technological barriers to low-energy injection, however, are high, so the majority of SR systems have used 100–200 MeV injection.

Unstable electron-beam motion can be suppressed by emitting SR. This is called radiation damping. The damping time τ_r is expressed as:

$$\tau_r\,[\mathrm{s}] = C[\mathrm{m}]\rho[\mathrm{m}]/(13.2JE^2[\mathrm{GeV}]), \tag{3.6}$$

where C is the circumference of the ring and J is a ring parameter in the order of unity. This equation indicates that the higher-energy electron beams are more stable, and this is why the higher-energy injection scheme is more advantageous for storing large beam currents. Stabilizing the beams is therefore the main problem in the medium- and, especially, low-energy injection schemes. Beam storing of more than 200 mA has been achieved with both injection schemes.

Ion trapping has a large effect on beam motion. Residual gas molecules in the ring are ionized when electrons collide with them. The ionized molecules are attracted towards the electron trajectory center and interact with electrons, resulting in reduced beam life or abrupt loss of beam current. Ion-cleaning electrodes are therefore attached to the inner side of the vacuum chamber. Applying a DC voltage of 100–1000 V to these electrodes removes the ions from the electron-beam trajectory.

There is, however, a positive aspect of the ion-trapping effect: the ions create multipole electric fields without narrowing the dynamic aperture, and these multipole fields stabilize electron oscillating motion. This is known as Landau damping. There is, therefore, a possibility that the electron motion can be stabilized by the fields generated by ion trapping. This stabilizing mechanism plays an important role in medium- and low-energy regions, where radiation damping is very weak. Taking advantage of this effect, it has been demonstrated that the superconducting compact ring Super-ALIS can store more than 1A of beam current.[12] In this experiment the beam was stable; that is, beam current was limited by RF power shortage and duct heating. This result implies that a machine which can store several amperes can be developed.

Electrons circulating in the ring are lost through collision with residual gas molecules. The circulating electron-beam current I_b is expressed as:

$$I_b = I_0 e^{-t/\tau}, \tag{3.7}$$

where I_0, is the initial current, t is the elapsed time after beam injection and τ is beam lifetime. Beam lifetime is proportional to the square of beam energy and inversely proportional to the vacuum pressure. Beam lifetime generally increases with decreasing beam current because the vacuum pressure decreases with decreasing beam current. For a given initial beam current, a higher-energy machine has a higher time-averaged beam current.

Designing a SR system requires knowledge of accelerator physics and also computer code analysis. Interested readers are referred, for example, to Wiedemann.[13,14] The computer code list is included in References 15 and 16 and can also be obtained over the 'Internet'.[17]

3.2.7 Practical compact SR rings

Several compact rings have been developed,[11,18,19,20,21,22] and the principal parameters of these rings are listed in Table 3.1. Both the AURORA[18] and the HELIOS[20] are commercially available. The structure and design concepts differ from machine to machine, and this section will briefly describe Super-ALIS[11] as an example of a compact ring.

Figures 3.14 and 3.15 show an outline drawing and a photograph of Super-ALIS, which in 1989 was the first compact superconducting SR ring. Its racetrack-type design uses two superconducting magnets with 180-degree bending angles. Three quadrupoles are installed in a straight section, where four horizontal steering magnets, four wobblers, a pair of sextupoles, and an octupole are also installed. For beam monitors, there are six button electrode position monitors, three profile monitors, and current transformers for measuring beam current. Shim coils wound in each quadrupole enable the measurement of betatron functions. Hall devices attached to the bending magnets measure vertical magnetic flux-density at the magnet center. The bending magnets are connected in series and are driven by one power supply. The inductance of the superconducting magnets changes from 1.75 to 0.3 H depending on the excitation current. Each straight section has a 50-mm diameter button-type ion-clearing electrode, and the bending magnets have long-plate-type ion-cleaning electrodes along the outer side of the electron orbit. The Super-ALIS has two inflectors used for low-energy- and high-energy-injection experiments.

Superconducting magnets have iron yokes. At full excitation, the yokes saturate magnetically and thus flux leakage becomes large, magnetizing neighboring magnets such as those used for steering. This magnetization makes it difficult to reproduce injection conditions.

Table 3.1 *Main parameters of compact storage rings.*

Ring name (Manufacturer)	AURORA (SHI)[a]	HELIOS (Oxford Instrum.)	LUNA (IHI)[b]	MELCO (Mitsubishi E)[c]	NIJI-III (SEI)[d]	Super-ALIS (NTT)[e]	
Ring type	circular	racetrack	rectangular	racetrack	rectangular	racetrack	
Bending magnet type	super	super	normal	super	super	super	
Energy (MeV)	575	700	800	600	600	600	
Injection energy (MeV)	150	100	45	600	100	15 (LEI)[f]	520 (HEI)[g]
Injector	Microtron	Linac	Synchrotron	Linac	Linac	Linac	Synchrotron
Critical wavelength (Å)	14.8	8.4	21.9	14.8	12.9	17.3	
Bending magnetic flux (T)	3.8	4.5	1.33	3.5	4.0	3	
Bending radius (m)	0.5	0.52	2.0	0.593	0.5	0.66	
Betatron number							
(horizontal)	0.8	–	1.75	1.38	2.37	1.565	
(vertical)	0.6	–	0.76	0.43	1.37	0.556	
Harmonic number	2	16	14	4	10	7	
Accelerating time (s)	450	130	42	0	–	120 (LEI)[f]	260 (HEI)[g]
Acceleration frequency (MHz)	190.86	499.7	178.5	130	158.7	125.855	
Accelerating voltage (kV)	120	~ 300	70	60	60	40	
Beam lifetime (h)	16.6 (300 mA)	35 (200 mA)	5	4 (200 mA)	13 (100 mA)	5.2 (500 mA)	

Vacuum pressure (Torr)							
(no beams)	2×10^{-10}	3×10^{-10}	1×10^{-10}	1×10^{-10}	1×10^{-10}	2×10^{-10}	
(with beams)	1×10^{-9}	1×10^{-9}	2×10^{-9}	2×10^{-9}	1×10^{-9}	5×10^{-10}	
Circumference (m)	3.14	9.6	23.5	9.2	18.9	16.8	
Beam ports	16	18	4	10	4	10	
Floor space (m^2)	9.1	13.2	46.2	18	33.3	22.0	
	(3.4ϕ)	(6.0×2.2)	(6.8 sq.)	(3×6)	(7.4×4.5)	(2.5×8.8)	
Beam current (mA)							
(achieved)	723	297	80	380	200	200 (LEI)[f]	1215 (HEI)[g]
(operating)	>300	200	50	220	150	150 (LEI)[f]	500 (HEI)[g]
Beam size (σ_x/σ_y) (mm)	1.3/0.14	0.5/0.5	0.9/0.5	0.8/0.4	0.25/0.25	0.8/0.6	
Total power consumption (kW)	215	296	300	–	300	390	
Status	operating	operating	operating	operating	operating	operating	

[a] Sumitomo Heavy Industries Ltd; [b] Ishikawajima–Harima Heavy Industries Co., Ltd, [c] Mitsubishi Electric Corp.; [d] Sumitomo Electric Industries Ltd; [e]Nippon Telegraph and Telephone Co.; [f] Low-energy injection; [g] High-energy injection.

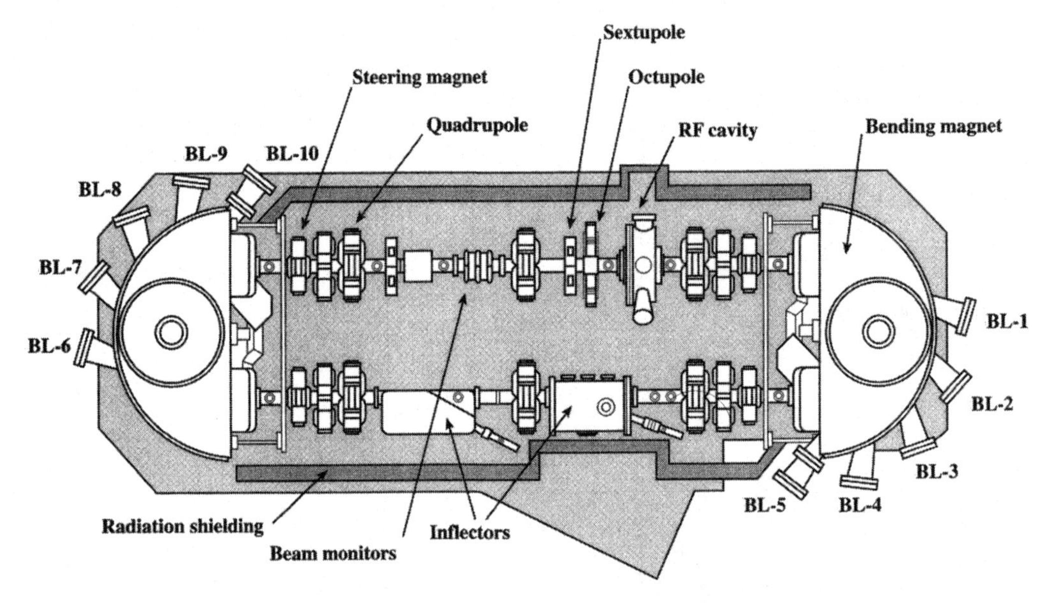

Figure 3.14 Outline drawing of basic structure of the Super-ALIS compact synchrotron; BL-1 to BL-10 are beamline positions.

Figure 3.15 Photograph of the Super-ALIS. Beamlines BL-1 to BL-5 are shown.

Therefore, in the low-energy injection scheme, all the magnets, except the bending ones, are degaussed before the injection process.

Figure 3.16 shows a layout drawing of the ring and the beamlines. Radiation shielding is installed around the straight sections of the Super-ALIS. Radiation intensity during electron-beam injection and that during the beam-storage stage at selected points one meter distant from the ring are shown inset in Figure 3.16. When a current of 500 mA is being stored, the radiation dose-rate one meter from the ring, including the background dose-rate, is less than 0.15 μSv/h. During electron-beam injection, the dose-rate becomes much higher and people have to be evacuated from the ring room. In the clean room, which will be occupied by people during beam-injection time, the integrated dose can be kept much lower than the value regulated by law.

3.3 **X-ray mask**

3.3.1 Introduction

As described in Section 3.1, X-ray image contrast is created by the photoelectric effect or by the Auger effect taking place in an X-ray absorber pattern. The photoelectric effect, however, also takes place in the mask substrate. To minimize X-ray absorption, the mask substrate is

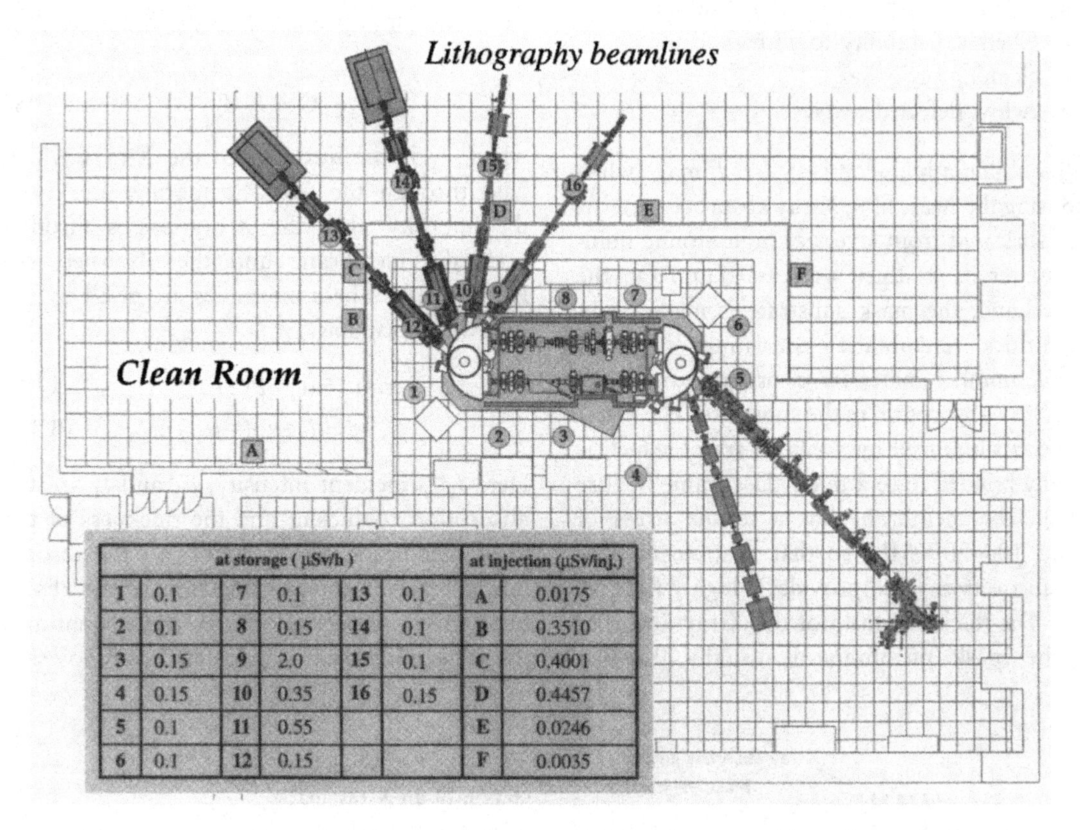

at storage (μSv/h)						at injection (μSv/inj.)	
1	0.1	7	0.1	13	0.1	A	0.0175
2	0.1	8	0.15	14	0.1	B	0.3510
3	0.15	9	2.0	15	0.1	C	0.4001
4	0.15	10	0.35	16	0.15	D	0.4457
5	0.1	11	0.55			E	0.0246
6	0.1	12	0.15			F	0.0035

Figure 3.16 Layout of the Super-ALIS and connected beamlines. The inset shows radiation intensity at selected points one meter distant from the ring.

made of a thin membrane consisting of materials with a low atomic number. This unique feature of X-ray masks means there are particular requirements that both the mask materials and the fabrication processes must meet to achieve the pattern-placement accuracy and CD control required for 0.1-μm-ULSI fabrication.

The principal requirements for X-ray masks used for 0.1-μm-ULSI fabrication are as follows:

- Providing appropriate mask contrast to soft X-rays,
- CD control within ±8 nm for the X-ray absorber pattern,
- Pattern-placement accuracy within 20 nm (3σ),
- X-ray resistance (long-term stability),
- Chemical stability to withstand the cleaning processes,
- A low defect-density.

To obtain a high-contrast X-ray image when exposing the resist film, X-ray absorber patterns are fabricated from layers of high-atomic number materials as thick as 0.4–0.5 μm. On the other hand, the mask substrate is made of 1–2-μm-thick membrane consisting of low-atomic-number materials in order to minimize the X-ray absorption in the substrate. The membrane is supported by a silicon frame which is usually bonded onto a thick glass frame (Figure 3.17). The membrane has a tensile stress of about 100–200 MPa so that it stretches over the silicon frame and provides a highly flat surface. The thickness ratio of the X-ray absorber to the mask membrane is usually 0.2–0.4, which is approximately 10^4 times as large as that of the reticles used for 4× or 5× optical steppers.

The aspect ratio of the X-ray absorber pattern for a 0.1-μm feature size will be as high as 4, which is about an order of magnitude higher than that of conventional reticles. Sufficiently controlling the critical dimensions in such absorber patterns is a key issue in XRL. Pattern-placement accuracy, X-ray resistance, and chemical stability are also important to achieve the required pattern-overlay accuracy. Practical techniques for achieving these requirements are described in this section.

3.3.2 Mask contrast

The contrast C produced by the X-ray mask is given by:

$$C = I_1/I_2 \tag{3.8}$$

where I_1 is the intensity of the X-rays which pass through the mask membrane and I_2 is the intensity after the X-ray passes through both the membrane and the absorber (see Figure 3.18). These intensities are given by the following formulas:

$$I_1 = I_0 \exp(-\mu_1 t_1), \tag{3.9}$$

$$I_2 = I_1 \exp(-\mu_2 t_2). \tag{3.10}$$

Here I_0 is incident intensity, μ_1 and t_1 are the absorption coefficient and the thickness of the mask membrane, and μ_2 and t_2 are the absorption coefficient and the thickness of the absorber pattern, respectively. From these equations, the mask contrast C is calculated as follows:

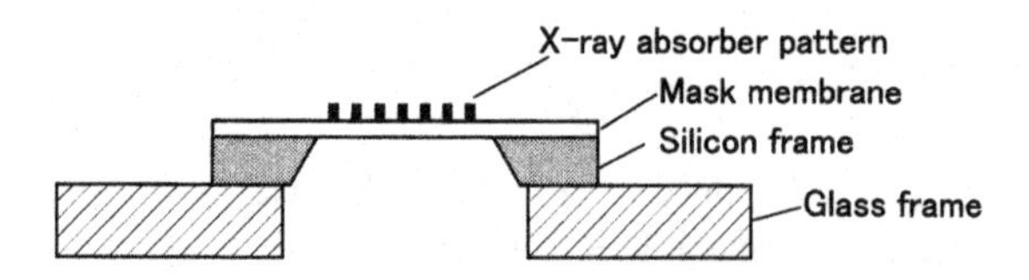

Figure 3.17 Schematic cross-section of an X-ray mask.

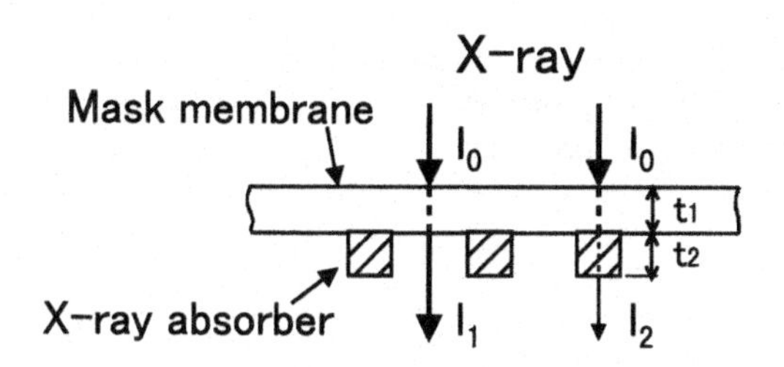

Figure 3.18 Illustration for X-ray mask contrast.

$$\begin{aligned} C &= I_1/I_2 \\ &= I_0 \exp(-\mu_1 t_1)/I_1 \exp(-\mu_2 t_2) \\ &= I_0 \exp(-\mu_1 t_1)/I_0 \exp(-\mu_1 t_1 - \mu_2 t_2) \\ &= \exp(\mu_2 t_2). \end{aligned} \tag{3.11}$$

Thus, X-ray mask contrast is determined only by the X-ray absorption coefficient and the thickness of the absorber pattern. (In the case of a broad-band X-ray spectrum like SR, however, mask contrast also depends on the material and the thickness of the mask membrane because the absorption coefficient changes as a function of wavelength.)

3.3.3 Mask-membrane materials

Linear X-ray absorption coefficients μ of the principal materials used for X-ray mask membranes are shown in Figure 3.19. It is clear that Si-compound materials have high transparency at the wavelength range 0.7–1.0 nm (7–10 Å). This is due to the high transparency of Si which has a K absorption edge at a 0.69-nm wavelength.

Among several Si-compound materials, SiN[23] deposited by LPCVD (low-pressure chemical vapor deposition) was the most popular material for mask membranes from the 1980s to the mid-1990s. However SiN was found to be damaged by X-ray irradiation, resulting in pattern-placement error.[24] On the other hand, SiC is more stable for X-ray irradiation.[25] In addition, SiC[26] is preferable because of its high Young's modulus which leads to high stability against the stresses generated in X-ray absorber patterns. For these reasons, SiN membranes have been replaced by SiC membranes.

SiC is usually deposited by LPCVD using reactant gases such as dichlorodimethylsilane ($(CH_3)_2SiCl_2$)[27] or dichlorosilane (SiH_2Cl_2)+ acetylene (C_2H_2).[28] To make smooth the surface of the SiC membrane, chemical–mechanical polishing (CMP) is carried out. The surface roughness of the SiC membrane can be made as small as 2 nm (R_{max}). Figure 3.20[28] shows the transmission of a 1.0-μm-thick polished SiC membrane to visible light. The transmission is improved because CMP decreases diffused reflection at the surface.

3.3.4 X-ray absorber materials

In the early days of XRL, the X-ray absorber pattern was usually made of gold (Au) and was formed by an electroplating process. This is called an 'additive process' because a Au pattern is fabricated after resist patterning by electroplating Au into the opening of the resist pattern. Because of the difficulty of controlling defects in the electroplating process, however, the additive process was replaced with a subtractive process. This process was based on the dry-etching of an X-ray absorber made of a refractory metal such as tantalum (Ta),[29] tungsten (W)[27] or their compound materials. Table 3.2 compares the principal characteristics of the additive and subtractive processes.

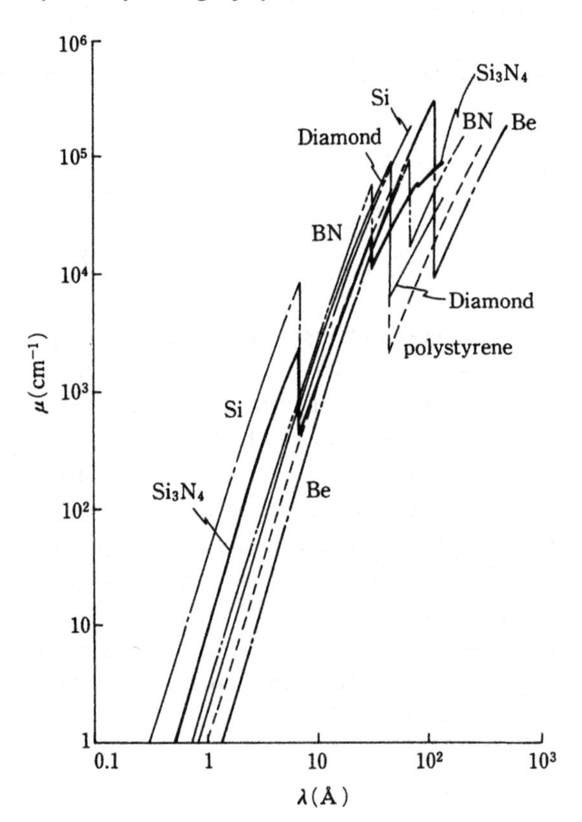

Figure 3.19 Linear absorption coefficient μ of principal materials used for X-ray masks.

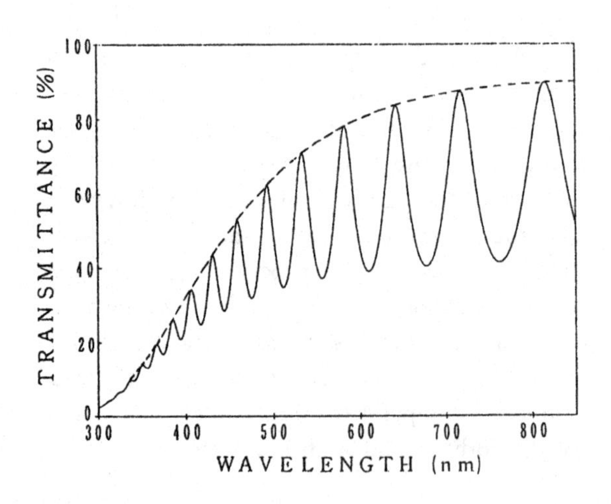

Figure 3.20 Optical transmission spectrum of 1.0-μm-thick polished SiC membrane.

Table 3.2 *Principal characteristics of the additive and subtractive processes for fabricating X-ray absorber patterns.*

Feature	Additive process	Subtractive process
CD control	Good	Good
Thickness control	Poor	Good
Thermal expansion coefficient	Poor	Good
Defect control	Poor	Good
Process simplicity	Poor	Good
Stress control	Good	Poor

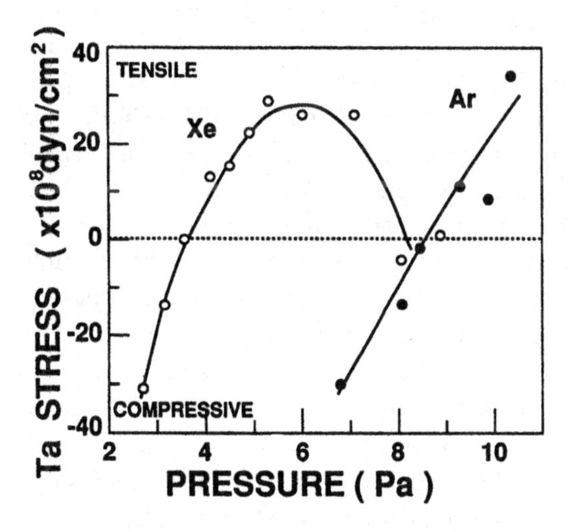

Figure 3.21 Stress in sputtered Ta films as a function of working-gas pressure.

Stress in the X-ray absorber affects the pattern-placement accuracy. Stress control for an X-ray absorber has thus been investigated for a long time. Generally, the X-ray absorber material is deposited either by RF-sputtering or by DC-sputtering. In both cases, stress in the absorber changes from compressive to tensile with increasing working-gas pressure. It has been found that the stress-transition vs gas-pressure relationship depends on the atomic-mass ratio of target material to working gas.[30] Tantalum has been the most widely used absorber material in the subtractive process because of its excellent dry-etching characteristics and relatively good stress-controllability.[29] Figure 3.21[31] shows the stress vs pressure curves for Ta films deposited by RF-sputtering using either Xe or Ar gas. It is clear that in the case of Xe gas, the stress transition takes place at much lower pressure than in the case of Ar gas. It was also found that Xe was not observed in the sputtered film, even though more than 1% Ar was observed when the Ta film was deposited with Ar gas. Sputtering with Xe in preference to Ar as a working gas has thus been used in order to obtain pure and high-density Ta absorber material.

It has been found, however, that the stress in a Ta film changes rapidly toward compression when the specimen is taken out of the vacuum chamber, then the stress continues to change gradually toward compression at room temperature. This transformation is thought to be due to oxidation along the grain boundaries.[32] A change in stress of both Ta and W films by X-ray irradiation was also observed.[33]

To make a more stable absorber, some compound materials such as Ta_4B,[34] WRe,[35] TaSiN,[36] TaGe,[37] and TaReGe[38] have been developed. All have been found to be amorphous, which is a favorable feature to prevent oxidation along the grain boundaries. It was reported that Ta_4B showed no measurable change in stress even after 16 kJ/cm^2 X-ray irradiation.[33]

Amorphous-alloy material is generally superior to metal in achieving smooth side walls through dry-etching. Figure 3.22[39] shows an example of a 0.1-μm line-and-space pattern fabricated using a TaGe absorber. Note that edge roughness is negligibly small.

3.3.5 X-ray mask fabrication processes

The conventional subtractive X-ray mask fabrication steps are shown in Figure 3.23. This process is called wafer processing because the patterning process is carried out on a bulk wafer (before back-etching). In the case of wafer processing, a systematic distortion (as shown in Figure 3.24) is induced in the back-etching process.[40,41] This distortion is caused by deformation of the mask frame due to a tensile stress in the membrane.

To eliminate mask distortion due to the frame deformation, a new fabrication process for X-ray masks has been developed (Figure 3.25). This process is called a membrane process because the pattern delineation is carried out on the membrane (after back-etching). In this method, the systematic deformation due to frame shrinkage will be negligibly small because the pattern delineation and dry-etching for the X-ray absorber are carried out after the back-etching process. The possible reasons for mask distortion in the membrane process are:

- Non-uniformity of stress in the X-ray absorber.
- Pattern-placement error in the electron-beam lithography (EBL) used for delineating the mask pattern.
- Damage of the membrane due to X-ray irradiation.

Figure 3.22 SEM photograph of 0.1-μm-L/S TaGe absorber pattern.

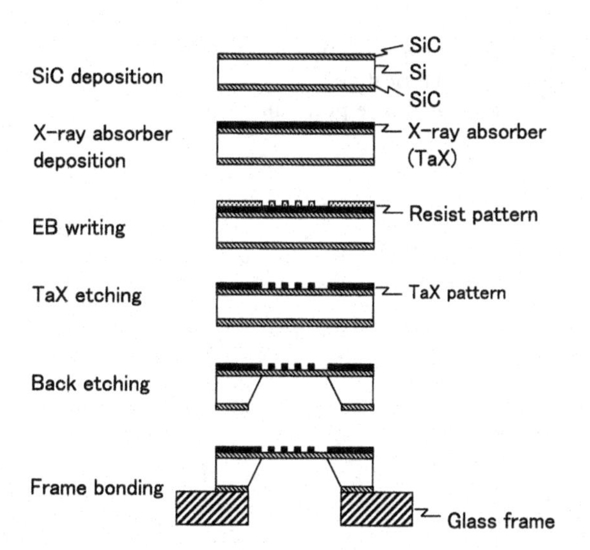

Figure 3.23 Conventional X-ray mask fabrication sequence.

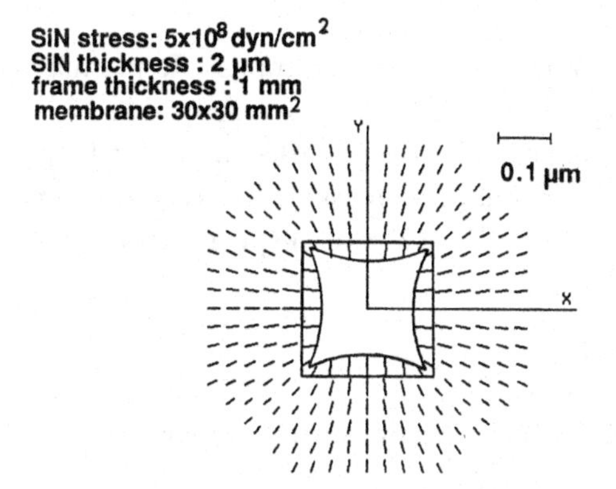

Figure 3.24 Si-frame in-plane distortion caused by membrane stress after a back-etching process.

All these problems have been carefully investigated, and there is room for further improvement in order to achieve 0.1-μm-design rule X-ray masks.

When the masks are used in the fabrication of devices with a minimum feature size less than 0.15 μm, the width of the X-ray absorber pattern should be controlled within less than 10 nm. The keys to making such absorber patterns are extremely precise EBL and dry-etching technologies.

In the EBL process for fabricating the X-ray mask, the mask substrate is covered with an X-ray absorber made of heavy metals which have a high emission coefficient of secondary electrons. Proximity effect correction is thus essential in this lithography in order to achieve the required CD control. High-acceleration-voltage (75–100-kV) EBL systems have been developed to achieve higher resolution with smaller beam spot-sizes.[42,43] Such systems are advantageous in achieving high resolution with a simplified proximity effect correction because the forward-scattering of electrons is suppressed and because back-scattered electrons from the substrate spread so widely that their effects can be

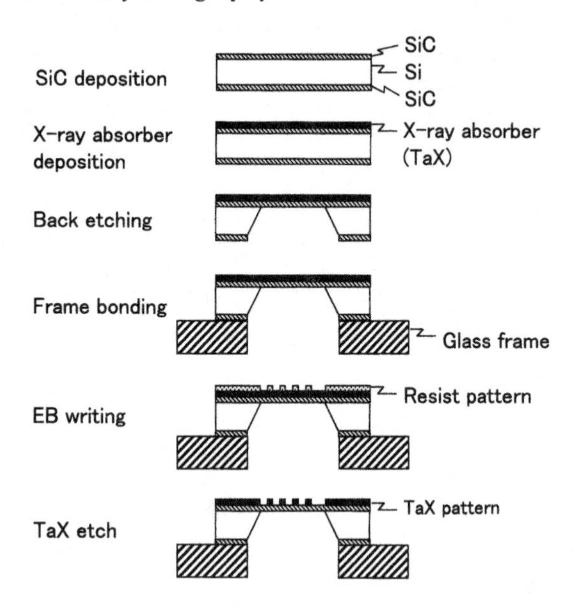

Figure 3.25 Low-distortion X-ray mask fabrication processes (membrane process).

estimated by considering only pattern density. As a result, a 75-nm line-and-space X-ray absorber pattern has been fabricated by electroplating of gold by using a 0.75-μm-thick PMMA resist pattern as a plating mask.[42]

Multiple-pass delineation technology[44,45] is effective in improving both stitching accuracy and image placement and in reducing pattern-width fluctuation.[44] Multiple exposures are carried out typically 3 or 4 times by using different main fields and sub-fields.

Dry-etching characteristics of the X-ray absorber are another key to CD control. The micro-loading effect (an etching-rate phenomenon) becomes increasingly serious with decreasing pattern width. The key points in suppressing this effect are to use high-density plasma, side-wall protection for the absorber pattern during the etching process, and a hard mask which enables a thinner resist to be used. ECR (electron cyclotron resonance) plasma-etching at low temperature has been one of the most popular methods for fabricating X-ray absorber patterns. A 0.1-μm line-and-space TaGe pattern (see Figure 3.22) was fabricated by using low-temperature ECR plasma-etching with SF_6 gas. An extremely thin (75-nm) CrN hard-mask was used as an etching mask.[39] To minimize mask distortion, stress in the CrN hard-mask was kept to less than 10 MPa.

3.4 X-ray mask alignment technology

In proximity XRL, overlay accuracy depends on the mechanical positioning of the mask and the wafer. Extremely accurate mask-to-wafer overlay, as well as precise control of the mask-to-wafer gap, is therefore necessary for taking full advantage of the excellent resolution of XRL. Position detection and mechanical positioning in advanced XRL systems require nanometer-order precision. In addition, high-speed positioning is required to achieve high throughput.

3.4.1 Theoretical study

If we assume the coordinate system shown in Figure 3.26, overlay errors in one-to-one proxi-

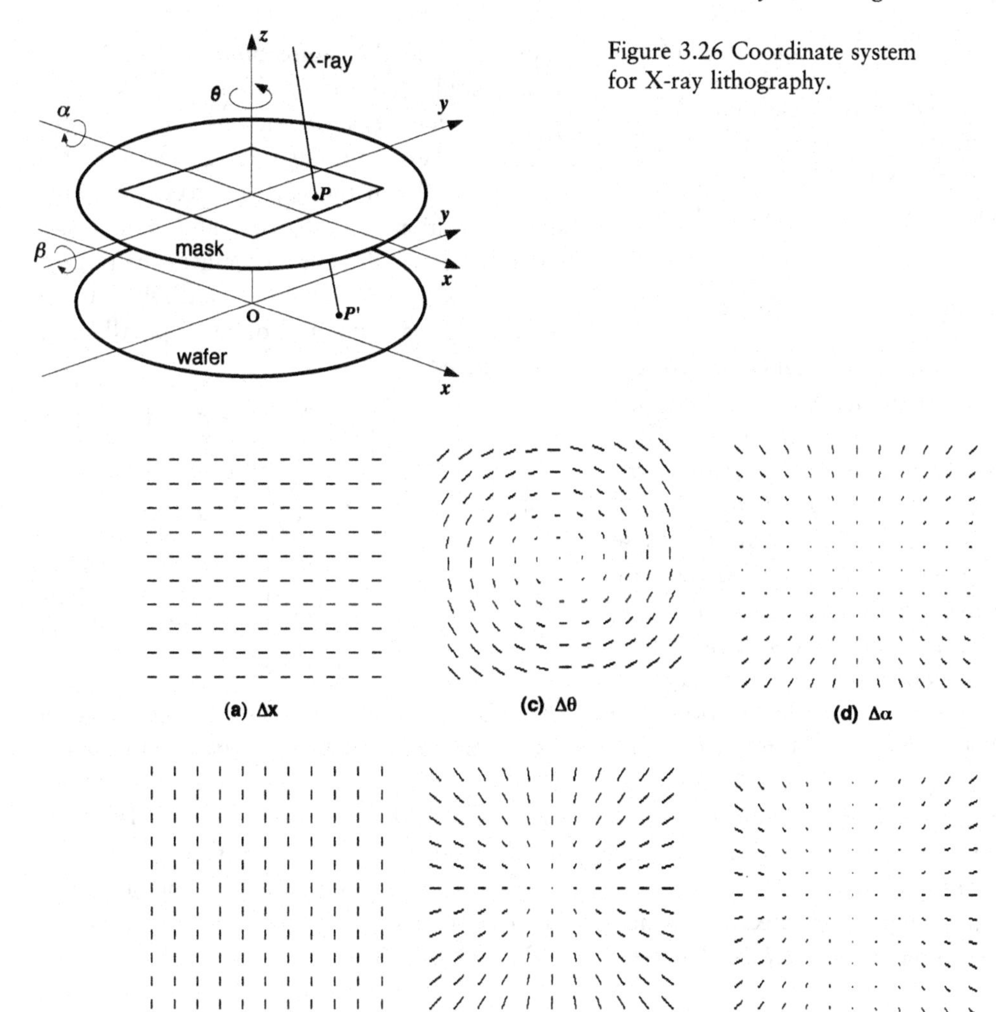

Figure 3.26 Coordinate system for X-ray lithography.

Figure 3.27 Overlay-error distribution caused by misalignment.

mity XRL can be calculated geometrically. The noteworthy errors are: Δx and Δy, the in-plane translations along the x- and y-axes; $\Delta\theta$, the whirlpool-like error due to rotational misalignment, which generally increases with distance from the center of the mask; Δz, the run-out error due to the mask-to-wafer gap change between exposures, that results in expanded or shrunken exposed patterns; and lastly, $\Delta\alpha$ and $\Delta\beta$, the run-out fluctuations distributed in the exposure field and due to tilting of the gap. These six kinds of errors are illustrated in Figure 3.27. The purpose of mask alignment is to determine these misalignments (Δx, Δy, Δz, $\Delta\alpha$, $\Delta\beta$, $\Delta\theta$) and to eliminate them before or during exposure.

The overlay-error vector $\Delta\varepsilon$ observed at the point P' on the wafer (see Fig. 3.26) is given by the following equation using six axes misalignments:

$$\Delta\varepsilon = \mathbf{A}(\Delta x, \Delta y, \Delta z, \Delta\alpha, \Delta\beta, \Delta\theta), \quad (3.12)$$

where **A** is a constant matrix determined by the location of point P in the exposure field and the incident angle of the X-rays. If the determinant of **A** is not zero, Equation (3.12) has the solution

$$(\Delta x, \Delta y, \Delta z, \Delta\alpha, \Delta\beta, \Delta\theta) = \mathbf{A}^{-1}\Delta\varepsilon. \quad (3.13)$$

To give the solution to Equation (3.13), the mask alignment system should thus be equipped with both (i) positioning apparatus along the above-mentioned six axes and (ii) an error-detection system for measuring the overlay errors of all six axes independently. For these measurements, sets of alignment marks are made on the mask and wafer. After detecting the marks, the machine drives each axis a based on the amount of $\Delta\mathbf{d} = (\Delta x, \Delta y, \Delta z, \Delta\alpha, \Delta\beta, \Delta\theta)$, which is calculated according to the measured error $\Delta\varepsilon$. To keep in-plane errors within the alignment margin, the factors Δx, Δy, and $\Delta\theta$ are much more important than the factors Δz, $\Delta\alpha$, and $\Delta\beta$; and each set of three axes is controlled separately.

The following simple and effective calculation can thus be applied in a practical alignment system. Along the z-direction, the mask and wafer should be set parallel but separated by a gap of a few tens of microns, and the gap should be kept constant during exposures for each layer. When the gap is detected at three sensor locations (h_1, h_2, h_3) and the aligner has three z-axis-driving mechanisms at (z_1, z_2, z_3) as shown in Figure 3.28, the driving stroke required for each point to obtain a parallel gap is calculated as:

$$\begin{bmatrix} \Delta z_1 \\ \Delta z_2 \\ \Delta z_3 \end{bmatrix} = \frac{4}{27} \begin{bmatrix} 1+2a & 1-a & 1-a \\ 1-a & 1+2a & 1-a \\ 1-a & 1-a & 1+2a \end{bmatrix} \times \begin{bmatrix} \Delta h_1 \\ \Delta h_2 \\ \Delta h_3 \end{bmatrix}, \quad (3.14)$$

where a is a parameter defined by the layout of the sensors and actuators as shown in the figure. With regard to the lateral alignment along the mask- and wafer surfaces, the mark layout is usually designed as shown in Figure 3.29. The set of driving amount (Δx, Δy, $\Delta\theta$) is calculated from the set of three relative displacements (ΔY_1, ΔX_2, ΔY_3) as

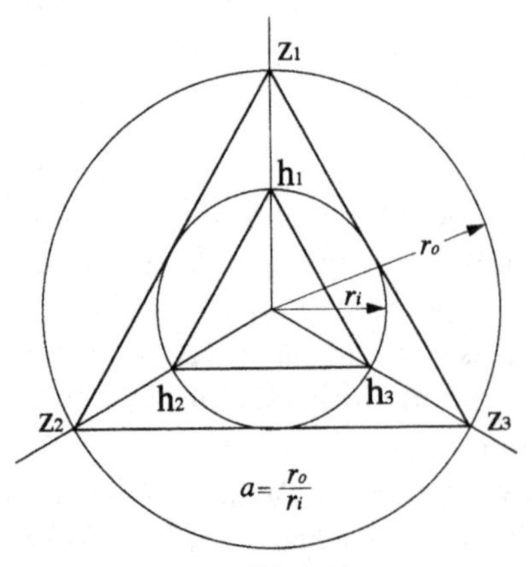

Figure 3.28 Gap-control arrangement.

Figure 3.29 Alignment control for the three in-plane axes.

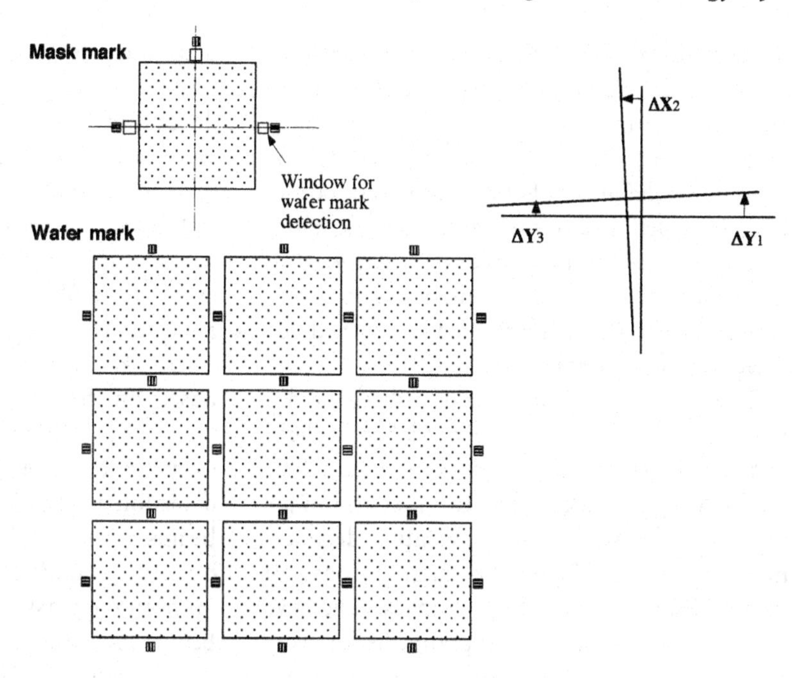

$$\begin{bmatrix} \Delta x \\ \Delta y \\ \Delta \theta \end{bmatrix} = \begin{bmatrix} 1 & 0 & 0 \\ 0 & 1/2 & 1/2 \\ 0 & 1/2R & -1/2R \end{bmatrix} \begin{bmatrix} \Delta X_2 \\ \Delta Y_1 \\ \Delta Y_3 \end{bmatrix}, \tag{3.15}$$

where R is the radius of the mark position from the center of the exposure field.

3.4.2 Mark-detection technology

Several methods for accurately detecting marks have been studied, and techniques for detecting marks in X-ray masks can be roughly classified into two groups. The first group of methods utilizes the reflection or scattering of white light from the alignment mark or its edges. Generally, a combination of white-light illumination, optical-microscope imaging, video-camera detection, and video-signal processing are used to define the mark position. To detect both mask marks and wafer marks, which are separated by a proximity gap, several techniques have been applied. The simplest way is to detect the mask- and wafer marks independently in a sequential operation by moving the objective lens along the optical axis of the alignment microscope.[46] This method, however, is time-consuming because the sequential operation takes a long time. There are also techniques for detecting the marks on the mask and wafer simultaneously. One technique utilizes dual-focus optics which can form both mark images on the same image plane. A dual-focus optical system utilizing chromatic aberration of two-wavelength illumination has been developed.[47] Another uses a Fresnel zone plate (FZP) for the mask mark, and alignment light through the mask mark is focused on a wafer mark. This makes it possible to measure the superimposing status between the mask- and wafer alignment targets.[48]

The signal for detecting the mark position is strongly affected by the reflectivity of the wafer surface. In device-fabrication processes, for example, the alignment-mark signal from an aluminum-wiring layer tends to be much brighter than that from silicon-based layers. This causes halation of the images and, consequently, signal saturation that leads to degradation of mark-detection accuracy. The key to these optical alignment techniques is to maintain sufficient and suitable image contrast for detecting mark edges.

The second group of methods utilizes the diffraction and/or interference of coherent light from gratings on the mask and wafer. The dual-grating method[49] was the first technique proposed for measuring the relative displacement of two gratings. The main purpose of using diffraction from the gratings is to achieve high position-detection resolution: the intensity of diffracted light is theoretically very sensitive to the relative position of the two gratings. For detecting intensity, however, changes in reflectivity in wafer gratings and the influence of gap fluctuation on the intensity of the diffracted light still pose serious problems for stable mark detection.

Optical-heterodyne interferometry with gratings can overcome both these problems simultaneously and detect marks highly sensitively and stably with nanometer-order precision. There are several alignment-system designs based on the optical-heterodyne scheme using linear or checkerboard gratings.[50,51,52] An example of a method for optical-heterodyne detection is shown in Figure 3.30. Two laser beams with slightly different wavelengths (frequencies) illuminate the linear gratings formed on both the mask and the wafer. The diffracted beams interfere with each other and the phase of the resulting beat signal is detected. The phase difference between the signals from the

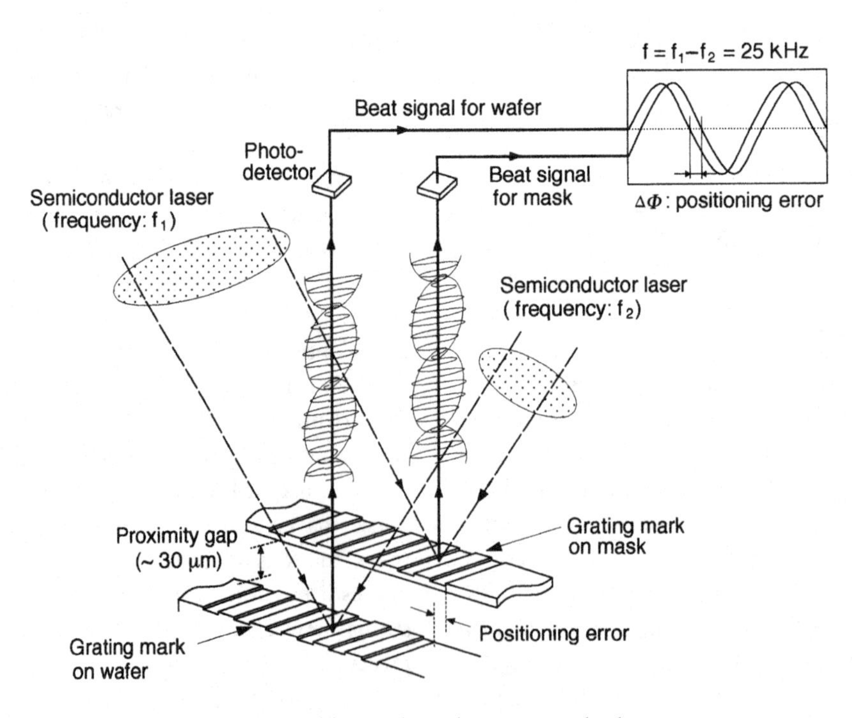

Figure 3.30 Optical-heterodyne detection method.

gratings on the mask and wafer is used to determine the relative position of the gratings. In theory, the phase difference $\Delta\phi$ between two beat signals is $2\Delta x/(p/2)$, where Δx is the lateral displacement between the mask- and wafer gratings and where p is the grating pitch. The relationship reveals that the position-detection characteristics are linear and periodic, repeating with the period of displacement, $p/2$.

This method is advantageous for achieving high sensitivity to the relative displacement between the mask- and wafer gratings. According to the above relationship, a phase difference of 0.5 degree for gratings with a 6-μm pitch corresponds to a displacement sensitivity of less than 5 nm. The symmetric optics prevents the detection signal from being affected by fluctuation of the mask-to-wafer gap. The accuracy of phase detection is not directly affected by changes in wafer-surface reflectivity due to the wafer processes. Moreover, because this scheme uses the same optical path for the mask and wafer, detection signals are not affected by laser wavelength fluctuations caused by the atmospheric environment. A technical concern raised by this method is how to deal with the influence of multiple reflections of light at the surfaces of the mask and wafer. One way to avoid this problem is to apply an anti-reflection coating (ARC) or opaque film to the surface of the masks.[52,53,54]

3.4.3 Positioning mechanism

Another important element for an X-ray mask alignment system is the mechanism for positioning the mask and wafer. The mechanism should provide two primary functions. The first is long-stroke x–y positioning for the wafer, that is, it should step the wafer in a vertical plane. The vertical x–y stage movement should be sufficiently accurate that the proximity gap can be made extremely small, and it should be fast enough to enable high-throughput exposure. To achieve high-speed movement with small pitch and yaw, the x–y positioning stage is usually guided by air bearings and drive by ball screws equipped with some gravity-compensation mechanisms. In addition, fine-positioning mechanisms for ultra-fine alignment are attached to the moving wafer table.[46,47,51,52]

The second function that the positioning mechanism should provide is multi-axis micro-positioning for three-dimensional alignment as described theoretically in Subsection 3.4.1. The device most often used for multi-axis micro-positioning is a piezoelectric actuator, which is very suitable for this purpose because of its compactness and high-precision motion. Many multi-axis micropositioners that use piezoelectric translators in combination with elastic support springs have been designed.[46,47,51,52,55]

One positioning mechanism is a fully air-lubricated x–y table driven by air-bearing lead screws without fine positioners. Air-lubrication is advantageous for a high-precision positioning table because the absence of solid friction eliminates both stick–slip and backlash. And, the error-averaging effect of the air film improves the motion accuracy of the positioning tables. The only disadvantage of air bearings is their rather low mechanical stiffness, which is most serious in a lead screw. Thus, if an air-bearing lead screw with sufficient stiffness can be produced, a lead-screw-driven positioning table can be used for both long-stroke x–y positioning and high-precision wafer alignment. Figure 3.31 is an outline drawing of a fully air-lubricated x–y stage driven by air-bearing lead screws[55,56,57] which provides the nanometer-order motion precision needed for aligning the X-ray mask.

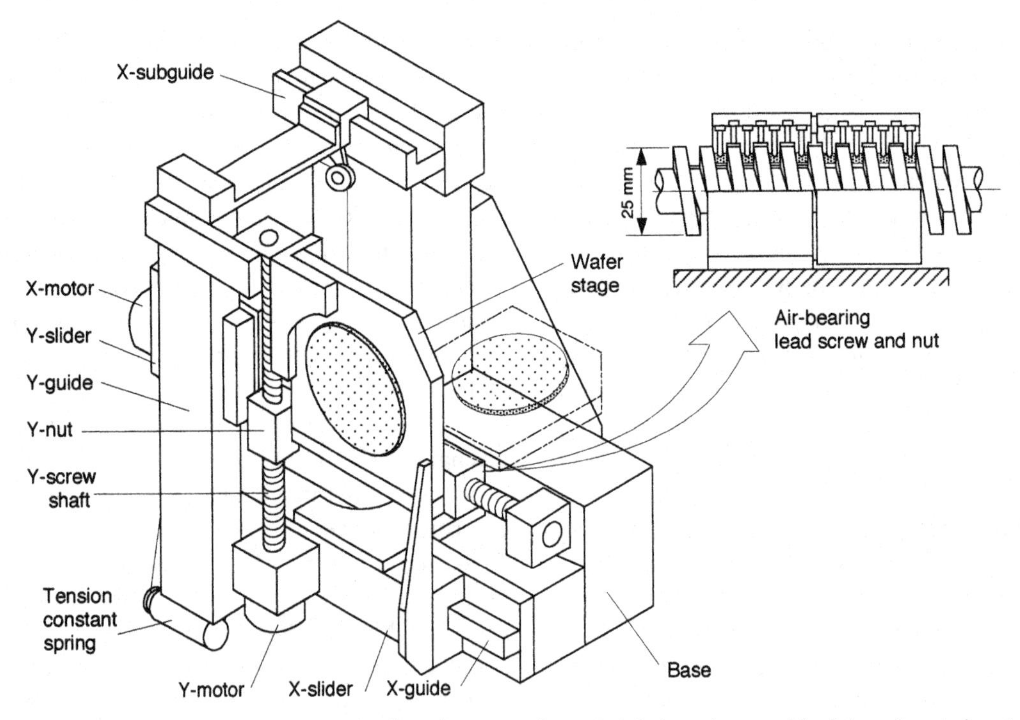

Figure 3.31 Outline drawing of an air-lubricated x–y table driven by air-bearing lead screws.

3.4.4 X-ray steppers and their performance

Several X-ray steppers to be integrated into synchrotron-based XRL systems have been developed and installed at SR facilities.[46,47,51,53,56] There are two types of alignment-control schemes: global; and die-by-die. In the global scheme, alignment-mark positions are measured before a series of exposure operations. Then exposures (without an alignment procedure) are repeated only by step-positioning according to the initial mark position measurements. On the other hand, the die-by-die alignment procedure is executed at each exposure field. The global alignment technique is useful in achieving high throughput, but the die-by-die scheme provides better overlay accuracy.

Figure 3.32 shows the results of a dynamic step-positioning and alignment operation of the X-ray stepper SS-1,[56] which is equipped with an optical-heterodyne alignment system and an air-lubricated x–y stage. After step-positioning with the laser interferometer for measuring the stage position, the positioning control is switched to a servo-alignment mode using an optical-heterodyne detection signal.[57] Both positioning operations are performed by a PID (Proportional–Integral–Differential) control system with optimum control parameters. As shown in the figure, 18-mm step-positioning and precise alignment within 10 nm can be performed in less than 1.5 seconds.

Overlay error in XRL has various causes: mask distortion, stepper error, wafer distortion which occur during device fabrication, etc. Alignment capability is usually evaluated by exposure experiments.[52,53,54,57,58] Figure 3.33

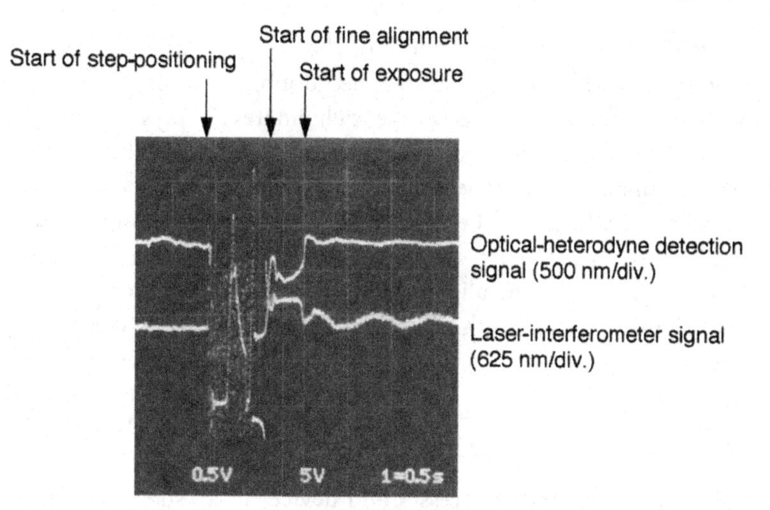

Figure 3.32 Dynamic positioning characteristics of the air-lubricated table.

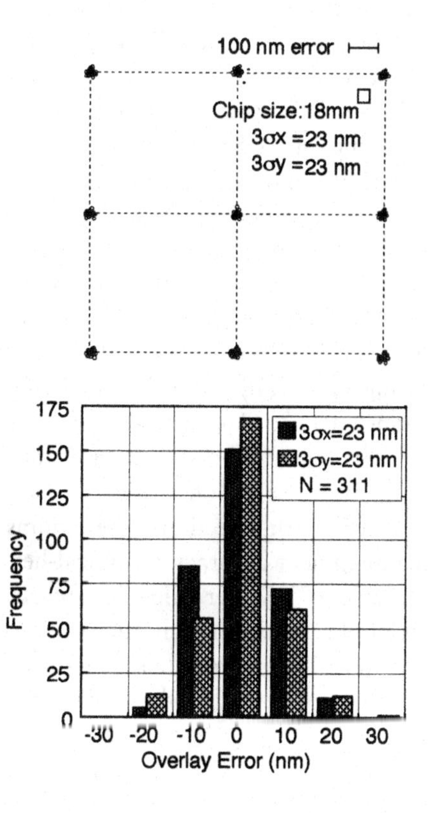

Figure 3.33 Alignment repeatability of SS-1 X-ray stepper evaluated by 'double-exposure' method.

shows the alignment repeatability of the X-ray stepper SS-1 evaluated by the so-called double-exposure method.[57] This experiment eliminates mask distortion and thus reveals the pure mechanical-alignment repeatability of the X-ray stepper, which can be represented by the standard deviation from the mean error. As shown in the figure, the alignment repeatability was 23 nm (3σ) for both the x- and y-direction.

3.5 X-ray lithography processes and device fabrication

3.5.1 X-ray lithography processes

The XRL processes described in this section have been carried out using the Super-ALIS, which is described in Subsection 3.2.7, as an X-ray source.

The beamline has two toroidal mirrors[59] as shown in Figure 3.1. The first mirror converges SR beams with a divergence angle of 1.8 degrees and at the same time removes the hard radiation. The second mirror collimates the beam horizontally and oscillates in order to enlarge the exposure area vertically. X-rays are extracted through three windows in the beamline.[60] The first window is a 1.3-μm-thick diamond film which seals an ultrahigh vacuum and blocks the ultraviolet light that would cause the Be window and X-ray mask to overheat. The diamond-membrane window is highly resistant to irradiation damage. The second window is a 15-μm-thick Be window which seals the vacuum, and the third window at the end of the beamline is a 1-μm-thick SiN membrane which seals the He gas filling the chamber where the alignment optics are installed. This window system enables the mask and wafer to be exposed in the open air and makes the mechanical operation of the stepper fast and easy. The wavelength of X-rays striking the wafer through the mask membrane ranges from 0.65 to 1.2 nm, and the peak-power wavelength is 0.7 nm. When the mirror scan-width is 25 mm, the X-ray power is about 0.7 mW/cm^2 per milliampere of storage current. This means that, at a storage current of 500 mA, a 1.4 s exposure is sufficient for a resist with a sensitivity of 50 mJ/cm^2. Recently, the possibility of a beamline which will offer an X-ray intensity as high as 75–80 mJ/cm^2 with $\pm 2\%$ uniformity has been suggested through simulation assuming a compact normal conducting synchrotron (AURORA-2S) as an X-ray source.[61] If such a high-efficiency beamline is realized, 1 s exposure for a 50-mm square field will be achieved using a 75-mJ/cm^2-sensitivity resist.

X-ray masks consisting of a Ta absorber and a 2-μm-thick SiN membrane were used for the experiments described below. Absorber thickness was typically 0.65 μm (contrast: 7) but was sometimes reduced to 0.4–0.3 μm (contrast: 2.5–3.5) to get better resolution for line-and-space patterns below 0.25 μm. A PAT method (previous analysis of distortion and transformation of coordinates) was used to improve the placement accuracy of the mask pattern. This method compensates for the complicated distortion induced during mask fabrication by modifying the EB-writing data.[62] An opaque film was coated on the mask marks and an anti-reflection film was coated on the mask windows for wafer mark-detection. These films improve the alignment accuracy of optical-heterodyne alignment.[63] Wafer alignment-marks were renewed on scribe-lines in each individual process in order to avoid mark deformation that would degrade the alignment accuracy.[64]

Both negative and positive resists have been used for XRL processes. The negative resist SAL-M8™ (Shipley),[65] which is a novolac-based three-component chemically amplified resist, was used for patterning layers for isolation, gates, and metallization in CMOS LSI fabrication. The positive resist CANI™,[66] which is a three-component chemically amplified resist based on poly(hydroxystyrene) (PHS), was used for patterning contact holes and through-holes. The sensitivity of both resists was approximately 100 mJ/cm^2 along with a high resolution in 0.2-μm regions. CANI™, however, is sensitive to airborne contamination during post-exposure delay. Poly-αmethylstyrene (PαMSt), which is removable by a xylene treatment in the development process, was over-coated on the CANI™ resist to stabilize the sensitivity and pattern size. Both SAL-M8™ and UV-cured CANI™ exhibited etching durability as good as that of conventional novolac-based optical resists.

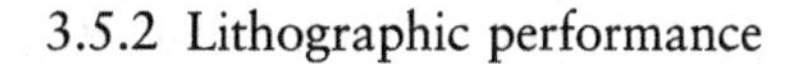

3.5.2 Lithographic performance

3.5.2.1 *Resolution and critical dimension (CD) control*

In XRL, Fresnel diffraction, photoelectron scattering, and acid diffusion in the resist all deform the pattern. Among them, Fresnel diffraction is the most dominant factor and is usually reduced by decreasing the proximity gap. The minimum mask–wafer gap in practical applications, however, is limited to about 10 μm (preferably 20 to 30 μm for the stepper operation), because a small gap causes dynamic out-of-plane distortion of the mask membrane during wafer stepping and results in contact between the mask and wafer.

It has recently been shown that a low-contrast mask (with a contrast of 2 to 4) can provide high-resolution line-and-space patterns in 100-nm regions with a proximity gap between 10 and 30 μm (which is relatively large compared with the gap used with a conventional mask). However, this mask reduces the expo-

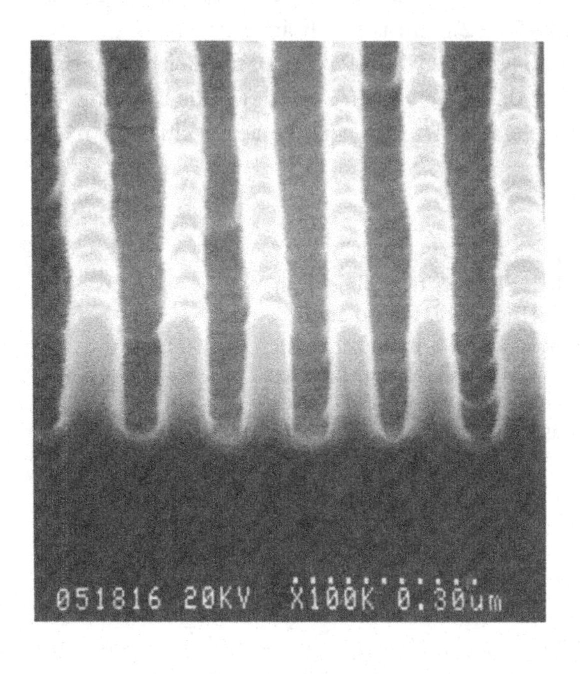

Figure 3.34 SEM photograph of 80-nm line-and-space patterns replicated onto a 0.3-μm thick ZEP520™ resist using a low-contrast (2.5) mask.

sure latitude for patterns larger than a quarter-micron.[67,68,69,70,71,72,73] This is because the interference between X-rays diffracted from the absorber edges and X-rays passing through the absorber patterns with a different phase shift increase the exposure contrast for fine patterns. For example, 80-nm line-and-space patterns have been resolved in a 0.3-μm thick ZEP520™ resist (Nippon Zeon Co.) with a mask contrast of 2.5 and a 10-μm gap (Figure 3.34). Pattern collapse is avoided by using an organic solution in the development process. Resolutions of 55, 70, 80, and 90 nm have been achieved with gaps of 10, 15, 20, and 30 μm, respectively. But when chemically amplified resists are used with alkaline-development and water-rinse processes, the resolution is degraded by 10 to 30%.

Resolution in the practical resist process is limited either by pattern collapse during development or by pattern deformation, by residue and scum over the unexposed area, when a negative resist is used. Pattern collapse is particularly common in the patterning of dense patterns such as line-and-space, and is mainly attributed to adhesion between adjacent patterns caused by the surface tension of the rinse water during the drying process.[74] The maximum slant of the resist pattern is inversely proportional to the Young's modulus of the resist material, is proportional to the surface tension of the rinse liquid and to the third power of the aspect ratio. The highest aspect ratio that can be obtained in the process using a water-rinse has been found to be about 5. This value can be increased by using a low-surface-tension liquid for rinsing. Using an organic rinsing solution, for example, the maximum aspect ratio has been increased by 40–50%. However, the pattern-collapse problem will be more serious with decreasing minimum feature size.

CD control was evaluated under standard wafer conditions for device fabrication.[64] As shown in Figure 3.35, XRL offers an exposure latitude exceeding 20% for ±10% CD control for all kinds of patterns as small as 0.2 μm, and the latitude remains good to 0.1-μm patterns. XRL also provides excellent CD control even when there are steps on the substrate. Figure 3.36 shows 0.2-μm resist patterns with 2-μm thickness replicated over 1-μm cross steps and 1-μm parallel steps. There is no influence from the substrate materials or topography. The CD error for various resist thicknesses ranging from 0.45 to 2.0 μm is less than ±10%.

Figure 3.37 shows the pattern-size variation for 0.2-μm gate across both (a) the chip and (b) the wafer. The same gate on each chip was mea-

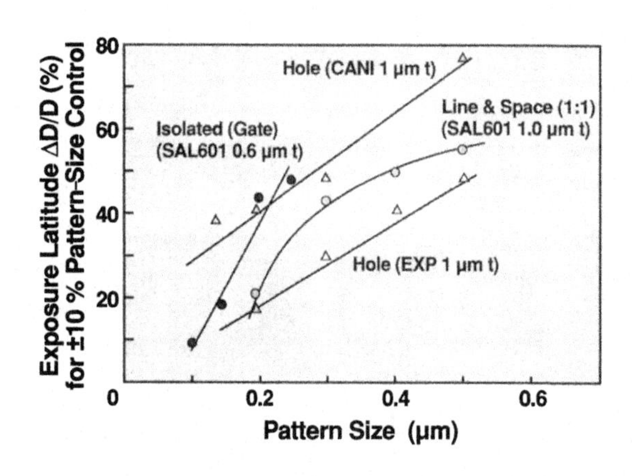

Figure 3.35 Exposure latitude for ±10% critical dimension (CD) control, for isolated, hole, and L/S patterns. SAL601, CANI, and EXP are negative resists of thickness *t*, as indicated.

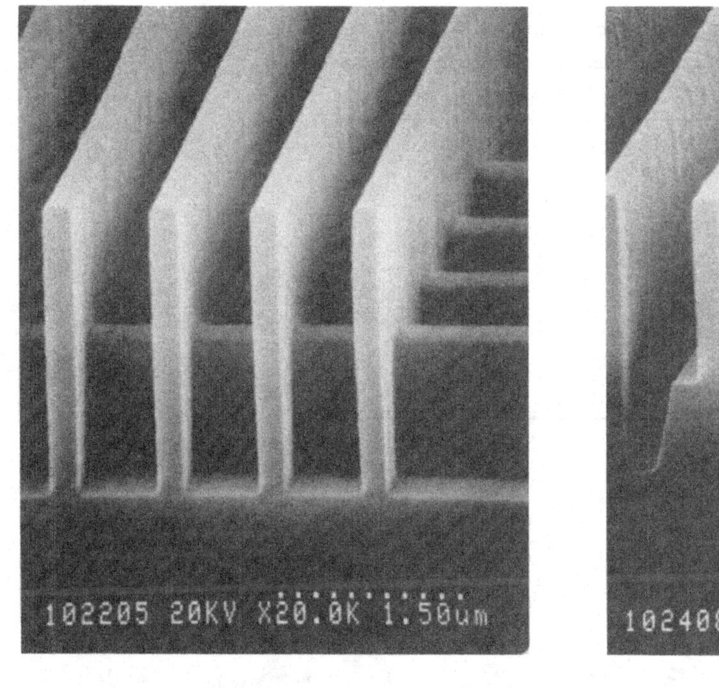

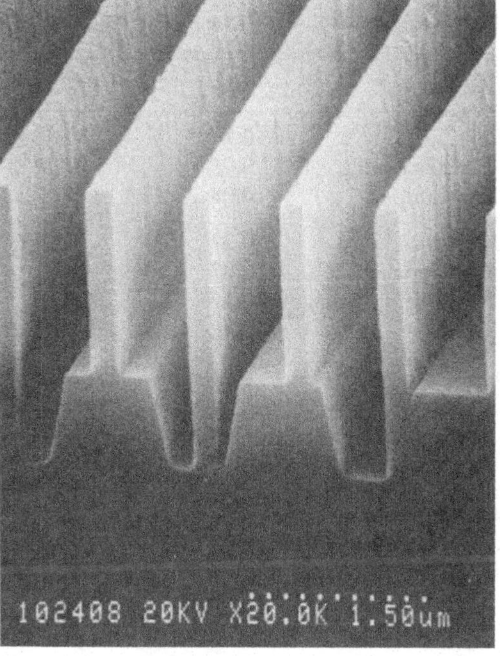

Figure 3.36 SEM photographs of 0.2-μm patterns replicated in a 2-μm thick resist over 1-μm (a) cross and (b) parallel steps.

sured for 37 chips on a wafer and for 25 wafers in the same lot. The pattern-size variation across the chips in a wafer was less than 20 nm for 3σ, and the overall pattern-size variation (including the measurement error) was about 27 nm. Taking the measurement error into consideration, the CD control with XRL is satisfactory for the requirement of less than $\pm 10\%$ in 0.2-μm regions. Mitsubishi[75] also showed that the fluctuation from the designed value of a 0.14-μm pattern was 14 nm (10%)

3.5.2.2 *Overlay*

Overlay accuracy was evaluated during actual device fabrication. Table 3.3 lists the trend of the overlay-error factors divided into two dominant factors: mask pattern-placement error; and alignment error. Offset errors could be removed by a dummy exposure using a bare Si substrate prior to device fabrication. This was possible because substrate-dependent offset errors (changes mainly caused by multiple reflections and cross-talk†) were eliminated by using both an opaque coating and an anti-reflection coating on the masks. The overlay error due to mark deformation (caused by several wafer processes such as the planarizing processes) is prevented by using renewal marks (Table 3.4; excluding mask errors).

† 'Noise' from the neighboring mark-detecting channel. In this case, a noise from a mask mark to a wafer mark or converse.

Table 3.3 *Overlay accuracy at the 3σ level. Unit is μm.*

Error factors	'92	'93–'94	'98
Mask	0.1	0.08	0.05–0.06
E-beam writing	0.06	0.04	0.03
Fabrication	0.08	0.07	0.05
Alignment	0.22	0.13	0.06–0.08
Repeatability	0.07	0.05	0.03
Mark deformation	0.17	0.10	0.05
Others	0.12	0.07	0.05
Overall	0.25	0.15	0.07–0.01

Figure 3.37 Pattern-size variation of 0.2-μm gates across (a) chip and (b) wafer.

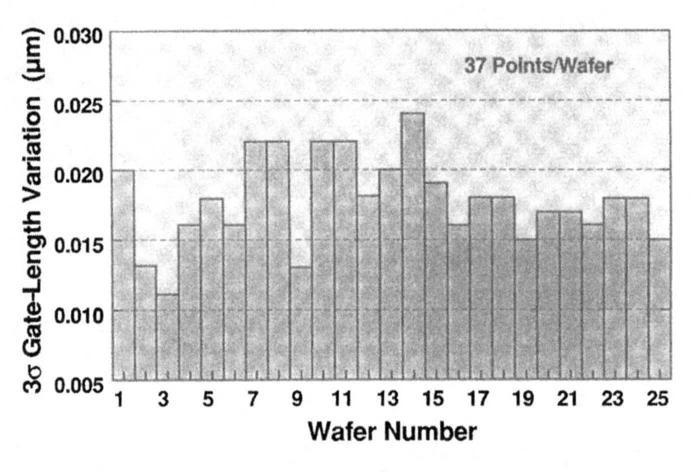

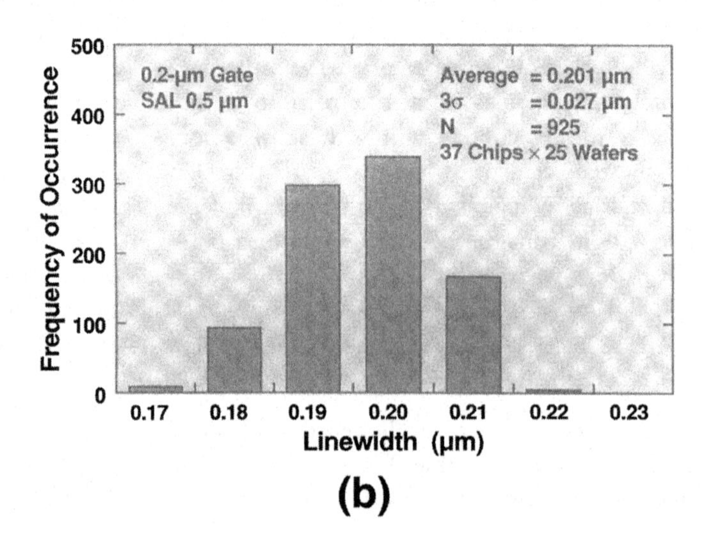

To achieve a total overlay accuracy which will satisfy ≦0.1-μm-ULSI fabrication, both mask accuracy and alignment accuracy must be improved. In addition, pattern displacement errors caused by wafer distortion due to high-stress film deposition and that caused by mark deformation should be reduced. Moreover, neither the in-plane distortion of the mask and wafer (due to temperature changes caused by X-ray irradiation) nor changes in flatness caused by the holding mechanism should be ignored. The linear distortion, however, can be compensated by using linear shrinkage/enlargement-feedback of the EB-writing of mask patterns. Magnification correction for X-ray masks during X-ray exposure has also been developed.[76] By reducing these errors, the total overlay accuracy should be improved by better than 30 nm to meet the requirements for 0.1-μm-ULSI fabrication.

3.5.2.3 *Throughput*

Throughput was evaluated using 6″ (150-mm) ϕ wafers,[77] and the optimum exposure dose and corresponding exposure time for one shot (an 18-mm-square exposure field) at a 400-mA storage current are listed in Table 3.5. The resist process conditions are also listed. The optimum exposure doses range from 1500 to 2500 mA · s for SAL-M8 resist and from 700 to 1500 mA · s for CANI resist. The optimum exposure dose for each exposure level changed in such a wide range to achieve the targeted CD control. The practical exposure time for a 400-mA storage current was about 2 to 6 s per field.

The total exposure time for a wafer is shown in Figure 3.38 as a function of storage-beam current. At a 400-mA storage current, the total exposure time for a 6″ (150-mm) ϕ wafer was between 2 and 4 min for SAL-M8 and between 1 and 2 min for CANI. The maximum throughput was 12 wafers/h, including the wafer loading/unloading time and stepping/alignment time of the stepper (typically about 2 min/wafer and 2 s/shot, respectively).

To reduce the exposure time, a high-power X-ray source using a compact synchrotron with a high-efficiency beamline[61] is being developed as described in Subsection 3.5.1. An X-ray stepper employing global-alignment technology in order to achieve high throughput is also under development.[78] The stepper is designed

Table 3.4 *Effects of substrate on alignment accuracy.*

Layer	$3\sigma x$ (μm)	$3\sigma y$ (μm)	N	Mark	Measurement
Active	0.096	0.107	1232	0-level	Optical-to-X-ray
Gate	0.111	0.100	1680	0-level	[Between the mark fabrication
S/D[a] contact	0.111	0.107	1120	0-level	level and each exposure level]
Gate contact	0.115	0.117	1344	0-level	
Metal 1	0.089	0.080	1480	Renewal	X-ray-to-X-ray
Through hole 1	0.153[b]	0.101[b]	740	Renewal	[Between the renewal mark
Metal 2	0.064	0.098	592	Renewal	exposure level and each exposure level]

[a] Source/Drain; [b] Without θ-axis compensation.

Table 3.5 *Optimum exposure dose and time for an exposure field of 18 × 18 mm² at a storage current of 400 mA. [1000 mA · s ≡ 100 mJ/cm².]*

Layer	Resist	Thickness (μm)	Dose (mA · s)	Exposure time (s)
Active	SAL-M8 (95 °C)	0.5	1500	3.8
Gate	SAL-M8 (90 °C)	0.5	2100	5.3
Gate contact	OverCoat/CANI	0.26/0.56	1000	2.5
S/D contact	OverCoat/CANI	0.26/0.56	1200	3.0
Metal 1	SAL-M8 (90 °C)	1.0	2300	5.8
Through hole 1	OverCoat/CANI	0.26/0.56	1500	3.8
Metal 2	SAL-M8 (90 °C)	1.0	2500	6.3

Sensitivity: SAL $D_{50} = 80$ mJ/cm² (90 °C PEB), 60 mJ/cm² (95 °C PEB); CANI $D_0 = 70$ mJ/cm² (85 °C PEB).

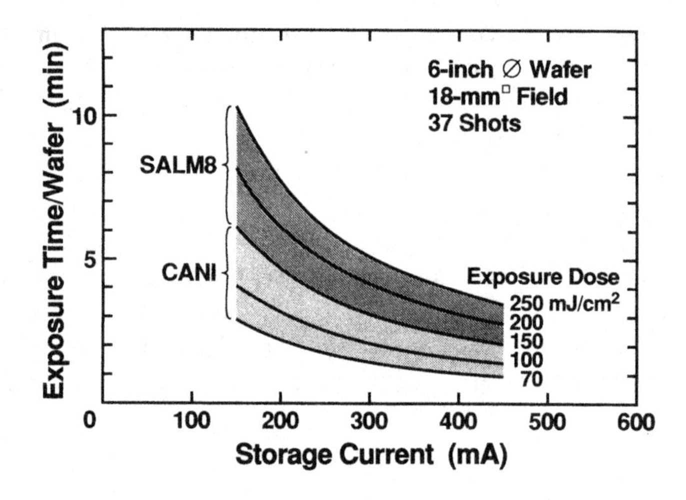

Figure 3.38 Exposure time versus storage current. The exposure time does not include wafer load/unload times and stepping/alignment times.

to achieve a throughput of 60 wafers/h when the exposure time is 1 s per field. If these systems are completed, an X-ray power of about 75 mW/cm² will be achieved. Therefore, if an X-ray resist with 75 mJ/cm² sensitivity can be developed, throughput of about 60 wafers/h will be achieved.

3.5.3 Application to LSI fabrication

To develop low-power and high-speed LSIs using thin-film silicon-on-insulator (SOI) structures for high-performance communication systems (such as high-frequency dividers, multipliers, prescalers, multiplexers, and demultiplexers) SR lithography was used either at several critical levels – the active, gate, source/drain contact, gate contact, and four metallization levels – or at the gate level only. Optical lithography was used for the other levels. An example of a finished chip is shown in Figure 3.39. An assortment of CMOS/SIMOX† gate-array LSIs with various minimum feature sizes and overlay tolerances ranging from 0.32 to 0.16 μm were designed on a 16-mm-square chip.

The device characteristics of gate-array LSIs (including operating frequency and functionality) were studied in order to evaluate various aspect of XRL such as CD control, overlay accuracy, and yield. The delay time of a two-input NAND†† with 0.2-μm gates is shown in Figure 3.40. The results show high-speed performance at a low-voltage supply. A 48 × 48-bit multiplier comprising 20.7 k gates with a 0.2-μm gate length was also fabricated with excellent and fully functional characteristics. A 12-kilobit-two-port SRAM was also operational. There was no clear difference in the chip yield of the SR processes and the 0.35-μm optical processes even though the gate length for XRL was much smaller than that for optical. The functionality has continued to increase as a result of efforts to fabricate defect-free masks by using a KLA SEM Spec™ inspection system and a Micrion M8000™ ion-beam repair system.[79]

The effects on hot-carrier reliability of the residual damage (in the gate oxide) caused by X-ray irradiation was also evaluated for

† SIMOX: separation by implanted oxide, one of the silicon-on-insulator (SOI) technologies.

†† NAND: A logical circuit, the inverse of AND (product of propositions).

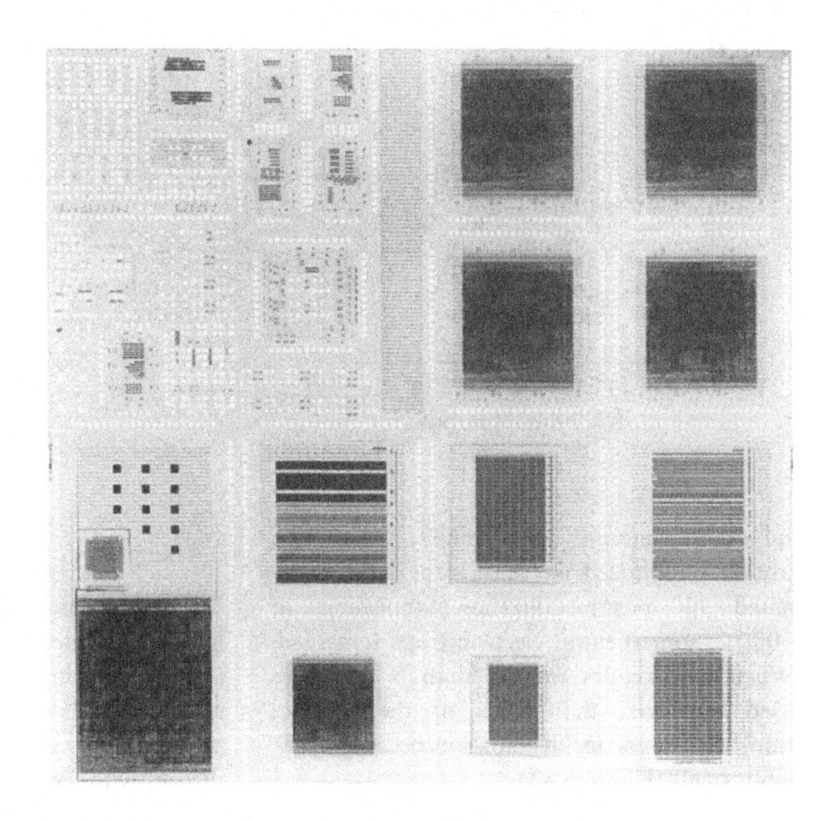

Figure 3.39 CMOS/SIMOX test LSIs. The chip size is 16-mm square.

Figure 3.40 Delay time of a two-input NAND with 0.2-μm gates fabricated using CMOS/SIMOX technology.

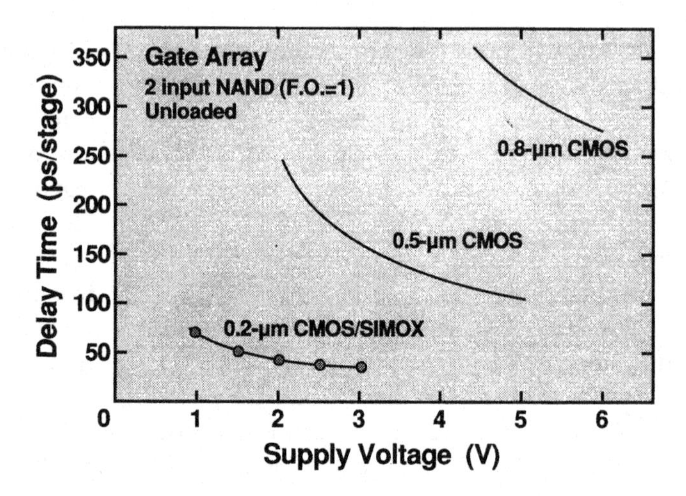

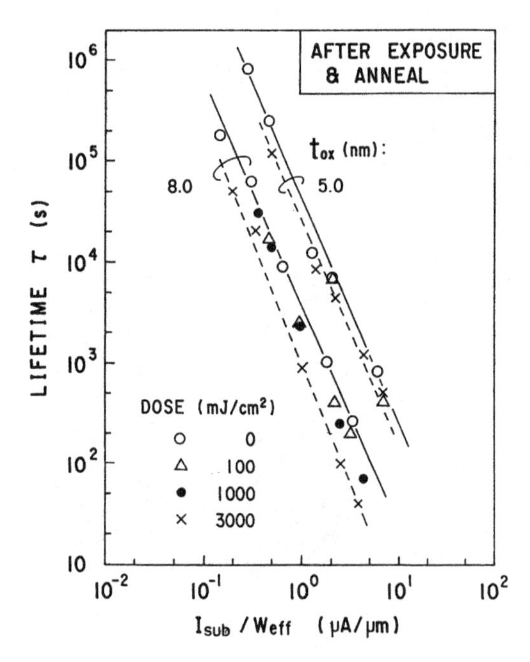

Figure 3.41 DC device lifetime τ vs substrate current normalized by channel width (I_{sub}/W_{eff}); gate length, $L_g = 0.2\,\mu m$.

sub-quarter-micron nMOSFETs. Although irradiation-induced interface traps were eliminated by post-metallization annealing at 400 °C, some neutral electron traps remained. When gate oxides thinner than 5 nm were used, however, the effects of the residual traps on hot-carrier degradation became negligibly small (Figure 3.41).[80]

3.6 Summary

Key technologies for XRL, such as synchrotron-based high-intensity X-ray source, high-precision 1× membrane masks and an X-ray mask aligner have been developed. All these components are now commercially available. Various devices with less than 0.2-μm minimum feature size have been fabricated using SR-based XRL at several organizations, and have demonstrated high-quality printing characteristics as well as a wide process latitude.

There are several issues remaining before XRL can be put to practical use. One is an overlay accuracy which will meet the requirements of sub-0.1-μm ULSI fabrication. This should be achieved by development of an extremely high-precision EB-lithography system and improving the membrane processes. A second issue is throughput. For achieving a practical throughput, the following improvements need to be applied: (1) boosting the beam current in the synchrotron; (2) improving the X-ray transmission efficiency in the beamline; (3) developing a high-sensitivity X-ray resist; (4) reducing the overhead time of the stepper; (5) developing a high-precision global-alignment technique; and (6) enlarging the exposure field. All these improvements are currently being carried out, and will be realized in the near future.

3.7 References

1. D. L. Spears and H. I. Smith, *Electronics Letters*, **8**, (No. 4), 102–4 (1972)
2. E. Spiller *et al.*, *J. Appl. Phys.*, **47**, 5450 (1996)
3. S. Ishihara, M. Kanai, M. Suzuki and M. Fukuda, *Microelectronic Eng.*, **14**, 141 (1991)
4. T. Watanabe, Housyasen to Genshi Bunshi, Edited by S. Shida, Kyoritu-shuppan, p.45 (1996)
5. N. Atoda, *Proc. 3rd European Particle Accelerator Conf., Berlin*, p. 77 (1992)
6. R. Feder, E. Spiller and J. Topalian, *Polymer Eng. Sci.*, **17**, 385 (1997)
7. T. Hosokawa *et al.*, *NTT Review*, **7**, (No. 4), 26 (1995)
8. T. Hosokawa *et al.*, *Proc. 1989 Intl. Symp. on MicroProcess Conf.*, p. 72 (1989)
9. B. L. Henke, E. M. Gullikson and J. C. Davis, *Atomic Data and Nuclear Data Tables*, **54**, (No. 2), 181 (1993)
10. http://www-cxro.lbl.gov/optical_constants/
11. T. Hosokawa *et al.*, *Rev. Sci. Instrum.*, **60**, (No. 7), 1783 (1989)
12. K. Yamada, M. Nakajima and T. Hosokawa, *Nucl. Instrum. and Meth.*, **A370**, 323 (1996)
13. H. Wiedemann, *Particle Accelerator Physics*, Springer–Verlag (1993)
14. H. Wiedemann, *Particle Accel-erator Physics II*, Springer–Verlag (1995)
15. J. S. Colonias, *Particle Accelerator Designs: Computer Pro-grams*, Academic Press, New York and London (1974)
16. LA-UR-86-3320, *Computer Codes used in Particle Accelerator Design*, Los Alamos National Laboratory (1987)
17. http://www.atdiv.lanl.gov/ (The Los Alamos Accelerator Code Group)
18. N. Takahashi, *Nucl. Instrum. and Meth.*, **B24/25**, 425 (1987)
19. K. Emura *et al.*, *Rev. Sci. Instrum.*, **63**, 1 (1992)
20. M. N. Wilson *et al.*, *IBM J. Res. Develop.*, **37**, (No. 3), 351 (1993)
21. M. Marushita *et al.*, *Proc. of the Asian Forum on Synchrotron Radiation*, p. 275 (1994)
22. T. Nakanishi *et al.*, *Rev. Sci. Instrum.*, **66**(2), 1968 (1995)
23. M. Sekimoto, H. Yoshihara and T. Ohkubo, *J. Vac. Sci. Technol.*, **21**, 1017–21 (1982)
24. H. Oizumi, K. Mochiji and T. Kimura, *Jpn. J. Appl. Phys.*, **29**, 2600 (1990)
25. T. Arakawa *et al.*, *Jpn. J. Appl. Phys.*, **31**, 4459–62 (1992)
26. M. Bieber, H. Betz and A. Heuberger, *Nucl. Instrum. and Meth.*, **208**, 281 (1983)
27. H. Luthje, B. Matthiessen, M. Harms and A. Bruns, *Proc. SPIE*, **773**, 15 (1987)
28. T. Shoki and Y. Yamaguchi, *Proc. SPIE*, **2254**, 313–19 (1994)
29. M. Sekimoto, A. Ozawa, T. Ohkubo and H. Yoshihara, Extended Abst. *16th Conf. Solid State Devices and Materials, Kobe*, pp. 23–6 (1984)
30. J. A. Thornton and D. W. Hoffman, *J. Vac. Sci. Technol*, **18**, 203 (1981)
31. M. Oda, A. Ozawa, S. Ohki and H. Yoshihara, *JJAP*

Series 4, Proc. 1990 Intl. MicroProcess Conf., pp. 96–9 (1990)
32. T. Yoshihara and K. Suzuki, *J. Vac. Sci. Technol. B.*, **11**, 301–3 (1993)
33. K. Ashikaga *et al.*, *Proc. SPIE*, **2793**, 204–10 (1996)
34. M. Sugawara, M. Kobayashi and Y. Yamaguchi, *J. Vac. Sci. Technol. B*, **7**, 1561–4 (1989)
35. S. Sugihara *et al.*, *Jpn. J. Appl. Phys.*, **34**, 6716–19 (1995)
36. W. J. Dauksher *et al.*, *J. Vac. Sci. Technol. B*, **13**, 3103–8 (1995)
37. T. Yoshihara, S. Kotsuji and K. Suzuki, *J. Vac. Sci. Technol. B*, **14**, 4363–5 (1996)
38. T. Yoshihara *et al.*, *J. Vac. Sci. Technol. B*, **16**, 3491–4 (1998)
39. S. Tsuboi, S. Kotsuji, T. Yoshihara and K. Suzuki, *J. Vac. Sci. Technol. B*, **15**, (No. 6), 2228–31 (1997)
40. S. Ohki *et al.*, *Jpn. J. Appl. Phys.*, **28**, 2074 (1989)
41. M. Oda *et al.*, *Jpn. J. Appl. Phys.*, **31**, (No. 12B), 4189–94 (1992)
42. M. A. McCord *et al.*, *J. Vac. Sci. Technol. B*, **10**, (No. 6), 2764–70 (1992)
43. R. Butsch *et al.*, *J. Vac. Sci. Technol. B*, **13**, (No. 6), 2478–82 (1995)
44. S. Ohki, T. Matsuda and H. Yoshihara, *Jpn. J. Appl. Phys.*, **32**, 5933–40 (1993)
45. D. M. Puisto *et al.*, *J. Vac. Sci. Technol. B*, **14**, (No. 6), 4341–44 (1996)
46. E. Cullmann, K. A. Cooper and W. Vach, *Proc. SPIE*, **773**, 2–6 (1987)
47. F. Sato *et al.*, *J. Vac. Sci. Technol. B*, **10**, (No. 6), 3235 (1992)
48. E. Kuono *et al.*, *J. Vac. Sci. Technol. B*, **6**, (No. 6), 2135 (1988)
49. A. Une, N. Takeuchi and Y. Torii, *J. Vac. Sci. Technol. B*, **8**, (No. 1), 51 (1990)
50. M. Suzuki and A. Une, *J. Vac. Sci. Technol. B*, **7**, (No. 6), 1971 (1989)
51. N. Uchida *et al.*, *J. Vac. Sci. Technol. B*, **11**, (No. 6), 2997 (1993)
52. K. Koga, I. Higashihara and T. Itoh, *J. Vac. Sci. Technol. B*, **10**, (No. 6), 3248 (1992)
53. M. Fukuda *et al.*, *NTT Review*, **7**, (No. 4), 33 (1995)
54. R. Hirano *et al.*, *J. Vac. Sci. Technol. B*, **12**, (No. 6), 3247 (1994)
55. S. Ishihara, M. Kanai, A. Une and M. Suzuki, *J. Vac. Sci. Technol. B*, **7**, (No. 6), 1652 (1989)
56. S. Ishihara *et al.*, *Microelectronic Eng.*, **14**, 141 (1992)
57. M. Fukuda, *J. Vac. Sci. Technol. B*, **12**, (No. 6), 3256 (1994)
58. C. J. Progler *et al.*, *J. Vac. Sci. Technol. B*, **11**, (No. 6), 2888 (1993)
59. T. Kaneko, Y. Saitoh, S. Itabashi and H. Yoshihara, *J. Vac. Sci. Technol. B*, **9**, 3214 (1991)
60. K. Kuroda, T. Kaneko and S. Itabashi, *Rev. Sci. Instrum.*, **66**, (No. 2), 2151 (1995)
61. E. Toyota, *J. Vac. Sci. Technol. B*, **16**, 3462 (1998)
62. S. Uchiyama *et al.*, *Jpn. J. Appl. Phys.*, **34**, (No. 12B), 6743–7 (1995)
63. M. Suzuki *et al.*, *Proc. SPIE*, **2254**, 329–36 (1994)
64. K. Deguchi *et al.*, *J. Vac. Sci. Technol. B*, **13**, 3040 (1995)
65. H. Liu, M. deGrandpre and W. E. Feely, *J. Vac. Sci. Technol. B*, **6**, 379 (1988)
66. H. Ban, J. Nakamura, K. Deguchi and A. Tanaka, *J. Vac. Sci. Technol. B*, **12**, 3905 (1994)
67. Y. Somemura, K. Deguchi, K. Miyoshi and T. Matsuda, *Jpn. J. Appl. Phys.*, **31**, 4221 (1992)
68. S. D. Hector, H. I. Smith and M. L. Schattenburg, *J. Vac. Sci. Technol. B*, **11**, 2981 (1993)
69. J. Z. Y. Guo *et al.*, *J. Vac. Sci. Technol. B*, **11**, 2902 (1993)
70. M. A. McCord, A. Wagner and D. Seeger, *J. Vac. Sci. Technol. B*, **11**, 2881 (1993)
71. J. Xiao *et al.*, *J. Vac. Sci. Technol. B*, **12**, 4038 (1994)
72. Y. Chen *et al.*, *J. Vac. Sci. Technol. B*, **12**, 3959 (1994)
73. Y. Kikuchi *et al.*, *Jpn. J. Appl. Phys.* **34**, 6709 (1995)
74. T. Tanaka, M. Morigami and T. Atoda, *Jpn. J. Appl. Phys.*, **32**, 6059 (1993)
75. Y. Nishioka *et al.*, *IEDM 1995*, 903 (1995)
76. N. Mizusawa *et al.*, *Proc. SPIE*, **3096**, 230–9 (1997)
77. K. Deguchi *et al.*, *J. Vac. Sci. Technol. B*, **13**, 3040 (1995)
78. K. Sentoku and T. Matsumoto, *J. Vac. Sci. Technol. B*, **16**, 3466 (1998)
79. I. Okada *et al.*, *Proc. SPIE*, **2437**, 253 (1995)
80. Tsuchiya, M. Harada, K. Deguchi and T. Matsuda, *IEICE Trans. Electron*, **E76-C(4)**, 506 (1993)

Electron-beam lithography

4

Takayuki Abe, Koichi Moriizumi,
Yukinori Ochiai, Norio Saitou,
Tadahiro Takigawa and Akio Yamada

4.1 Direct writing

4.1.1 Gaussian electron-beam lithography

4.1.1.1 *Introduction*

Gaussian electron beam is a basic electron-beam technology which is used not only for lithography but also for a scanning electron microscope (SEM), scanning transmission electron microscope (STEM), and several metrology techniques. Gaussian electron-beam lithography is a very important technology, especially in fine-pattern lithography used for basic research, and in mask writing for optical, X-ray and other projection lithographies. The tungsten thermal electron gun and the LaB_6 gun have been used for electron-beam systems. The Schottky thermal field emitter (TFE) has also been used in practice because of its small beam diameter with a high beam current. A beam size smaller than 5 nm is obtained, so a nanometer-scale resolution is achieved. The Gaussian beam, as previously mentioned, is used as a mask writing tool because it allows accurate positioning and high resolution. This subsection describes Gaussian electron-beam lithography for direct pattern writing at 0.1 μm or less.

4.1.1.2 *Gaussian electron-beam apparatus*

The Gaussian electron-beam apparatus has a basic and simple optical column compared with a variably shaped electron-beam system or the systems used for high-throughput lithography. The Gaussian apparatus is widely used for fine-pattern lithography, especially at less than 0.1 μm, to fabricate small quantities of devices and/or for pre-production, and in the development of quantum devices. In nanometer-scale lithography, problems with a system are how to obtain a small spot-size and a stable beam. With the development of the thermal field emitter (TFE), the beam spot-size was dramatically decreased. The TFE gun is composed of a tungsten needle covered with Zr/O to reduce the Schottky barrier height and it is operated at around 1500 K to migrate the Zr/O and activate the emitter surface. Electron-beam exposure systems using the TFE gun have been reported since 1981.[1,2,3,4,5]

Figures 4.1 and 4.2 are a diagram and a cross-section of the beam-optics column of a 50-kV TFE electron-beam system (the JBX-5FE or JBX-6000FS made by JEOL).[6] The specifications for the system are described in Table 4.1. It is commonly used and the specifications are almost standard for nanolithography. The system is designed to operate at 50 kV, and uses a Zr/O/W TFE gun. The pressure in the gun chamber is in the order of 10^{-10} Torr. The field size with a $5 \times 5\,\mu m^2$ subfield is $80 \times 80\,\mu m^2$. The electron beam is deflected by an electrostatic deflector and is digitally controlled. The minimum beam step-size is 2.5 nm and the maximum beam scanning frequency is 12 MHz. The system efficiently exposes a 6-inch (150-mm) wafer and an 8-inch (200-mm)

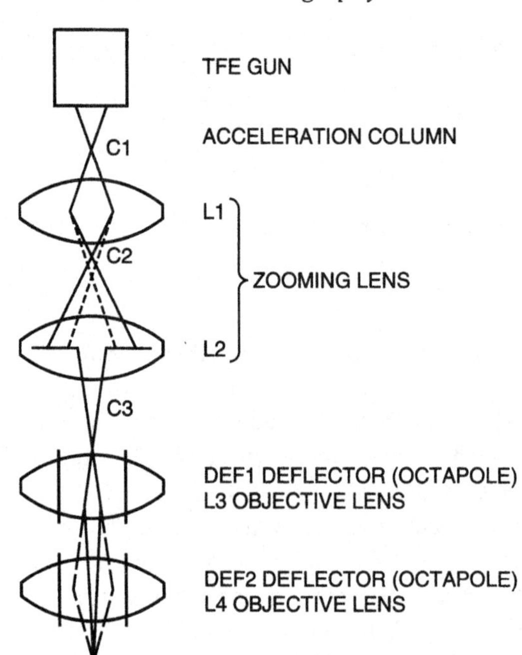

Figure 4.1 Schematic diagram of electron optics of 50-kV electron-beam exposure system using a thermal field emission (TFE) gun (JEOL: JBX-6000FS).

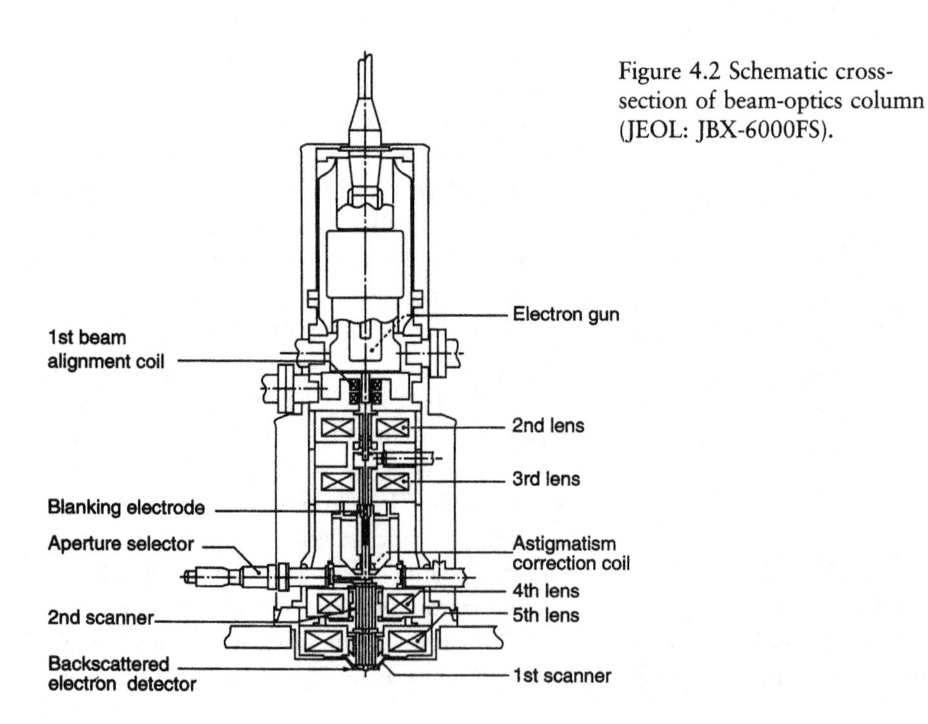

Figure 4.2 Schematic cross-section of beam-optics column (JEOL: JBX-6000FS).

Table 4.1 *Specifications of 50-kV electron-beam exposure system using a thermal field emission gun. (JEOL: JBX-6000FS.)*

Acceleration voltage	50 kV
Electron-beam type	Gaussian
Electron source	Zr/O/W thermal field emitter (TFE)
Scan field	$80 \times 80\,\mu m^2$
Minimum step size	2.5 nm
Scan frequency	12 MHz (max.)
Wafer size	8 inch [200 mm] (max.)
Wafer movement	step-and-repeat

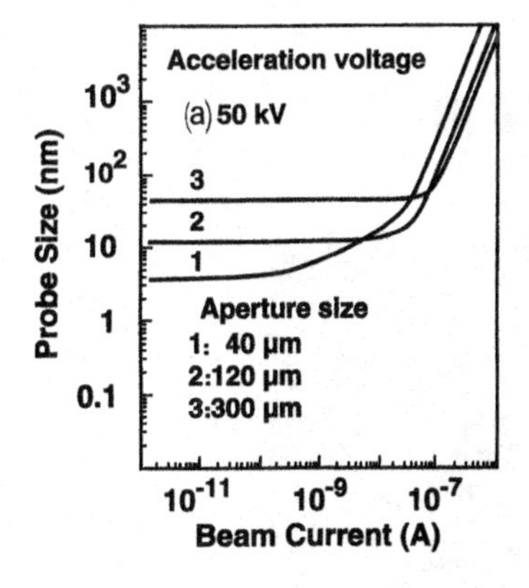

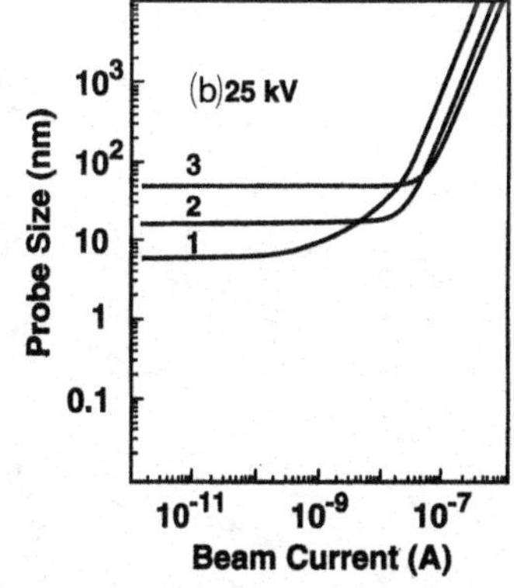

Figure 4.3 Calculated beam diameter (probe size) versus beam current as a function of objective aperture size and acceleration voltages of (a) 50 kV, and (b) 25 kV. For assumptions used in calculation, see text.

wafer. The calculated beam diameter (probe size) when using a 40-μm aperture and a beam current of 100 pA is 8 nm at 25 kV and 5 nm at 50 kV as Figure 4.3 shows, where the calculated beam diameter for different aperture sizes and beam currents at 25 and 50 kV is plotted. Calculation conditions are: a beam-energy spread ΔE of ± 1 eV, an extraction voltage V_e of 5 kV, an angular current intensity of $j_\alpha = 200\,\mu A/sr$, and beam convergence angles α_1, α_2, α_3, of 2.37, 7.05 and 11.80 mrad, respectively. This system has been operated at 50 kV and the measured beam diameter using the knife-edge method was about 5 nm, as Figure 4.4 shows.[7] The Ni-mesh edge was used as the knife-edge. In this measurement, beam diameter is defined as a width on beam-current change, which utilizes 70% of the total beam current. By using a 50-kV, 5-nm-diameter electron beam (100 pA), 10-nm-width lines were fabricated on a PMMA resist (thickness 50 nm) as shown in Figure 4.5.

The electron-beam diameter decreases with increasing acceleration voltage. Therefore, the pioneers in this field did lithography under high-energy electron beams using TEM or STEM[8] and they developed high-energy electron-beam exposure systems. A 100-kV electron

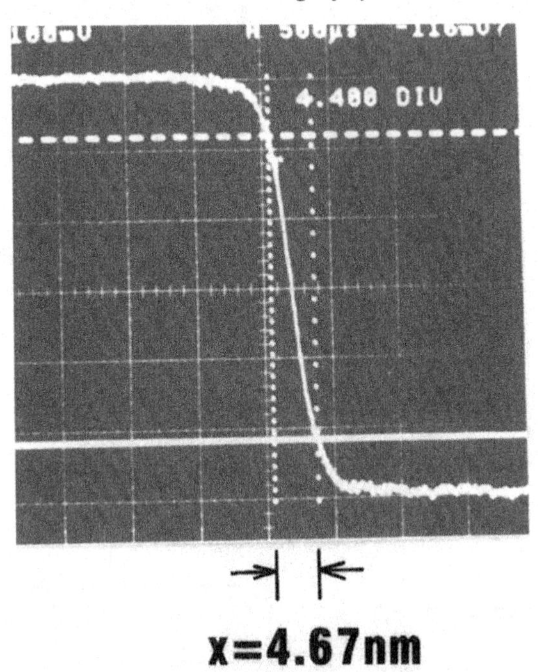

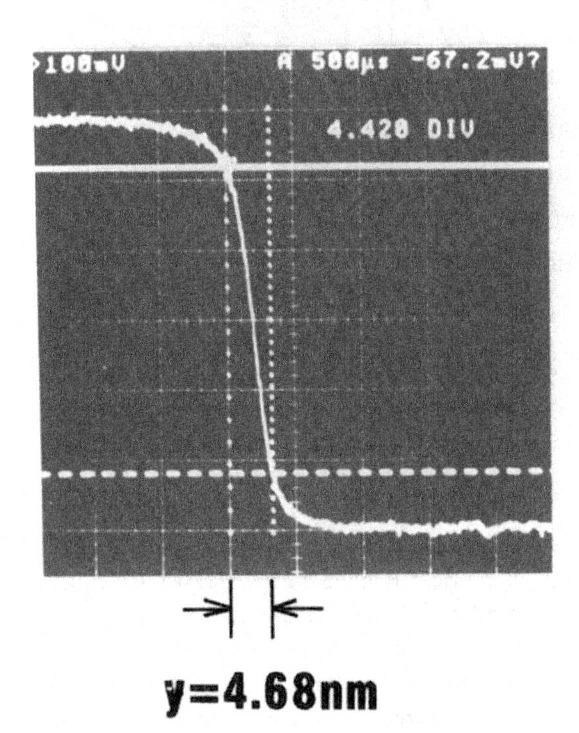

Figure 4.4 Beam diameter measurement by a knife-edge method (Ni wire).

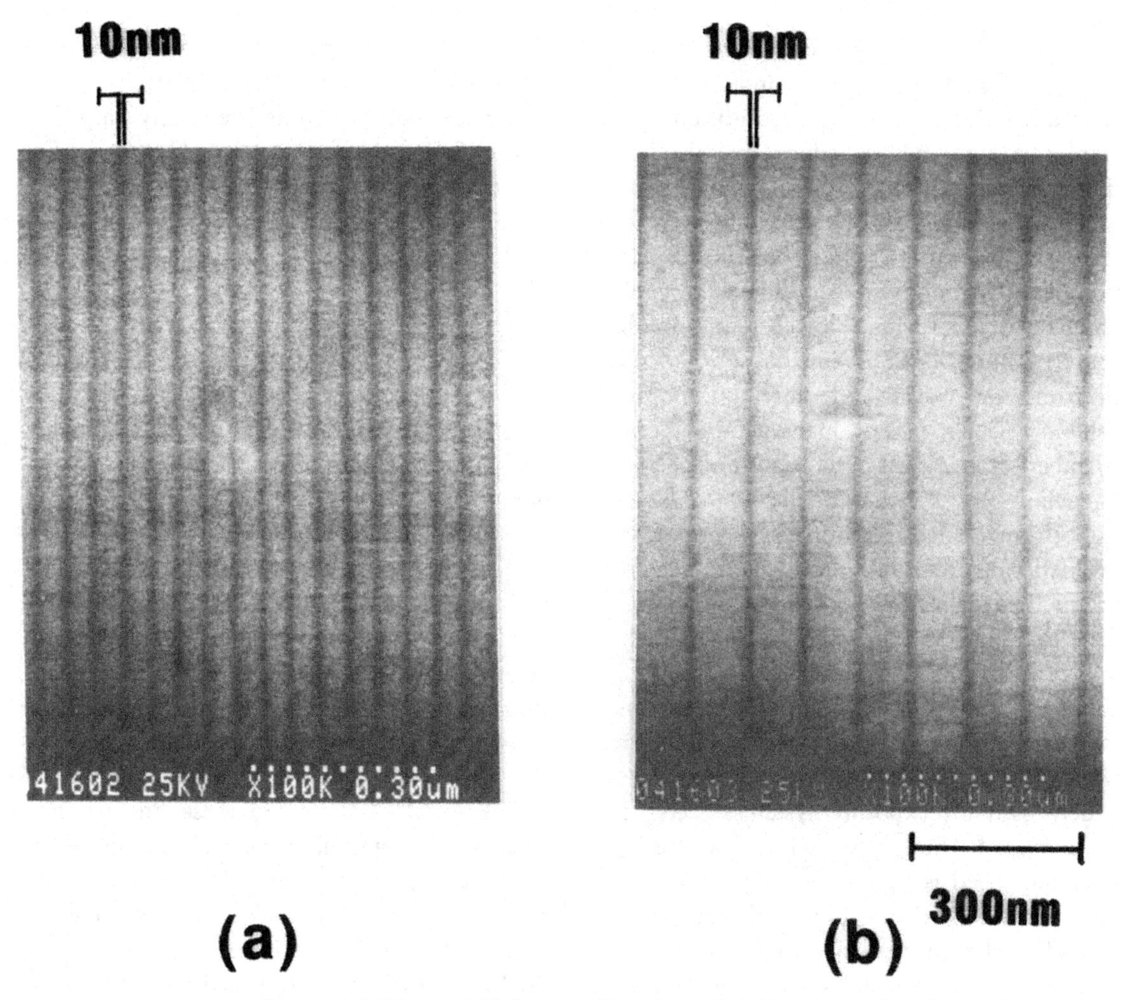

Figure 4.5 10-nm-width lines replicated on PMMA using 50-kV electron beam.

optical column using the LaB_6 gun was constructed.[9] This system used new concepts such as a swinging objective immersion lens (SOIL) which is based on a variable axis immersion lens (VAIL) and a swinging objective lens (SOL). The target is inside the objective immersion lens. The SOL has several advantages. These are: (1) it is possible to reduce the working distance, thereby decreasing spherical aberrations; (2) the scanned field can be increased without significantly increasing deflection aberrations; (3) it permits easy alignment of the column; and (4) an inherent attribute of the SOIL is shielding from the extraneous magnetic fields applied to the target. Using this technique, a deflection field size of 250 μm can be achieved at 100 kV. The target for the beam diameter is 4 nm at 100 kV. The exposure results on PMMA at 60 kV indicate that a linewidth of 15–20 nm can be obtained at all points within a 250 × 250 μm^2 field. Several 100-kV electron-beam systems using the TFE (Zr/O/W) gun have been developed.[10,11,12,13] High-energy electron beams are useful for nanolithography and X-ray mask fabrication. High-voltage electron-beam lithography is expected to reduce the proximity effects associated with forward scattering and backward scattering of electrons.[14] In the early stages of developing the 100-kV electron-beam system, the beam diameter was not as small as sub-10 nm. However, a 3-nm-diameter beam has been obtained in a 50-kV TFE electron-beam system by modifying the optical column, which has a high-vacuum sample chamber for investigation of beam-induced phenomena.[15]

Both an electrostatic force and an electromagnetic force can be used to deflect electron beams. The former force is generated by using a deflection plate and the latter by using a deflection coil. In SEM, TEM and STEM, an electromagnetic deflection system is used, because magnetic deflection is cheaper than electrostatic. However, electrostatic deflection is widely used for exposure systems, except for those made by Etec Systems Inc., because high scanning speed is easy to obtain. However, a disadvantage of electrostatic deflection is that it tends to be affected by noise deriving from its own electronic circuits and from the surrounding environment. On the other hand, magnetic deflection is resistant to noise, because a magnetic deflection coil itself works as a noise filter.

In recent Gaussian electron-beam systems, a digital circuit controls the beam. Figure 4.6 shows the relationship between scan frequency and scan step as a function of electron doses. The scan step is usually set at one-half an electron-beam diameter. When using a 5-nm-diameter beam, it is necessary to scan the beam with a step of 2.5 nm. As can be seen from the dashed line on the figure, for a dose of 300 $\mu C/cm^2$, which is the same as the sensitivity of a PMMA resist at 50 kV, the necessary scan frequency is about 10 MHz. Therefore, high-sensitivity resists, such as chemically amplified resists (CAR), cannot be exposed correctly. So, a high-frequency deflection system is required for high-sensitivity-resist exposure, and also for high throughput.

4.1.1.3 *High-resolution lithography using inorganic and organic resists*

Only a Gaussian electron beam can delineate fine patterns smaller than 0.1 μm. The resolution of electron-beam lithography mainly depends on the beam diameter and the resolution of the resist. The pattern is usually delineated on organic resists and inorganic resists.

PMMA is a well-known high-resolution positive resist and it is used to evaluate exposure systems.[16,17] Figure 4.7 shows an example of a PMMA resist pattern produced by exposure with a 100-kV electron beam emitted from a TFE gun.[18] To develop these fine patterns with 5–7-nm width, ultrasonic agitation was used during development. Figure 4.8 shows the pattern of 10-nm-wide AuPd metal wires on Si produced by the lift-off method using a PMMA resist.[19] Because PMMA has low etching durability, other high-resolution and high-durability resists have been used, such as the ZEP series resist.[20] ZEP is a non-chemically amplified resist and has a fairly high resolution of about 10 nm as Figure 4.9 shows.

In attempts to obtain higher resolution, the use of many kinds of inorganic resists has been investigated. The exposure mechanism for inorganic resists is self-development, that is, the composition dissociates by energy-beam irradiation, and decomposed materials evaporate or migrate from the irradiated region. For example, NaCl, LiF, and AlF have been delineated to obtain a few nanometers of

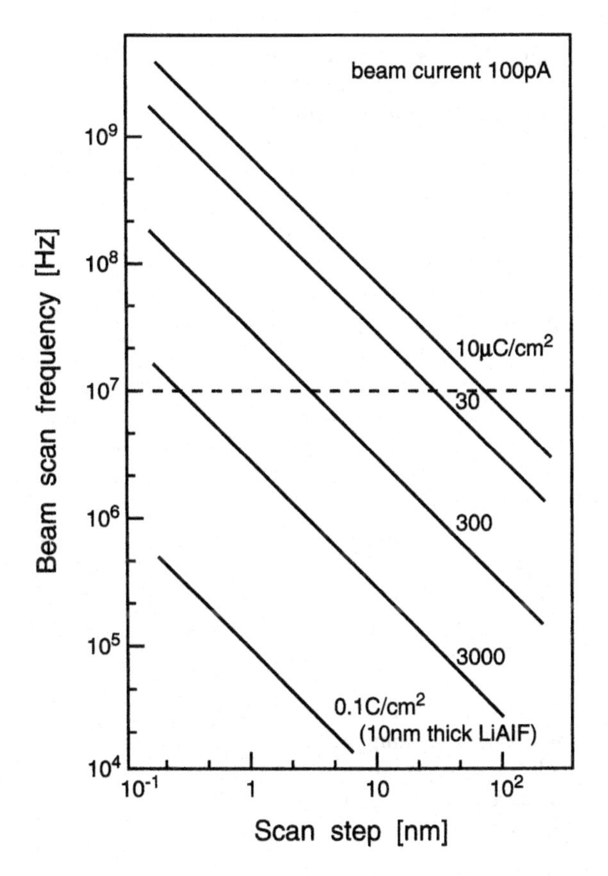

Figure 4.6 Beam scan frequency vs scan step as a function of resist sensitivity (μC/cm²) for a beam current of 100 pA.

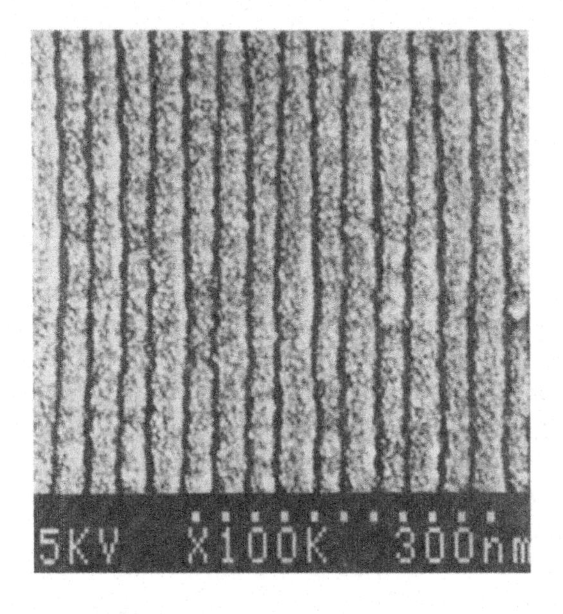

Figure 4.7 A 5–7-nm-width PMMA resist pattern produced by 100-kV electron-beam exposure using TFE gun.

Figure 4.8 10-nm-width AuPd metal wires on Si substrate produced by a lift-off method.

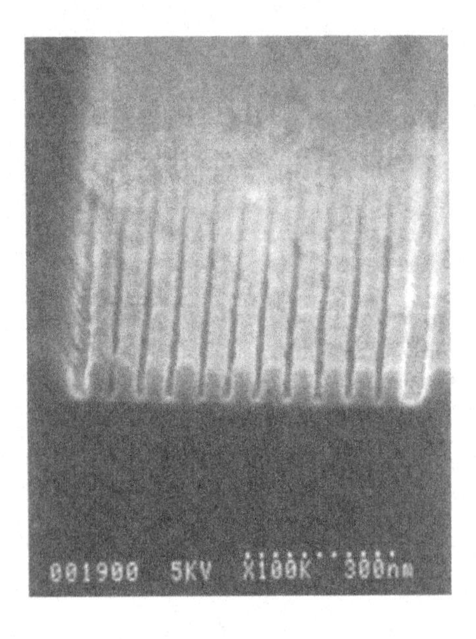

Figure 4.9 10-nm positive non-chemically amplified (ZEP-520) resist pattern produced by electron-beam exposure.

resolution.[4] In these experiments, high-energy electron beams (greater than 100 kV) have been used. There are some problems in that there are some grains with a diameter as large as 10 nm in these inorganic resists and, in addition, the sensitivity of the materials is very low for electron-beam exposure. To obtain a finer grain-size and higher sensitivity, a mixture of these inorganic resists was investigated.[21,22] LiAlF film was deposited on a Si substrate by an ion-beam-sputter apparatus using LiF and AlF targets. Figure 4.10 shows SEM photographs of a LiF film and a $(Li_{0.9}Al_{0.1})F$ film.[23] The grain size was dramatically reduced in the $(Li_{0.9}Al_{0.1})F$ film. The exposure characteristics of $Li_{1-x}Al_xF$ film with various concentrations of Al under 50-keV electron-beam irradiation are shown in Figure 4.11. For pure LiF film, self-development did not occur. However, with increasing Al content, the $(Li_{1-x}Al_x)F$ film can be developed up to 25 nm at $x = 0.1$. So, the inorganic resist has acceptable sensitivity at a voltage of 50 kV. However, the depth of the irradiated region decreased again for doses higher than 0.6 C/cm^2. This was due to a contamination from the vacuum and the residue of decomposed material such as Li and Al. Figure 4.12 shows a high-resolution $(Li_{0.9}Al_{0.1})F$ resist pattern. Lines with a width of 5 nm have been delineated with a period of 60 and 30 nm.

In general, positive resists have a higher resolution than negative resists, because the exposure mechanism is chain scission of resist resin (high-molecular-weight organic material) for positive resists but cross-linking for negative resists. The cross-linking reaction increases the molecular weight of the irradiated portion of the resist. Nevertheless it has been reported that 10-nm-order lithography can be achieved by using a calixarene resist, which has a negative tone.

Calixarene[24] is a general term for specific cyclic phenol resins. A calixarene resist is a cyclic oligomer consisting of 6-phenol having

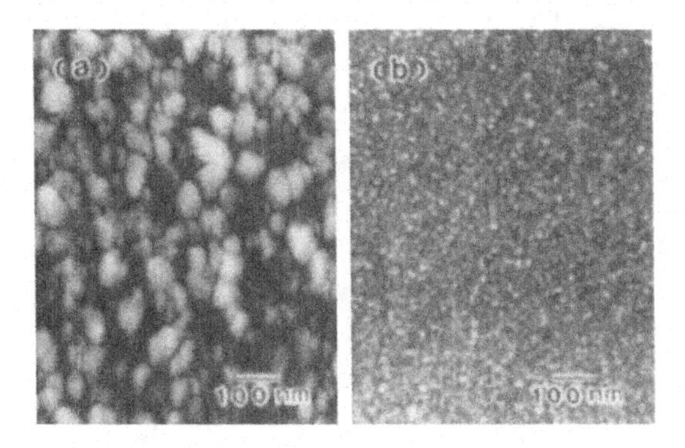

Figure 4.10 SEM photographs of (a) LiF film, and (b) $(Li_{0.9}Al_{0.1})F$ film, ion-beam sputtered as inorganic resists.

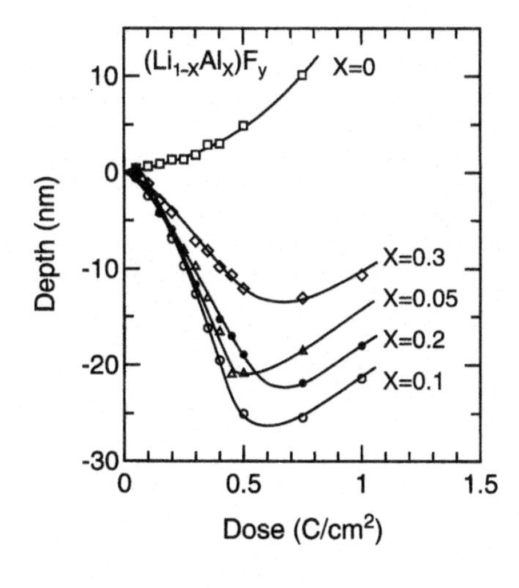

Figure 4.11 Exposure characteristics of $Li_{1-x}Al_xF$ film with various Al concentrations exposed by 50-keV electron-beam irradiation.

a molecular diameter of 1 nm (Figure 4.13(a)).[25] The molecule in Figure 4.13(a) is a calixarene derivative, hexa-acetate *p*-methylcalix[6]arene (MC6AOAc), while that in Figure 4.13(b) is hexachloromethohexa-methoxycalix[6]arene (CMC6AOMe). Their respective molecular weights are 972 and 996. Most calixarene derivatives are poorly soluble in organic solvents, but these calixarene molecules are soluble in organic solvents such as *o*-dichlorobenzene or monochlorobenzene, and they have high heat resistivity up to 320 °C.[26] Therefore, calixarene films are easily made by using a spin-coating method similar to that used in conventional resist processes. These calixarene films work well as negative

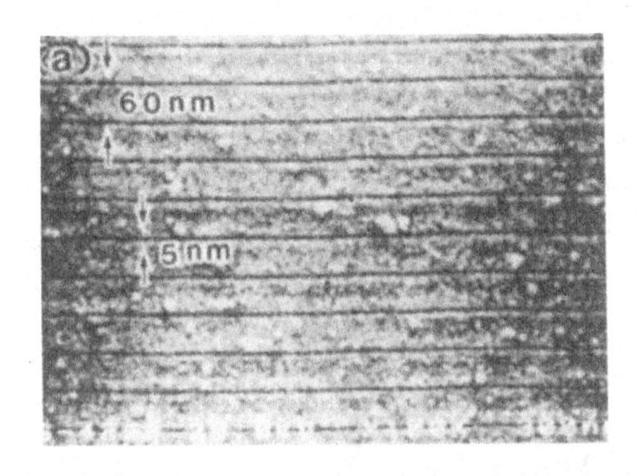

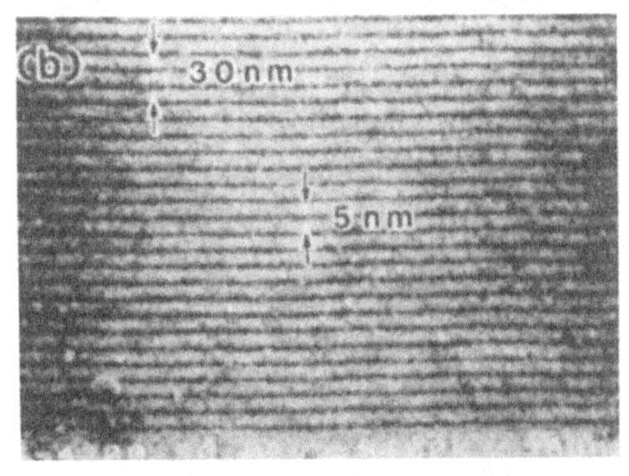

Figure 4.12 High-resolution $(Li_{0.9}Al_{0.1})F$ positive resist pattern produced by electron-beam exposure.

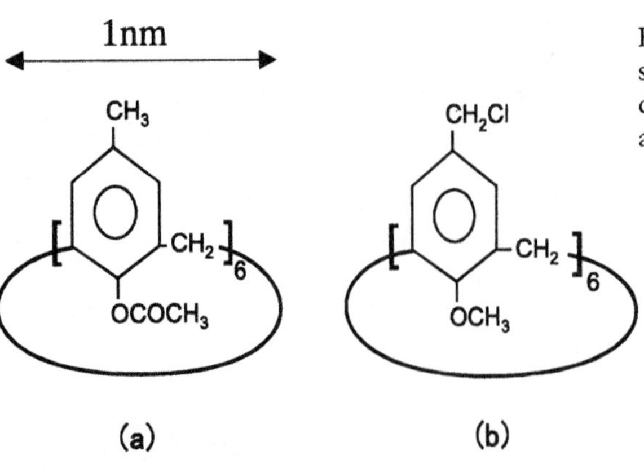

Figure 4.13 Molecular structure of calixarene derivatives: (a) MC6AOAc, and (b) CMC6AOMe.

electron-beam resists, and they have ultra-high resolution and high durability under halide plasma-etching.[27]

Figure 4.14 shows the electron-beam exposure characteristics of these calixarene films. 1 wt% of calixarene solution was spin coated on a Si wafer at 3000 rpm for 30 s to prepare a 30-nm-thick film. For MC6AOAc, the sensitivity was about 7 mC/cm^2, which is almost 20 times that for PMMA and almost 100 times that for a SAL601™ chemically amplified negative resist. The chloromethylated calixarene CMC6AOMe is about 10 times more sensitive than MC6AOAc. Substituting Cl atoms in the methyl groups improves sensitivity. The reason for this is that since the C–Cl bonding energy is low, their Cl bonds easily decompose. Activated Cl can affect Cl bonds at other sites and reduce their binding energy. These Cl bonds can then be broken by low-energy deposition. The resist contrast γ is about 1.6 for both these calixarene resists.

Electron-beam lithography using calixarene has been performed to evaluate resist resolution by exposing dot arrays. The dot pattern is fabricated by irradiating with the Gaussian electron beam onto a certain site for a certain time. Figure 4.15 is a SEM photograph of a dot array produced by two types of calixarene on a Si substrate. Figure 4.15(a) shows an MC6AOAc dot array with a 15-nm diameter and a 35-nm pitch and Figure 4.15(b) is a CMC6AOMe dot array with a 25-nm diameter and a 50-nm pitch. The electron beam used was 50 keV and 100 pA (JEOL: JBX-5FE), and the beam diameter was estimated to be about 5 nm. The typical exposure spot-dose was about 1×10^5 electrons/dot (16fC/dot). If we assume that the beam intensity distribution within a beam spot is constant, a spot-dose for a 25 nm^2 (5-nm square) spot-size corresponds to an area-dose of 64 mC/cm^2. Figure 4.16 shows a calixarene line pattern with a thickness of 30 nm and the transferred Ge pattern using the calixarene pattern as a mask.[28] A 10-nm-width calixarene line was delineated on 20-nm-thick Ge film on a Si substrate. The pattern was transferred by a plasma etching apparatus, using CF_4 gas at 5 Pa, a microwave power of 50 W with a 200-V DC-electrode bias for 1 min. A 7-nm-width Ge line pattern could be fabricated due to side etching during pattern transfer.

To investigate why a high-resolution negative resist pattern was obtained, polystyrene

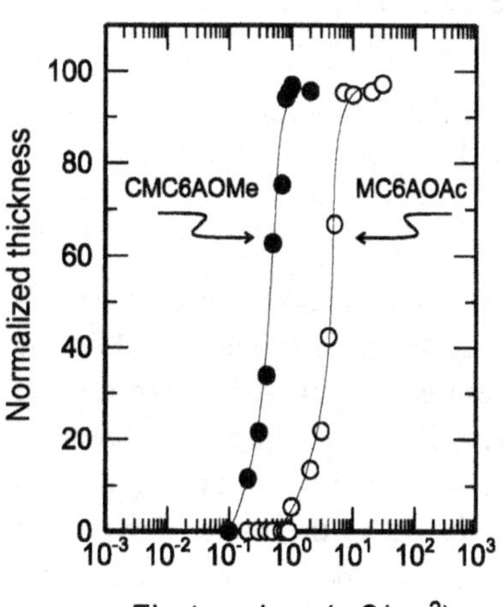

Figure 4.14 Electron-beam exposure characteristics of the calixarene resists shown in Figure 4.13.

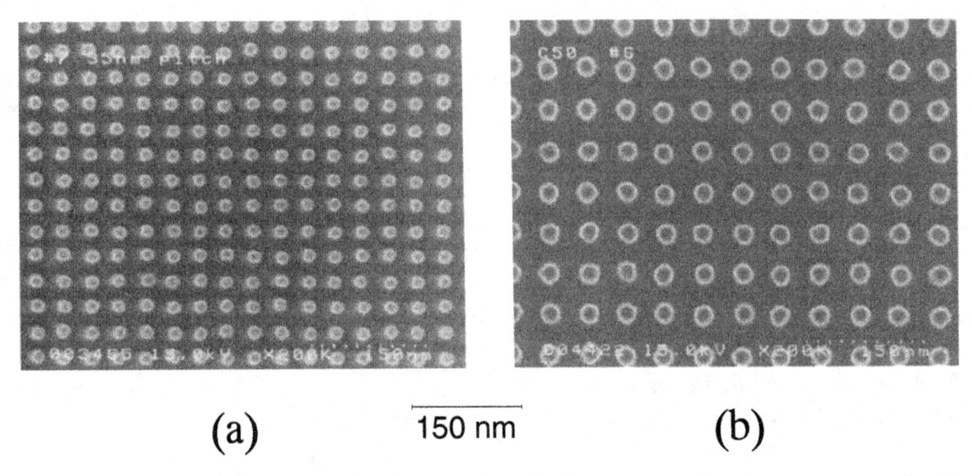

Figure 4.15 SEM photographs of a dot array made using two types of calixarene on a Si substrate: (a) MC6AOAc with 15-nm diameter and 35-mm pitch, and (b) CMC6AOMe with 25-nm diameter and 50-mm pitch.

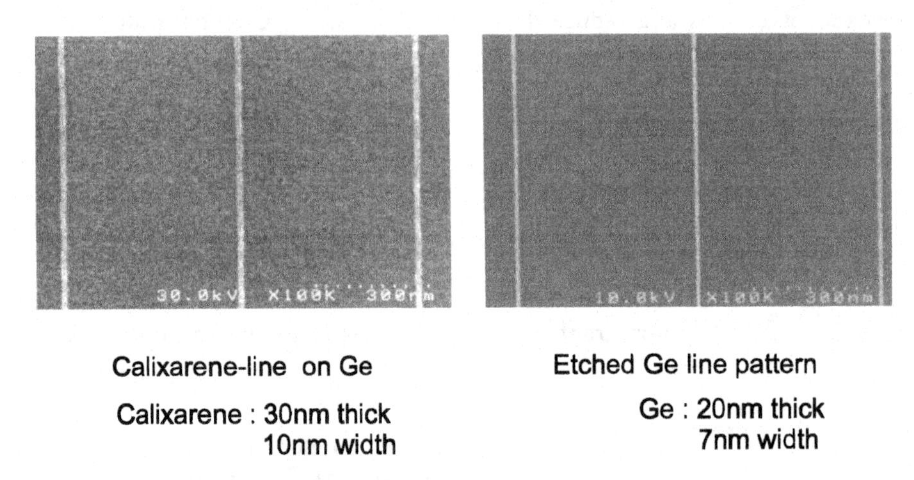

Figure 4.16 10-nm-width calixarene line pattern with a thickness of 30 nm and its transferred line pattern in Ge. Ar ion etching: 2×10^{-4} Torr.

negative resists with various molecular weights were investigated. Because polystyrene has a simple chain molecular structure with no additives to increase its sensitivity, it is suitable for such investigations. The molecular size for a molecular weight of 1000 is almost 1 nm assuming the polystyrene resist is piled up. The relation between exposed-and-developed resist dot diameter and electron dose as a function of molecular weight is in Figure 4.17. The experimental results for calixarene are also in the same figure. Dot-pattern diameter decreased with decreasing electron dose and the minimum dot diameter reduced as the molecular weight reduced. A 10-nm diameter dot pattern was obtained on a polystyrene resist with a molecu-

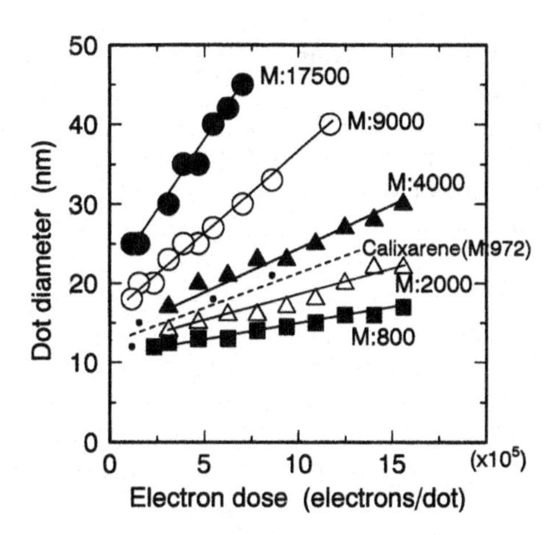

Figure 4.17 Relation between polystyrene-resist dot diameter and electron dose as a function of molecular weight of polystyrene resist. Experimental results for calixarene resist (see text) are also shown.

lar weight of 800. Thus, the resolution of negative resists closely depends on their molecular weight, that is, their molecular size.

4.1.1.4 *Low-energy lithography*

There is another way to achieve high-resolution lithography by using a low voltage to reduce the proximity effect.[29,30,31] Figure 4.18 shows the energy deposition profiles obtained by Monte Carlo simulation for PMMA resist of various thicknesses as a function of electron-beam energy.[32] We can see that there is an optimum incident energy for the exposure of the resist. For the 50-nm (500-Å)-thick resist, Figure 4.18(a), 1 keV is enough for the exposure and the energy is sufficiently low to minimize back-scattered electrons. Figure 4.19 shows how to reduce the proximity effect by using low-energy exposure.[33] 'Dose factor' is the exposure dose in arbitrary units. At energies of 10 keV and higher, the proper doses for an isolated line and an isolated space are very different. However, at energies lower than 3 keV, the required dose is equal to or much lower than that for higher energy. Another advantage of low-voltage exposure is that it is highly sensitive to resist exposures. Figure 4.20 shows the required dose as a function of beam energy for 70-nm-thick PMMA. This result means that the sensitivity becomes higher for lower incident energy.

To obtain a small beam diameter at low acceleration voltage, retarding optics are useful.[30,34] Figure 4.21 is a schematic view of some retarding optics.[35] The sample stage is isolated from the column and is maintained at high voltage to retard the incident electron beam just in front of the sample surface. Then, the electron beam irradiated onto the sample is at the desired low energy. By using variable-energy exposure, the cross-section of the resist can be modulated. Figure 4.22 is an inclined view of a resist pattern produced by exposure at three different energies of 1, 3 and 5 keV.[36] This variable-energy technique allows the desired cross-section or three-dimensional shape to be obtained.

4.1.1.5 *Application of Gaussian electron beam to device fabrication*

(a) Sub-0.1-μm MOSFET gate fabrication using chemically amplified resist[37]

The Gaussian electron beam has been used to fabricate not only quantum devices but also Si–

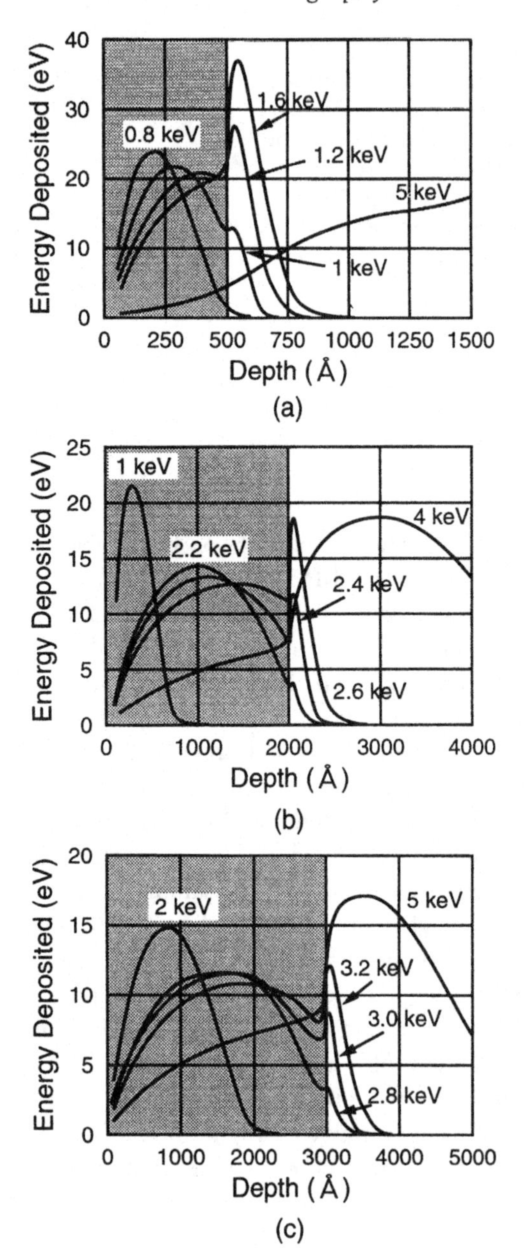

Figure 4.18 Energy deposition profiles obtained by Monte Carlo simulation as a function of incident electron-beam energy for various PMMA resist thicknesses.

MOS devices. Because of the need to increase the density of memory devices, research on Si–MOS devices has reached the sub-0.1-μm region. In this regime, electron-beam lithography is the only means available for delineation. This subsection describes the application of electron beams in the fabrication of gates for 40-nm gate MOSFETs.

Figure 4.23 illustrates the fabrication process for nanometer-order MOS transistors. A 3.5-

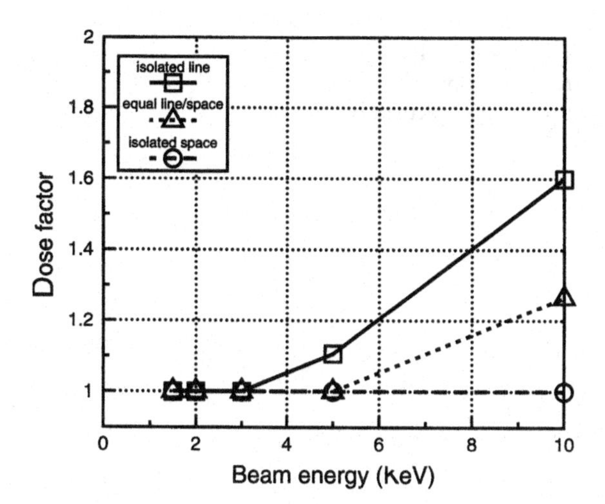

Figure 4.19 Proximity effect for various incident-electron-beam energies.

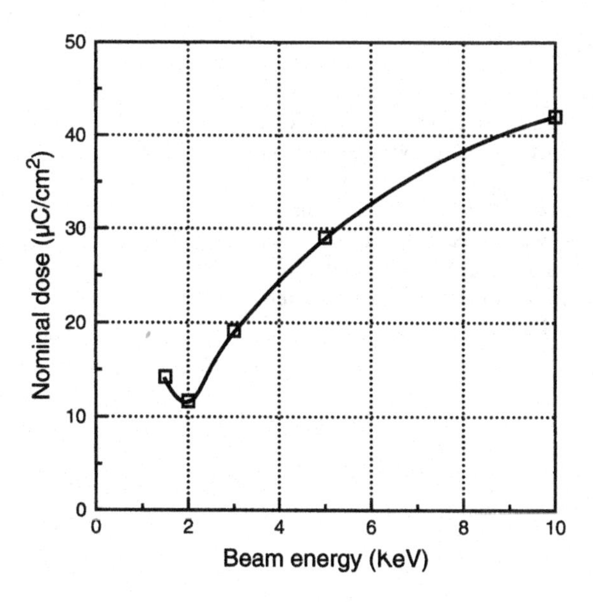

Figure 4.20 Exposure-dose dependence on beam energy for 70-nm-thick PMMA resist developed 1 min in 1 : 2 MIBK : IPA developer. The dose was calibrated such that isolated spaces developed to their nominal linewidth. The increased required dose at the lowest energy is a result of incomplete penetration of the resist by the electron beam.

nm-thick gate oxide is used to obtain high current drivability. The polysilicon thickness was 150 nm. A single-layer resist was used as a mask to simplify fabrication. A SAL601™ (Shipley Ltd) chemically amplified negative resist with a thickness of 200 nm was coated on a 6-inch (150-mm) ϕ Si wafer. The resist thickness was determined based on an etching selectivity of five for the polysilicon over the resist in order to obtain high-aspect-ratio polysilicon-gate patterns.

The resist was spin coated and pre-baked (PB) at 120 °C for 2 min. After electron-beam (EB) exposure, the wafer was post-exposure baked (PEB) at 100 °C for 2 min. Acid was produced in the resist during PEB, and was diffused. Such high PB and low PEB temperatures relative to conventional conditions are suitable for higher resolutions,[38,39] because this process decreases the diffusion length of the acid. After PEB, the resist was developed for 1 min in a TMAH(tetramethylammoniumhydroxide)

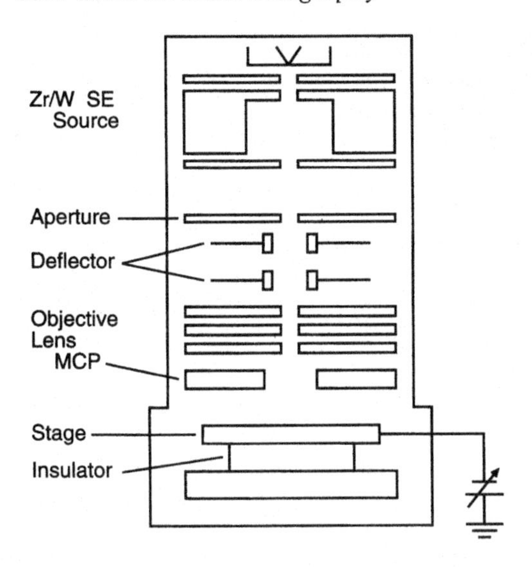

Figure 4.21 Schematic of retarding optics to obtain a low-energy electron beam.

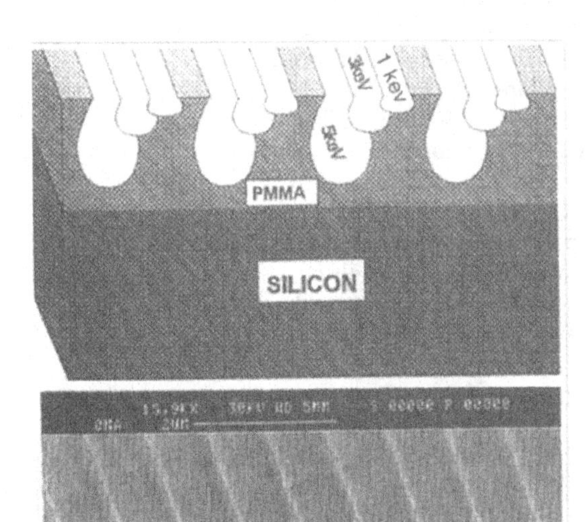

Figure 4.22 Inclined view of a resist pattern exposed at three different energies 1, 3, and 5 keV. *Upper*: schematic cut away; *Lower*: SEM photograph of exposed-and-developed resist pattern after cleavage.

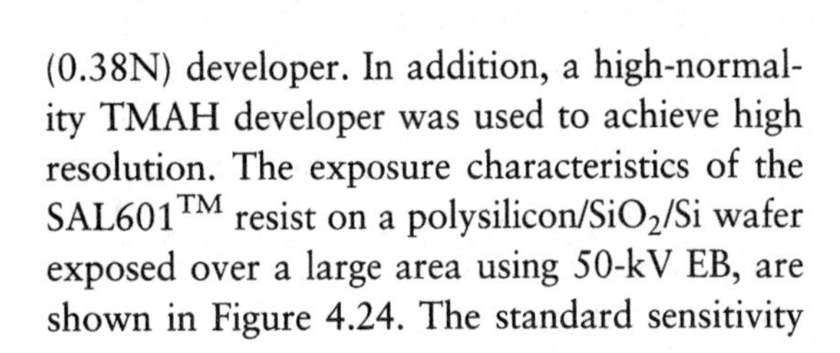

(0.38N) developer. In addition, a high-normality TMAH developer was used to achieve high resolution. The exposure characteristics of the SAL601^{TM} resist on a polysilicon/SiO_2/Si wafer exposed over a large area using 50-kV EB, are shown in Figure 4.24. The standard sensitivity was 50 $\mu C/cm^2$, which is about one-half that of conventional processes. The residual resist thickness was about 95%. However, the contrast ($\gamma = |\log(D_0/D_m)|^{-1}$) was greater than 4, thereby ensuring fine patterns with a high aspect ratio. Here D_0 is the projected dose

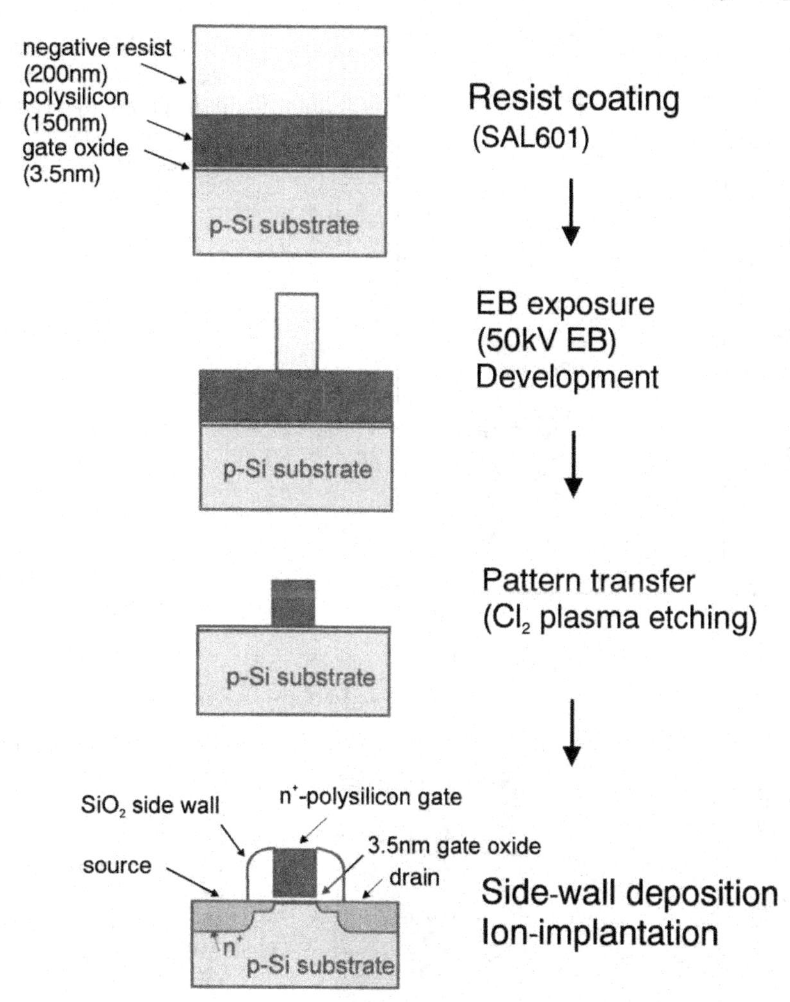

Figure 4.23 Fabrication process for sub-0.1-μm gate Si–MOS transistors.

required to remain at the initial resist film thickness and is determined by extrapolating the linear portion of the thickness–dose plot to a value of 1.0 (normalized film thickness remaining). D_m is the minimum dose required for any detectable resist layer to be formed.

Figure 4.25 shows SEM photographs of a 40-nm negative resist pattern and the transferred polysilicon gate pattern of a residual resist. The highest resolution for the SAL601^{TM} resist is 30 nm with a resist height of 200 nm at an exposure dose of 600 μC/cm^2.

It is necessary to fabricate FETs of various gate lengths on a wafer. Generally, to obtain an exposed resist line width which is the same as the designed line width, the EB dose should be varied, because of the inter-and intraproximity effect. To correct the proximity effect, a double-Gaussian-type energy intensity distribution (EID) function has been used, as shown in Figure 4.26. The experimental correction parameters used are: α (forward scattering) = 25 nm, β (backward scattering) = 12 000 nm, and η (backward-scattering coefficient) = 0.8.

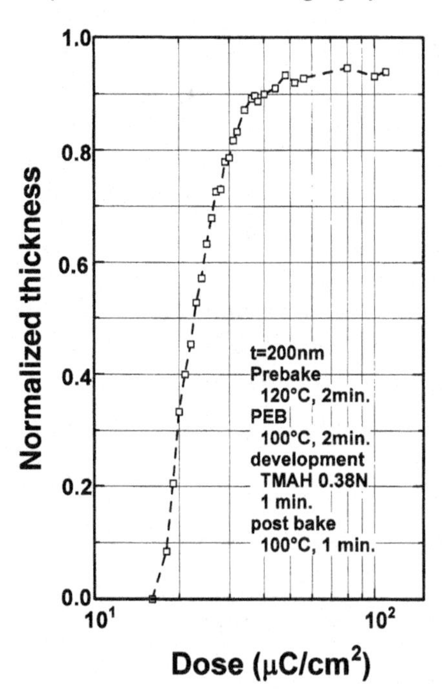

Figure 4.24 Exposure characteristics of SAL601^{TM} resist on a polysilicon/SiO_2/Si wafer exposed over a large area using 50-kV EB.

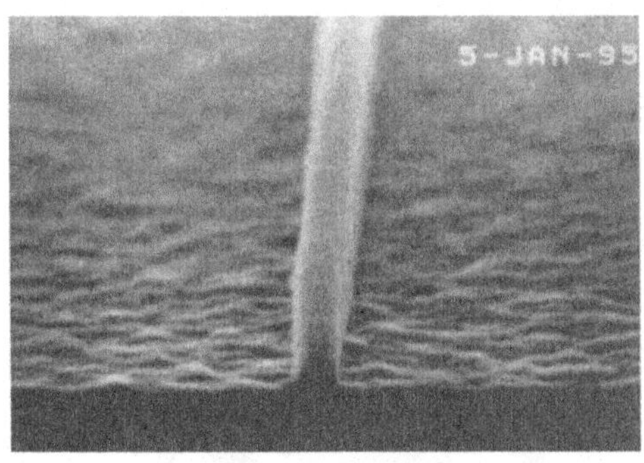

Figure 4.25 SEM photographs of a 40-nm-width SAL601^{TM} resist pattern (*upper*) and the polysilicon gate pattern (*lower*) fabricated by EB exposure and dry-etching. Designed width: 60 nm; dose 350 $\mu C/cm^2$; etching time: 1 min 45 s.

In this equation, $p(x)$ is the deposited energy in a resist due to forward- and backward-scattering electrons, and x is the distance from the electron-irradiated point. It should be noted that the forward-scattering parameter, α, is small due to the high acceleration voltage. This means that the high-energy EB lithography has the potential use for line widths down to 30-nm. Figure 4.26 shows SEM photographs of exposed resist lines. Allowing accurate correction, the resist patterns in Figure 4.26(b) were obtained, which are the same as the designed ones. The results confirmed that proximity effect correction using double-Gaussian EID works well, even for line widths of less than 100 nm.

n-MOSFETs with various gate lengths exceeding 40 nm on 6-inch (150-mm) ϕ wafers have been fabricated. Figure 4.27 shows the I_D–V_D (drain current vs drain voltage) charac-

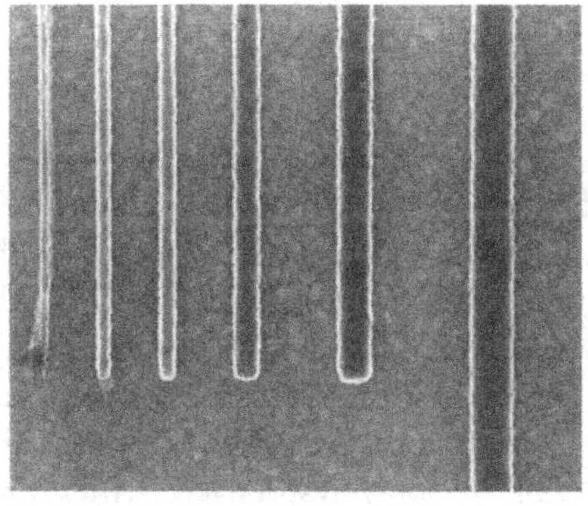

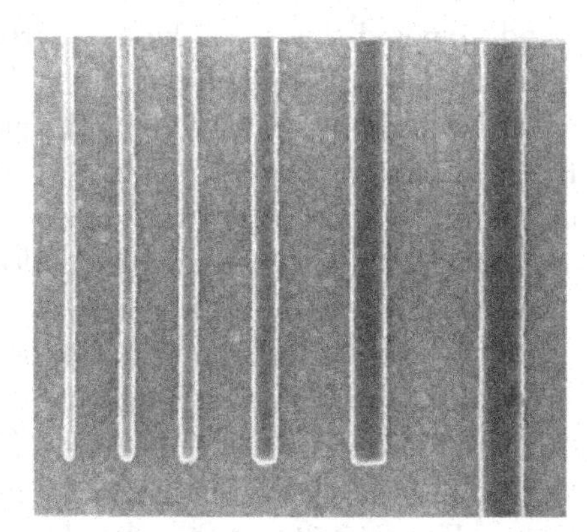

$$p(x) = \frac{1}{\pi(1+\eta)}\left\{\frac{1}{\alpha^2}\exp\left(\frac{-x^2}{\alpha^2}\right) + \frac{\eta}{\beta^2}\exp\left(\frac{-x^2}{\beta^2}\right)\right\}$$

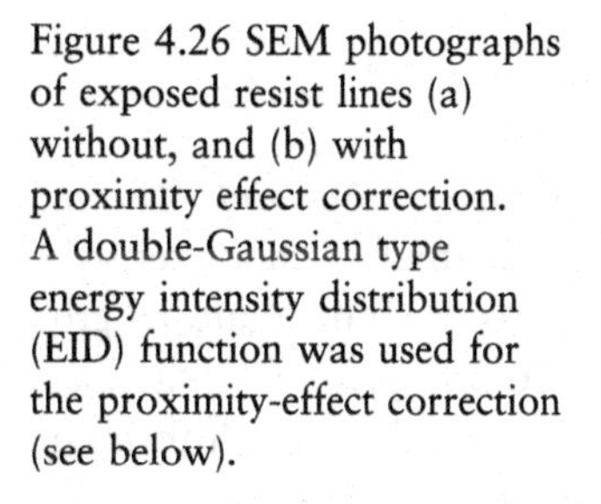

Figure 4.26 SEM photographs of exposed resist lines (a) without, and (b) with proximity effect correction. A double-Gaussian type energy intensity distribution (EID) function was used for the proximity-effect correction (see below).

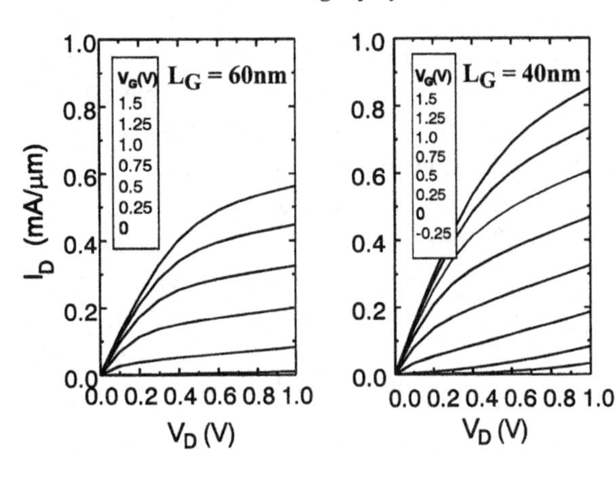

Figure 4.27 I_D–V_D characteristics for gate lengths of (a) 60, and (b) 40 nm. V_G is the gate voltage.

teristics for gate lengths of 60 and 40 nm. Well-behaved short-channel characteristics were obtained down to a 60-nm gate length. The operation of a 40-nm gate FET was also confirmed, although weak punch-through occurred. Maximum transconductance $[(I_D/V_G)_{max}]$ at $V_{DS} = 1$ V was 580 mS/mm for the 40-nm MOSFET.[40] According to these results, it was confirmed that 40-nm gate MOS devices can be used practically by improving other device parameters such as junction depth, gate-oxide thickness and carrier density in the substrate.

(b) Fabrication of a nanometer-scale gate using calixarene resist

Calixarene resist has been used for the fabrication of devices with the nanometer dimensions. Electrically variable shallow-junction MOSFETs (EJ-MOSFETs) were proposed and fabricated by Kawaura *et al.*[41] to obtain good ultra-narrow gate length MOSFET performance. These devices have a double gate; a lower gate which acts as a conventional gate and an upper gate which is an additional gate. A positive upper-gate bias induces inversion layers that act as source/drain regions on the silicon surface. They are extremely shallow, typically 5-nm deep.

Figure 4.28 is a schematic diagram of the fabrication process for EJ-MOSFETs.[42] The EJ-MOSFET was fabricated in a similar manner to conventional Si–MOSFETs. Boron and As were implanted for the substrate and source/drain regions, respectively. After gate oxidation (thickness = 5 nm) and polysilicon (thickness = 40 nm) deposition for the gate, calixarene resist to a thickness of 20 nm was coated and exposed by a 50-keV electron beam. After the resist was developed, the polysilicon was etched by RIE with CF_4 gas through calixarene as a mask. Then, a 20-nm-thick intergate oxide layer was grown by CVD and the upper gate and source/drain electrodes were formed by Au/Al evaporation. Figure 4.29[42] is a cross-sectional TEM photograph of an ultra-short gate surrounded by a gate oxide and intergate oxide. A 14-nm-width polysilicon gate can clearly be seen in the oxide layer. Figure 4.30 shows the I_D–V_D characteristics of EJ-MOSFETs.[43] As there was no I_D saturation in the measurement, it was confirmed that the lower-gate voltage (V_{LG}) could control the I_D. For further optimization of the fabrication process, it is expected

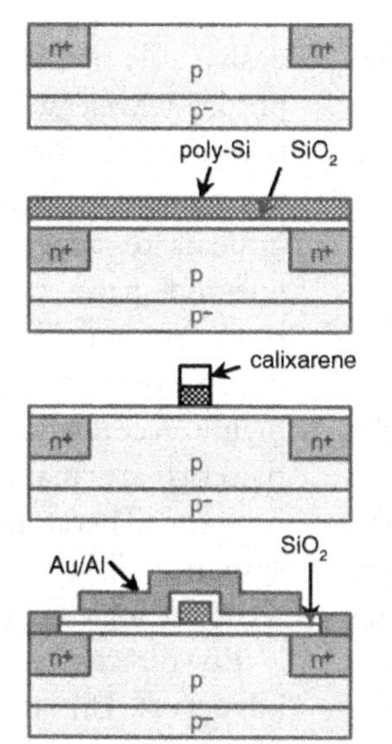

Figure 4.28 A schematic diagram of the fabrication process for an EJ-MOSFET (electrically variable shallow-junction metal-oxide-semiconductor field effect transistor). © 1997 IEEE.

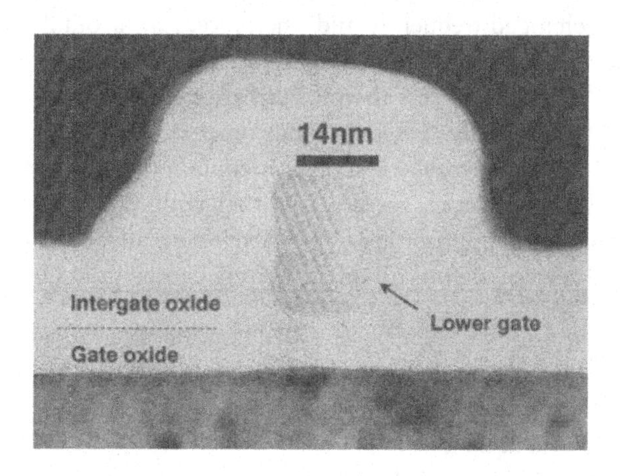

Figure 4.29 A cross-sectional TEM photograph of a 14-nm polysilicon gate surrounded by a gate oxide and intergate oxide layer. © 1997 IEEE.

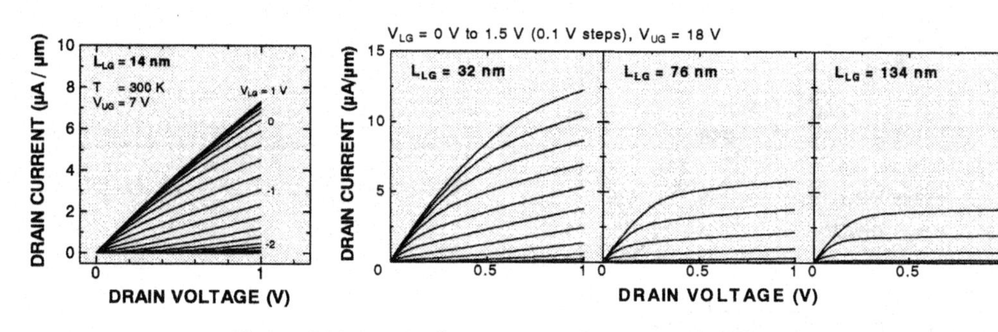

Figure 4.30 I_D–V_D characteristics for various gate-length (L_{LG}) EJ-MOSFETs. V_{LG} and V_{UG} are the lower and upper gate voltages, respectively.

that a 10-nm gate will be able to be produced by using calixarene.

4.1.2 Shaped electron-beam lithography

4.1.2.1 *Introduction*

Lithography with a resolution of 0.1 μm will be required for cost-effective LSI production in the next century. One of the most likely solutions will be a combination of optical lithography and EB direct writing.[44] In this method, critical layer patterns such as contact holes and gates that are hard to form by optical lithography could be written by EB. In particular, variably shaped beam (VSB) technology in conjunction with the projection method is appropriate for the combination of optical lithography and EB direct writing.

As shown in Figure 4.31, the gate length for logic devices (MPU – microprocessor unit) is being miniaturized at a greater rate than that for DRAM. Moreover, the critical dimension (CD) of the gate length should be controlled more severely than that of other layers. As the design rule of LSIs is decreasing below the exposure wavelength, CD control in optical lithography is becoming more difficult because of the limited resolution. In particular, it is very difficult to accurately form various kinds of line-and-space patterns with different duty ratio by optical lithography. The inherently high resolution of EB direct writing results in highly accurate CD control. Logic-device gate patterns are mainly composed from random patterns. Therefore, EB direct writing by VSB systems will be a solution to form the gate patterns for logic devices.

EB direct writing is required to develop advanced LSI technologies in the research stage when an appropriate duplication system is not available. The design of a test device is changed frequently and, therefore, turn-around time must be short. Highly accurate masks for pattern sizes less than 0.2 μm are expensive and it presently takes a long time to make an advanced mask. Test devices contain many random patterns, so direct writing with VSB systems is appropriate when developing advanced devices. Before EB direct writing can be used in

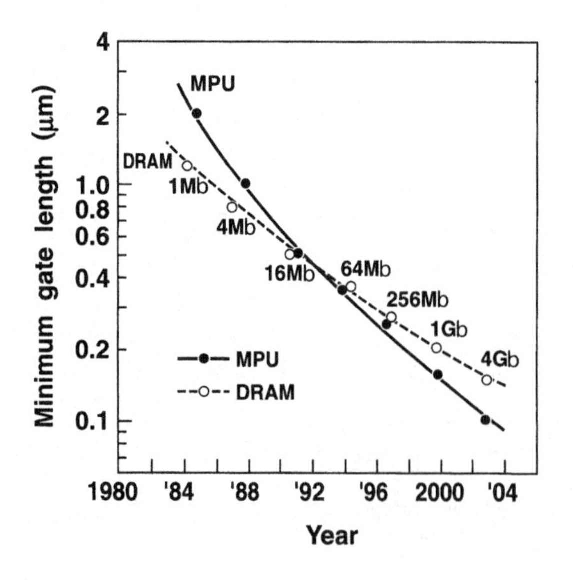

Figure 4.31 Miniaturization trend in the gate length of logic devices (MPU) and DRAMs.

production lines, the technologies related to throughput, reliability, accuracy, and handling have to be further developed.

4.1.2.2 *Beam shape*

The most challenging task concerning EB direct writing is to increase the throughput, and various kinds of writing strategies are therefore being studied and developed. As shown in Figure 4.32, beam shapes have been changed from round (or spot) Gaussian beams, through variably shaped beams,[45,46,47] to projection beams (cell projection,[48] block exposure,[49] and character projection[50]). Projection beams are considered to be advanced-shaped beams. Throughput is improved by increasing the beam area. Although projection systems have higher throughput than conventional VSB systems, they can only write a repeated pattern using an aperture mask, as shown in Figure 4.32, and they cannot be used to write random patterns. VSB systems, however, can generate any pattern and therefore are essential for writing the random patterns of logic devices, the peripheral region patterns of DRAMs, and mask patterns.

4.1.2.3 *Stage control*

There are two kinds of stage-control methods: the step-and-repeat (S&R) stage[51] and the continuously moving stage (Figure 4.33).[52] In S&R systems, writing is carried out when the stage is stopped. After writing is finished, the stage moves to the next position and writing begins again. Because this stage-moving results in lost time, the deflection field has to be enlarged in order to reduce the number of stage-steppings. Deflection aberration, however, increases drastically with increasing field size. Correction of deflection aberration is therefore essential.

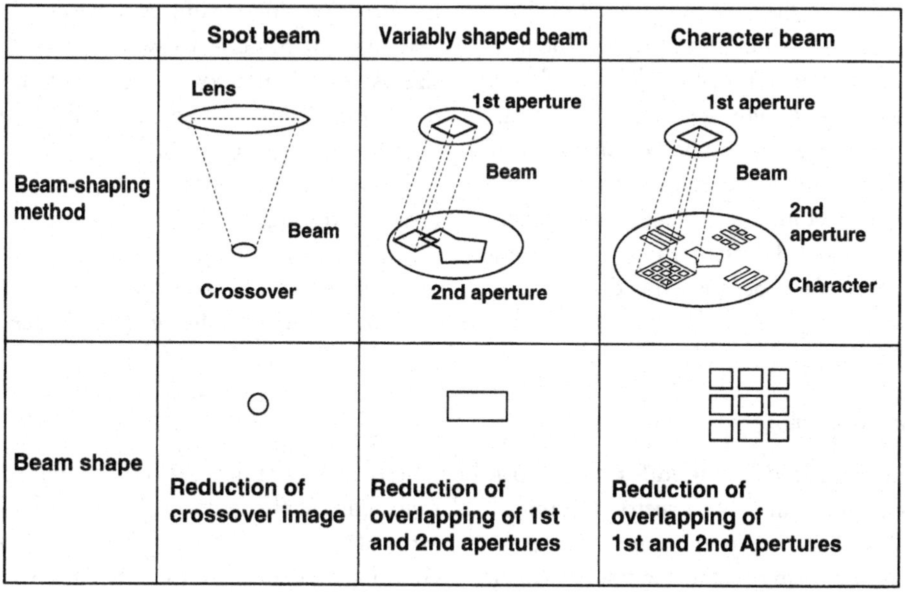

	Spot beam	**Variably shaped beam**	**Character beam**
Beam-shaping method	**Lens** **Beam** **Crossover**	**1st aperture** **Beam** **2nd aperture**	**1st aperture** **Beam** **2nd aperture** **Character**
Beam shape	**Reduction of crossover image**	**Reduction of overlapping of 1st and 2nd apertures**	**Reduction of overlapping of 1st and 2nd Apertures**

Figure 4.32 History (left to right) of electron-beam shape for lithography.

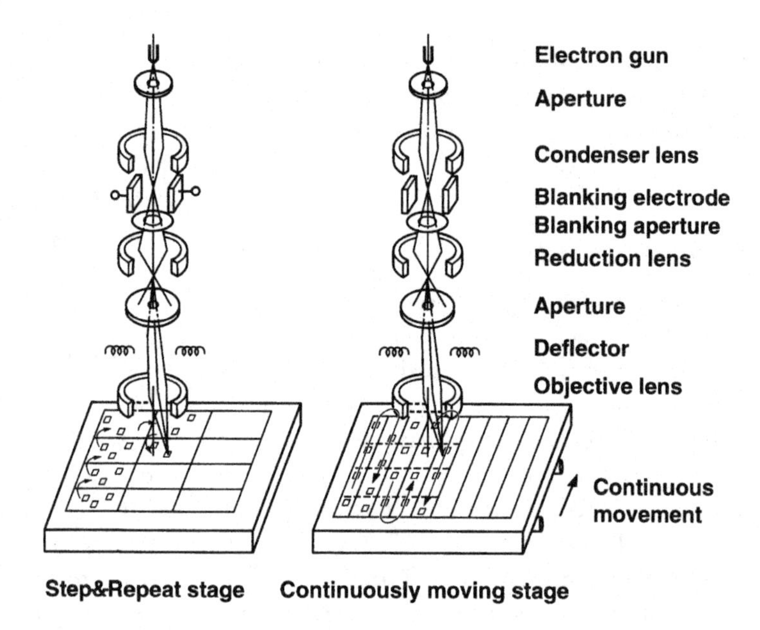

Figure 4.33 Stage-control methods for electron-beam lithography.

The continuously moving stage method, on the other hand, has less lost time because writing is carried out while the stage is moving in one direction. The deflection field is small, so the deflection aberration problem tends to be less serious than in the case of the S&R system. However, the field-stitching error problem is serious because the number of field stitchings increases with decreasing field size. A continuously moving stage is complicated in terms of system control and stage mechanics, but the electron optical column, especially the deflection system, is simpler than that of a system using an S&R stage because the deflection is smaller.

4.1.2.4 *Beam scanning*

Two scanning methods are used for beam positioning: raster scanning[53] and vector scanning[54] (Figure 4.34). Both types of scanning can be used in a system using round Gaussian beams. In the case of shaped beams, however, the vector scanning method is used. Beams are positioned for exposing the resist for a very short time and then moved to the next position.

4.1.2.5 *Acceleration voltage*

EB direct writing is subject to several problems that make it difficult to write a pattern whose size is less than about 0.25 μm, but most of these problems are ameliorated by using a high acceleration voltage.[55]

Beam-edge sharpness (beam resolution) and beam scattering in the target limit resolution in EB lithography. Deterioration due to factors originating from the resist process is not considered here. Beam resolution R is expressed as:

$$R = \sqrt{(\Sigma A_i^2 + \Delta^2 + \Sigma \varepsilon_i^2)}, \qquad (4.1)$$

where A_i is aberration, Δ is beam blur due to the space-charge effect, and ε_i is residual correction error of focus, astigmatism, beam blur caused by charging problems, and so on. The space-charge broadening (blur) Δ is proportional to $IL/V^{3/2}$, where I is beam current, L is optical path length,

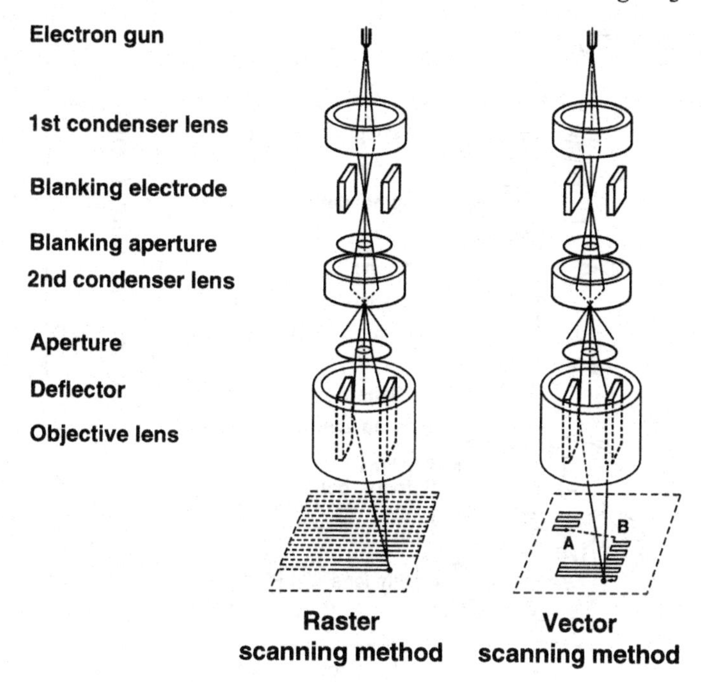

Figure 4.34 Beam-scanning methods for electron-beam lithography.

and V is acceleration voltage. A larger beam current I is better for increasing throughput, but beam resolution decreases with increasing beam current. In a VSB system, beam current, which is proportional to beam area, changes dynamically during writing. Fluctuation of the beam current changes the focus position and beam resolution. Changing the focus allows dynamic correction of the space-charge effect.[56] Figure 4.35 shows equipment for correcting the dynamic space-charge effect by refocusing. The space-charge effect is only partly corrected by this method; that is, a residual space-charge effect remains and is not removed by refocusing. This effect is currently the most serious problem in shaped-beam technology.

Since resist sensitivity is roughly inversely proportional to acceleration voltage, the required beam current is also proportional to the acceleration voltage V. Therefore, beam blur (Δ) caused by the space-charge effect decreases with $V^{1/2}$ if the optical path length is constant.

Chromatic aberration is one of the main factors reducing beam resolution and is proportional to the relative spread of the electron energy, $\Delta V/V$. Under the operational condition of the electron gun of VSB systems, energy spread ΔV is broadened by the Boersch effect and increasing acceleration voltage V reduces chromatic aberration.

As shown schematically in Figure 4.36, electron beams are broadened in the resist as a result of forward scattering. The forward-scattered electron range decreases with increasing acceleration voltage and decreasing resist thickness, so resolution is improved by increasing acceleration voltage or decreasing resist thickness. When resist thickness is limited by the surface topography of devices, high acceleration voltage is effective for improving resolution. Device surface topography can now be made

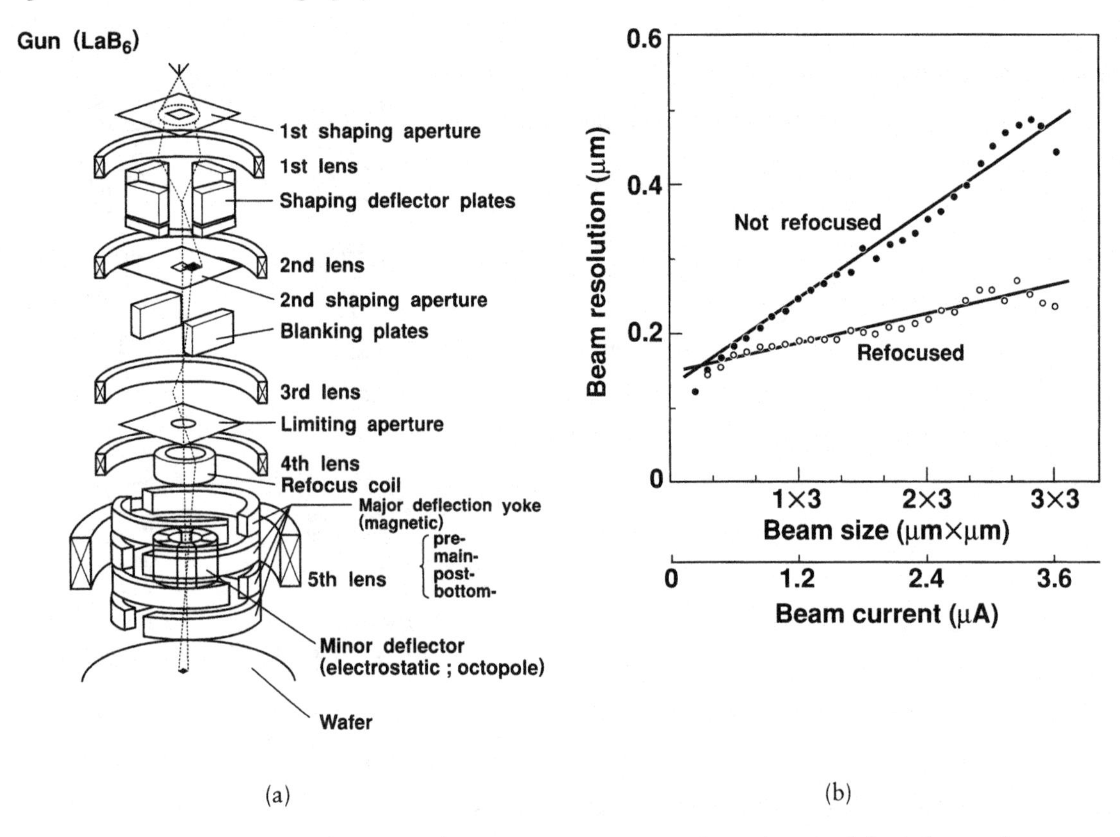

Figure 4.35 (a) Electron optical column with a refocus coil for reducing the space-charge effect.[56] (b) Dependence of beam resolution on beam current with and without refocus.[56]

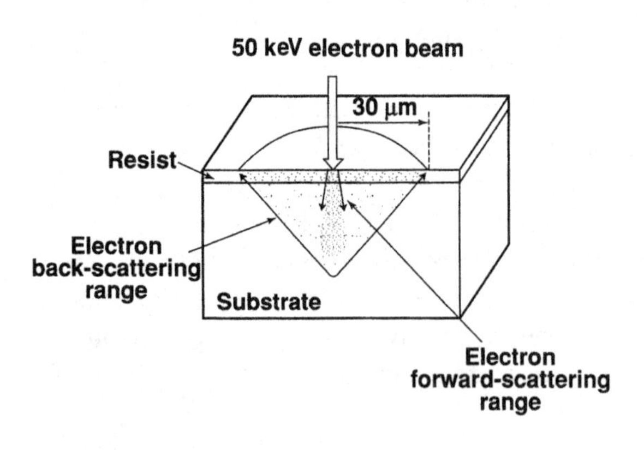

Figure 4.36 Electron-beam scattering in a solid.

flatter as a result of CMP (chemical–mechanical polishing) technology. High resolution will be obtained by using a thin resist as well as a high acceleration voltage.

It is easy to correct for the proximity effect when the acceleration voltage is high. Because the effect of forward-scattered electrons is negligibly small at high acceleration voltage, we need consider only the effect of the back-scattered electrons.[57] Moreover, at high acceleration voltage, back-scattered electron density is considered to be almost constant within a small area, and LSI patterns are replaced by one representative figure (the representative figure method).[58] This method enables proximity effect correction to be calculated very quickly.

Back-scattering electron signals are used for aligning the electron beam with marks on the substrate, and the error ε in detecting the alignment-mark position is given by

$$\varepsilon = Nd/S, \tag{4.2}$$

where N is a signal noise, S is signal amplitude, and d is the signal slope width (Figure 4.37(a)). The values of N, S, and d for several kinds of conditions using concave ($\vee$) and convex ($\wedge$) Si marks have been measured experimentally, and the theoretical mark-detection error ε has been calculated.[59] The dependence of the mark-detection error on the acceleration voltage is shown in Figure 4.37(b). The detection error decreases with increasing acceleration voltage. Good alignment accuracy was obtained at high acceleration voltages, even when the mark was covered with thick overlayers.

On the other hand, if resist thickness is reduced to less than 0.1 µm, EB direct writing with a low acceleration voltage will be attractive for fabricating fine patterns.[60] The influence of back-scattered electrons then becomes negligibly small and radiation damage can be greatly reduced. Since an electron beam is stable when the acceleration voltage is high, a retarding-field electron optical system (see Figure 4.21) is appropriate for direct writing. Ishii *et al.* demonstrated the use of this kind of column using a 50-keV electron beam to write a 0.6-µm line-and-space pattern.[61]

4.1.2.6 *Variably shaped beam system*

An example of an EB system with a VSB, a Toshiba EX-8D, is shown in Figure 4.38.[62] The system uses the VSB mode in addition to the character projection mode. The VSB mode is used for device development and for X-ray mask patterning where random pattern writing is required, and the character projection mode is used for repeated pattern writing. The EX-8D is used in the study of the system itself, in X-ray mask writing, and in pattern writing of the most advanced devices. Its main specifications are listed in Table 4.2.

Figure 4.39 shows schematically the electron optical system. The electron beam emitted from a LaB_6 gun illuminates the 1st shaping aperture. The beam current-density is changed by a condenser lens system. This system has a zoom function and forms the 1st crossover image on the deflection center position of the blanking deflectors. As a result, the crossover position is maintained in the optical path at the center of the blanking deflector. The first shaping aperture image is cast on the second (character) aperture plate by two projection lenses, and the shaping deflector changes the overlapping of the two apertures. The projection lens forms the 2nd crossover image on the center position of the shaping deflector. This keeps the current density on the target from being reduced when the beam size is changed. The aperture image is finally projected onto a wafer by using a reduction lens and an objective lens. The beam current is 0.3 µA, and the convergent semi-angle of the objective lens is 5 mrad. Beam blur is kept

Table 4.2 *Main specification of the EX-8D.*[62]

Item	Specification
Acceleration voltage (kV)	50
Current density (A/cm^2):	
CP + VSB mode	10
VSB mode	40
Beam-edge resolution (μm)	0.04
Character aperture field (mm)	2.0
Maximum beam size (μm):	
CP	2.5
VSB	1.25
Main field (μm)	500
Sub-field (μm)	20
Address size (nm)	2.5
Field-stitching accuracy (μm)	0.03
Overlay accuracy (μm)	0.07

Notes: CP = cell projection; VSB = variably shaped beam.

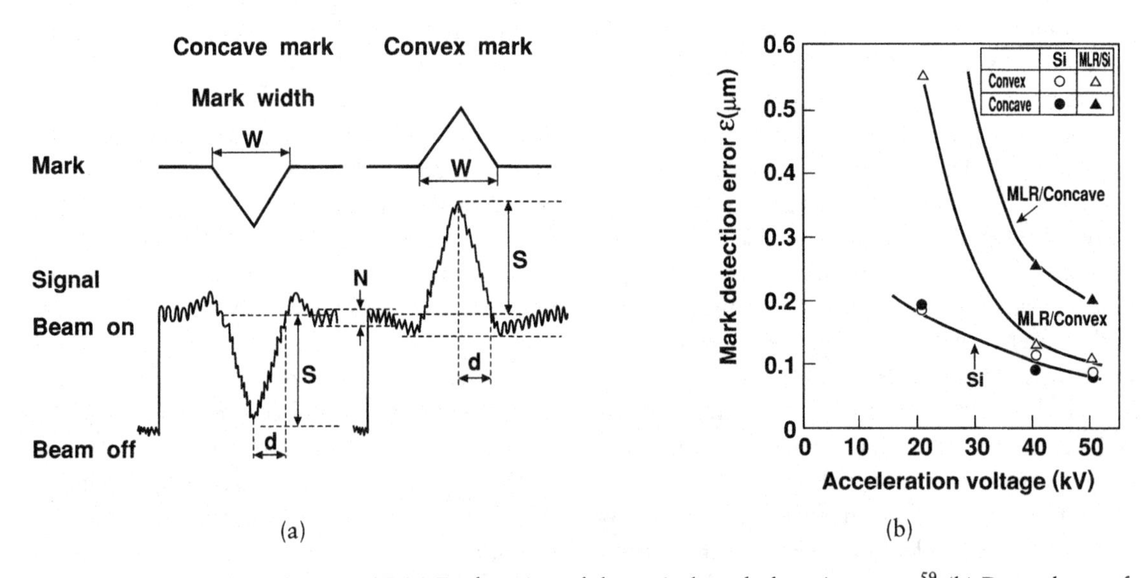

(a)

(b)

Figure 4.37 (a) Explanation of theoretical mark-detection error.[59] (b) Dependence of mark-detection error on acceleration voltage. Bare Si marks and marks covered with a multi-layer resist (MLR) were used in this experiment.[59]

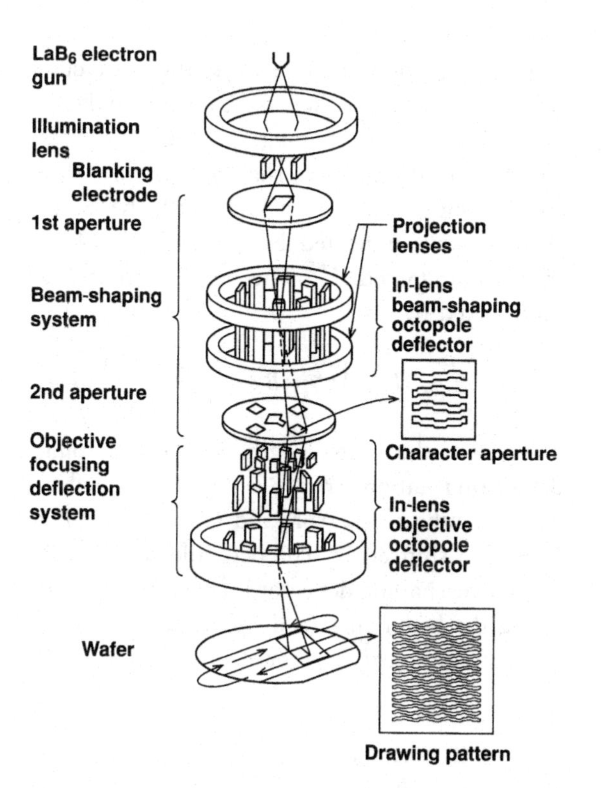

Figure 4.38 Outline of Toshiba's EB direct writing system, EX-8D.[62]

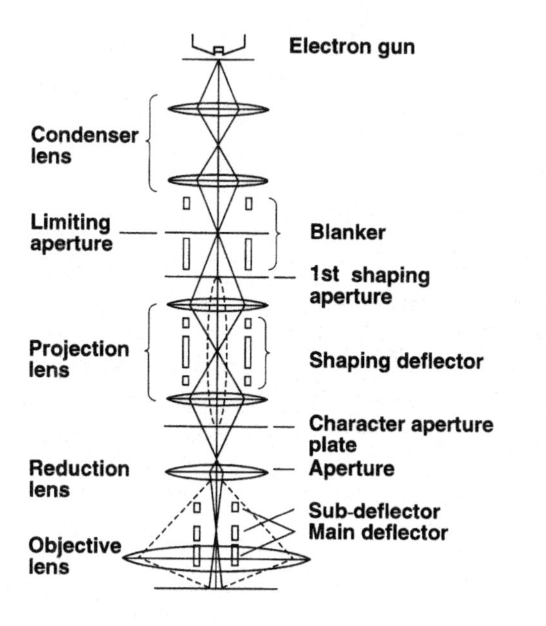

Figure 4.39 Electron optical system of the EX-8D.[62]

below 0.04 μm in a main field size of 500 μm. The electron beam is scanned over a fine gold particle placed on a Be plate.[63] The beam shape is observed using back-scattered electrons. The high acceleration voltage (50 kV) improves the beam-shape image quality.

As shown in the upper part of Figure 4.40, when a shaped beam is rotated inadequately, the line profile corresponding to the shaped-beam side edge inclines whilst the line profile corresponding to the shaped-beam interior remains uniform.[64] The aperture rotation must then be adjusted so that the line profile of the side edge becomes uniform. The rotations of the first and second shaping apertures can be controlled to within 6.6 mrad, which gives a beam rotation error of only 0.0066 μm for a 1-μm beam.

Changing the beam size varies the beam current on the substrate. When the deflection direction of the first shaping-aperture image is inclined to the second shaping-aperture side edge as shown in Figure 4.41 (*left*), the beam current curve indicated by (a) in Figure 4.41 (*right*) is obtained.[64,65] The beam current I is represented by:

$$I = Ax^2 + Bx + C \qquad (4.3)$$

where x is the beam length, A is the shaping deflection rotation error, B is the shaping deflection magnification error, and C is the shaping deflection shift error. The octopole electric circuit of the shaping deflector is corrected by analyzing the beam current data so that the beam current non-linearity error A and shift error C

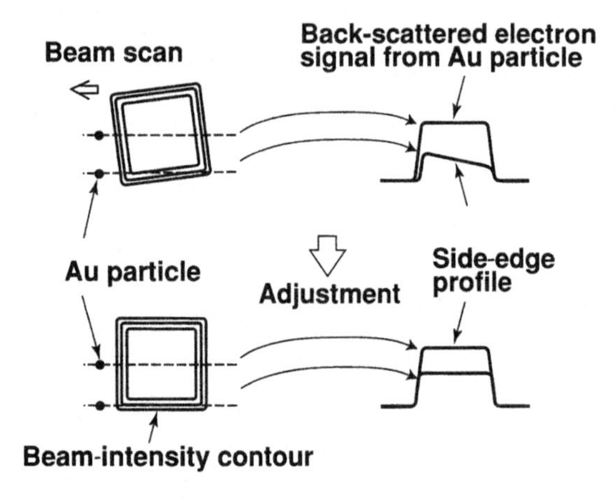

Figure 4.40 Rotation correction for shaped beams.[64]

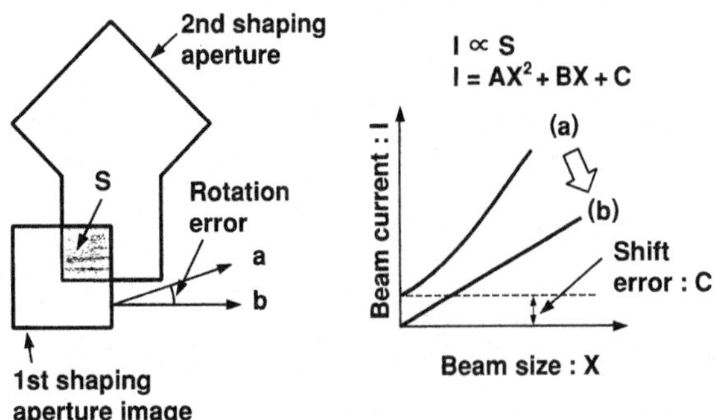

Figure 4.41 Beam-size calibration method using beam current.[64,65]

are below values corresponding to ±0.0005 μm. The magnification error B is corrected by measuring the sensitivity of the shaping deflector. The size of the shaped beam can thus be controlled to within ±0.0005 μm for rectangular beams and ±0.001 μm for triangular beams.

In making a 0.1-μm pattern, the edge roughness must be less than 0.01 μm. In Figure 4.42 the edge roughness caused by beam-size deviation from the calibrated value is plotted against this deviation. The open circles are for a PMMA resist and filled circles are for an SNR™ resist. For both resists an edge roughness of less than 0.1 μm was obtained under the condition of a beam-size control error of 0.005 μm.

Contamination is a serious problem in making LSIs with a design rule of 0.15 μm. The EX-8D wafer-loader system (Figure 4.43), therefore, excludes human operation and processes the wafer by computer control.

Temperature control is important, especially in the making of X-ray masks. The temperature of the wafer is controlled to within ±0.01 °C by a heater in the load chamber. This control compensates the temperature drop due to adiabatic expansion during quick evacuation. Cooling water controls the wall temperatures of the load chamber and the writing chamber.

4.1.2.7 *Throughput*

If we assume that a contact-hole layer contains 1×10^9 contact holes and that the hole pattern is written by one shot with a beam of current density 50 A/cm^2 on a resist with a sensitivity of 5 μC/cm^2, the total exposure time for 100 chips is more than 1.0×10^4 seconds. This is not an acceptable time for mass production. If an EB system is to be used in production lines for gigabit-class DRAMs, throughput will have to be improved by a factor of at least 100.

4.1.2.8 *Writing accuracy*

Writing errors in an EB system using a VSB are classified into three kinds: stitching error, image-placement error (pattern-placement deviation from the ideal grid position), and overlay error (Figure 4.44).

There are three kinds of stitching errors. Beam-shot stitching error is caused by errors in beam rotation, beam size, and beam position-

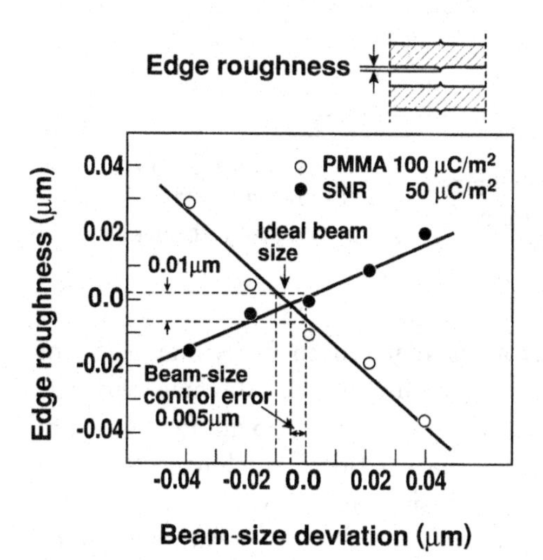

Figure 4.42 Dependence of edge roughness on beam-size deviation from the ideal value.[64]

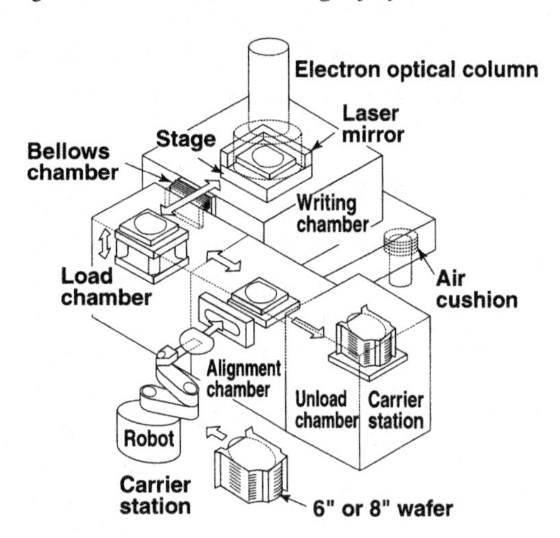

Figure 4.43 Wafer-loader system of the EX-80.[62]

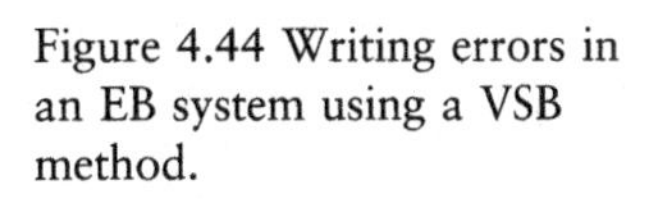

Figure 4.44 Writing errors in an EB system using a VSB method.

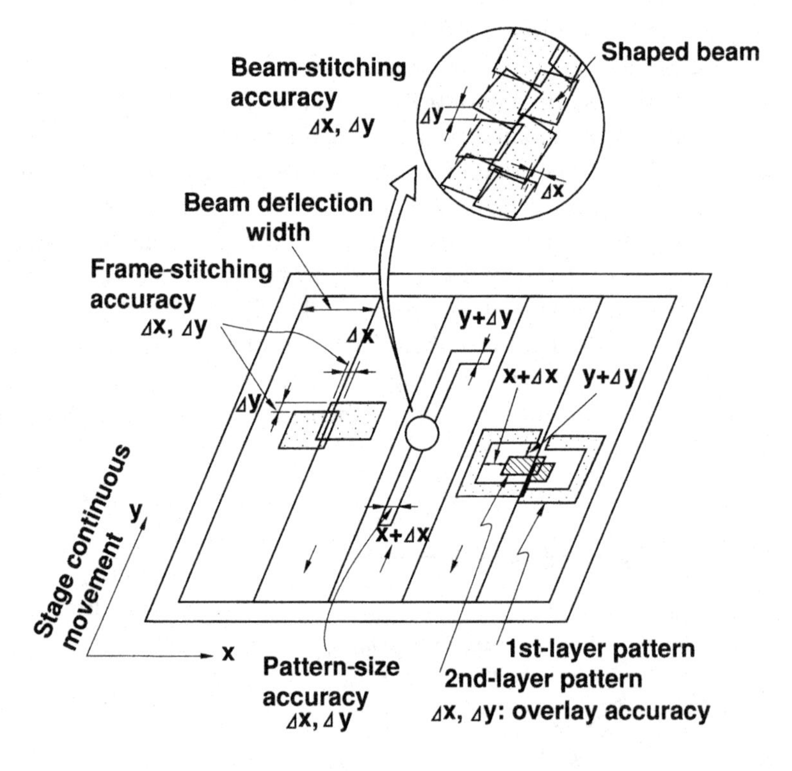

ing. Sub-field and main-field stitching errors are caused by errors of field rotation, field positioning, field distortion, beam vibration, and beam drift. Pattern-size error is greatest when the three kinds of stitching errors are overlapped. When the pattern-size error including the stitching errors is required to be within ±7% of pattern size, each individual stitching error has to be no more than about ±5 nm for a 0.15 μm pattern. In mask writing, stitching accuracy is improved by using multiple exposures. In direct writing, however, multiple exposures have to be avoided because they greatly reduce throughput.

Image-placement error is caused by target clamping and wafer process-induced stress, beam drift, wafer-placement drift, and positioning measurement error. The reduction of the image-placement error is especially important in X-ray mask writing.

Overlay error is a result of alignment error (the difference between the beam position and the alignment mark), beam drift, and process-induced stress. This kind of error is important in direct writing.

In efforts to improve accuracy, various error factors have been studied.[66,67,68] Some of these factors are: mechanical vibration,[69] beam vibration caused by stray magnetic fields,[70] and beam drift caused by contamination charging.[71]

4.1.2.9 *Future of shaped electron-beam lithography*

Electron-beam systems using the VSB method are adequate for mask making, small-scale production, and device development for which high throughput is not required. The throughput of an EB system using the simple VSB method, however, is not sufficient for mass production. Throughput has to be improved by at least two orders of magnitude. This may be possible using the projection concept if 100 contact-hole patterns are written at once. A mix-and-match between optical and EB direct writing will be attractive. In this combined system most pattern layers are formed by optical lithography and the critical layer, which is difficult to form by optical lithography, is written by EB direct writing. As yet, however, writing accuracy is insufficient and the factors related to accuracy need further study.

4.1.3 Electron-beam projection lithography

4.1.3.1 *History of development of high-throughput EB lithography*

EB direct writing lithography does not present a problem with regard to resolution, overlay accuracy, or pattern-generation capability. Many direct writing systems[62,72,73,74,75,76] for developing ASICs (application-specific integrated circuits) and advanced devices with nanometer dimensions have been reported, but the limited throughput and complicated process of direct writing remain major obstacles to their use in high-volume production. The throughput and overlay accuracy of the EB system have been improved during the past thirty years by drastically changing its beam shape and stage movement as shown in Figure 4.45, in which the historical development of the EB system is presented. Stage movement has changed from the step-and-repeat mode to a continuous-movement mode; and the Gaussian point (or spot) beam used in the 1970s was replaced by the variably shaped (VS) beam in the 1980s.[45,77] Then the more complex cell projection (CP) method appeared in the 1990s.

Four kinds of CP systems have been developed,[73,74,75,76] and their throughput has been increased about a thousandfold in the last 20 years. The effective throughput is approaching 100 wafers/h, and EB systems have become practical in quick turn-around time (QTAT) applications.

4.1.3.2 *Characteristics of the CP method and its throughput*

The pattern writing time T is expressed as

$$T = N_s(T_{ex} + T_d) + T_{oh} \qquad (4.4)$$

where $T_{ex} = S/J$ (S: resist sensitivity, J: EB current density), N_s is total number of exposure shots, T_{ex} is resist exposure time, T_d is settling

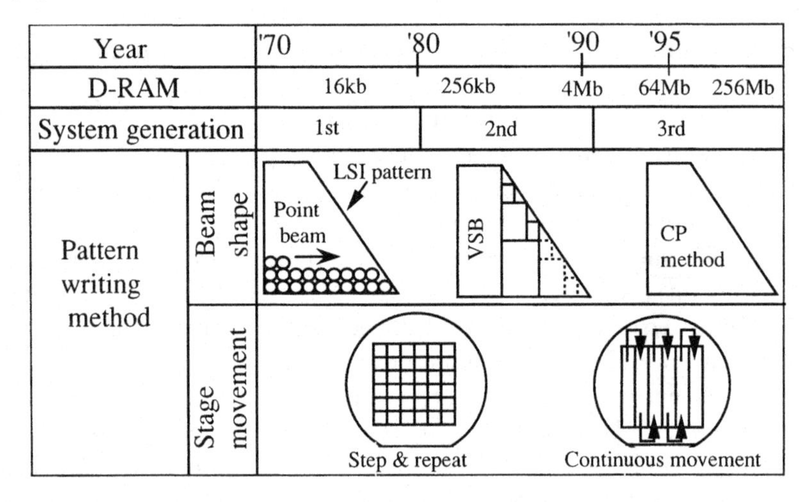

Figure 4.45 Electron-beam system development with regard to beam shape and stage movement.

time for EB deflection, and T_{oh} is overhead time. Overhead time includes the times required for stage travelling, loading/unloading wafers and aligning the beam. High resist sensitivity, small shot numbers, and reduction of overhead time are the technological objectives in improving the throughput of EB systems. The sum of $T_{ex} + T_d$ in Equation (4.4) is called the shot-cycle time and the minimum shot-cycle time is now about 200 ns. As we move beyond the 256-Mb-DRAM era, the minimum pattern size will become 0.2 μm or less. Even if a VS beam is used to delineate these patterns, the total shot number N_s will be greater than 10^{10} in an 8-inch (200-mm) wafer. The first right-hand term in Equation (4.4) (that is, N_s times shot-cycle time), will be longer than 2000 s. This means the throughput is less than 2 wafers/h. In the CP method, the number of shots is determined by the size of the cell beam. In an 8-inch wafer, there are about 10^9 shots if a 5-μm-square CP beam is used. Then the first term becomes 200 s. The CP method thus reduces the shot number to one-tenth to one-hundredth VSB and the throughput will be improved.

The original cell projection method was IBM's (International Business Machines') character projection method,[45] which used only character patterns. The CP method has recently been combined with the VSB method in order to delineate irregular patterns. In the variably shaped beam method, the apertures in both the first and second masks are square. In the recent CP method, multiple cell apertures and square apertures are arranged in the second mask as shown in Figure 4.46. The pattern writing of the CP method is shown in Figure 4.47. Periodic patterns are exposed by using the specially spaced cell beam, and peripheral patterns that have no regularity are exposed by using the conventional variably shaped beam.

In addition to providing high throughput, the CP method offers such advantages as data compaction and good pattern quality. The CP method can improve the pattern quality because the reduction ratio of the optics is high and because the accuracy of the cell mask is also high. In the variably shaped beam method, the beam-size correction error deteriorates the stitching accuracy. To delineate oblique patterns, a lot of VS beams are necessary. As shown in Figure 4.47, a CP beam can write such a pattern in one shot with high accuracy. This means the CP method results in good pattern quality.

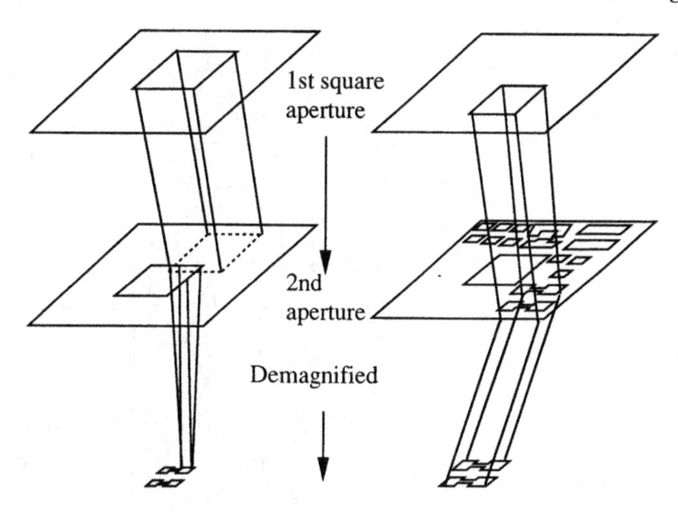

Figure 4.46 Comparison of variably shaped beam (a) and cell projection (b) methods.

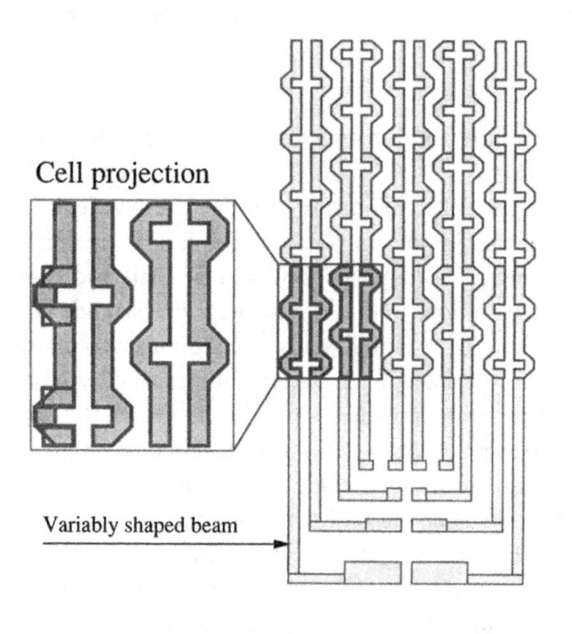

Figure 4.47 Cell projection lithography combining with the variably shaped beam method.

4.1.3.3 *Cell projection optics*

An example of the optical column in a CP system, the HL-800D, is shown in Figure 4.48 (*left*). It consists of an electron gun, beam-shaping lenses, reduction lenses, and an objective lens. The deflection system consists of one VS deflector and dual CP deflectors. The second mask consists of twenty-five 5×5 aperture groups, each of which has five cell apertures and one square aperture. The VS deflector defines the beam size on the second square aperture. The CP deflectors select the 5-cell apertures corresponding to the repeated cell pattern. The second-mask image is demagnified with a 1/25-demagnifying optical column and

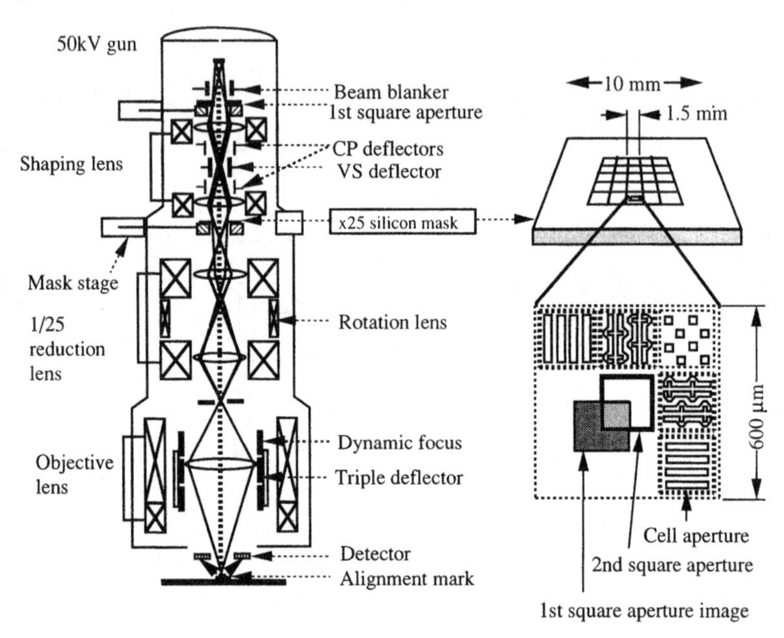

Figure 4.48 HL-800D cell projection optics.

exposed on a wafer. The maximum beam size on a wafer is 5-μm square for CP or VS beams. The beam current-density is 10 A/cm^2

A thermionic cathode is suitable for producing a CP beam, and the electron gun uses a LaB_6 thermionic cathode. The electron beam is accelerated to 50 keV in a vacuum of 10^{-5} Pa. the beam size and its current-density are limited by the electron gun, the objective lens, and Coulomb interaction. The current density J is expressed by the following three equations:

$$J \leqq B\pi(E/l)^2/2 \tag{4.5}$$

$$J \leqq B\pi\alpha^2 \tag{4.6}$$

$$\delta \geqq 10^4\, Jl^2 L/\alpha V^{3/2} \tag{4.7}$$

Here B and E are, respectively, the brightness and emittance of the gun, l is beam size, α is beam semi-angle, δ is beam blur due to Coulomb interaction, and L is the column length between the second mask and the wafer. Emittance E is the product of the cross-over size and the emission angle of the electron gun.

The deflection distance of the objective lens is another important parameter. If it is as small as 1 mm, the stage speed would be high and the acceleration and the deceleration would be large. This would cause the optical column and specimen chamber to vibrate and would deteriorate the beam positional accuracy. To obtain the large deflection length of 5 mm, a small beam semi-angle of 2.5 mrad has been adopted for the objective lens system.

Equations (4.5)–(4.6) are shown plotted in Figure 4.49, where practical values for the gun and optical parameters are substituted into them. The maximum current densities for a field emission gun, a LaB_6 cathode, and a W hair-pin cathode, are respectively 10^3, 10, and 0.1 A/cm^2. The applicable maximum beam sizes for a field emission gun, a LaB_6 cathode, and a W hair-pin cathode are respectively 0.1-, 10-, and 50-μm square. In the HL-800D optical column, a 10-μm-square beam using a LaB_6

Figure 4.49 Current density vs beam size.

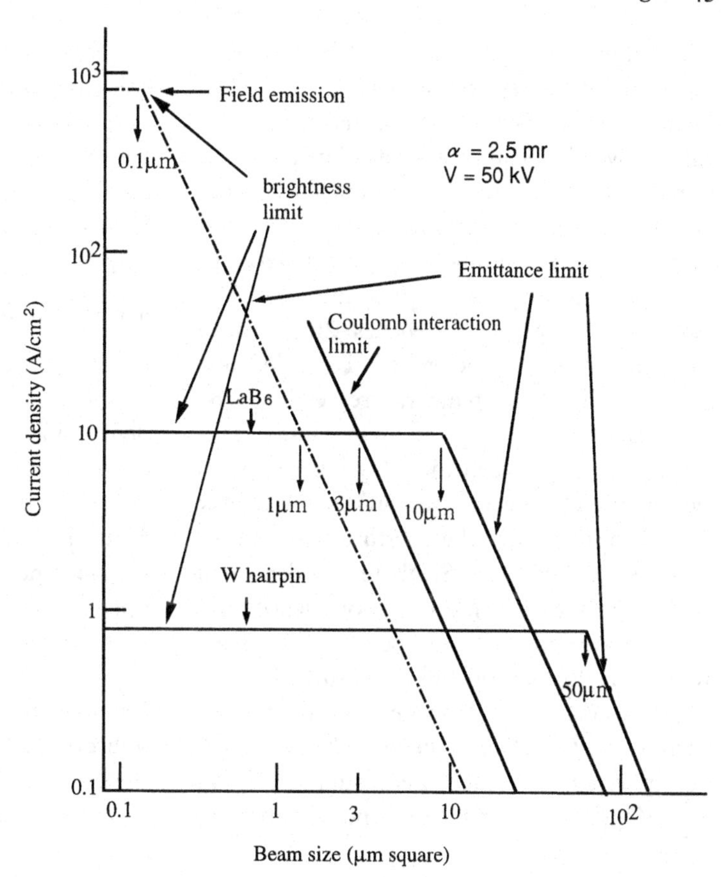

cathode could be an applicable maximum. However, if the beam blur should be less than 0.1 µm, the maximum beam size would be limited to 3-µm square because of the Coulomb interaction limit.

4.1.3.4 *Cell mask technology*[78]

The arrangement of the second (silicon) mask for a projection system is shown in Figure 4.48 (*right*). The mask is 10-mm square and 500-µm thick. The patterned areas are only 20-µm thick and are located in a 5 × 5 matrix. This 20-µm thickness is necessary to stop a 50-keV electron beam. To give a good thermal conductivity for irradiation of 50-keV electron beams, each thin area is restricted to 600-µm square and surrounded by a rib structure. Another function of the rib structure is to increase the mechanical strength of the mask. Each thin area is arranged with the pitch of 1.5 mm and has a square aperture for the VS beam and five CP apertures. These apertures are selected by a VS deflector and a CP deflector. The apertures of different thin areas are selected by moving a mask stage mechanically. In this way, 25 square apertures and 125 shaped-cell apertures can be easily selected. When this structure is used there are no problems due to thermal instability such as melting and positioning-drift of the mask. The temperature rise during EB exposure is less than 20 °C.

The fabrication processes used in making a silicon mask are shown in Figure 4.50. The starting material is a silicon-on-insulator (SOI) wafer in which a 1-μm-thick SiO_2 layer is sandwiched between a pair of Si wafers.[78] First, the aperture patterns are delineated from the front surface using EB lithography. After development, the patterns are etched into the SOI wafer. In this process, trenches 20-μm deep and 5-μm wide must be etched. To make such deep trenches, low-temperature etching by microwave plasma using SF_6 gas is used. This dry-etching stops when the SiO_2 layer appears because the etching rate for SiO_2 is extremely low. Then, to protect the aperture pattern during back-etching a thin Si_3N_4 film is deposited by low-pressure chemical vapor deposition. After deposition, window patterns for back etching are formed on the back surface. Back etching using a KOH–water solution stops when the SiO_2 layer appears. The Si masks are separated from each other after the Si_3N_4 and SiO_2 films are removed and all aperture patterns are opened. In the final step, the mask is mounted on a metal holder with adhesive. All fabrication errors in aperture width are within ±0.25 μm, which corresponds to ±0.01 μm on the wafer. Figure 4.51 shows a SEM image of one of these masks.

4.1.3.5 *Tuning of electron optical column*

In CP optics, the beam size is determined only by the mask dimensions and the projection-lens demagnification. The rotation and the magnification therefore need to be automatically corrected quickly and accurately. To correct the rotation and magnification precisely, a twin-beam method (shown in Figure 4.52) was developed.[79] The twin beams are generated by a twin-hole aperture and demagnified by 1/25 as shown in Figure 4.52. The twin beams are two 1-μm-square beams 4 μm apart for both *x*- and *y*-directions. The twin beams are scanned across the cross mark for *x*- and *y*-directions. The distances *X* and *Y* are calculated from the four

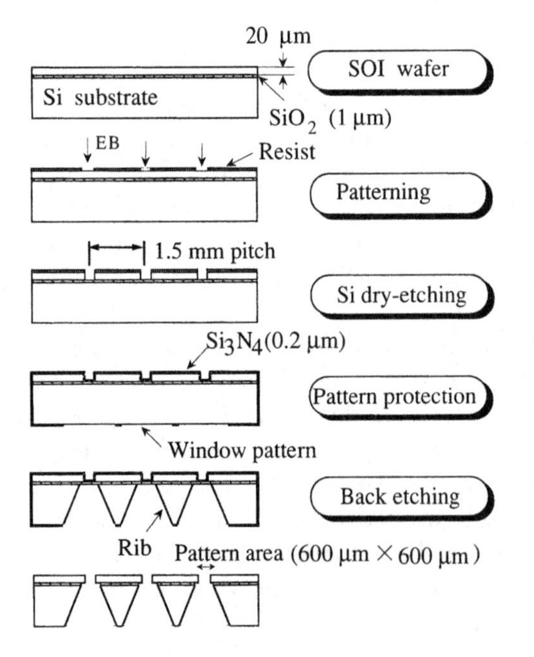

Figure 4.50 Fabrication process of silicon mask for cell projection lithography.

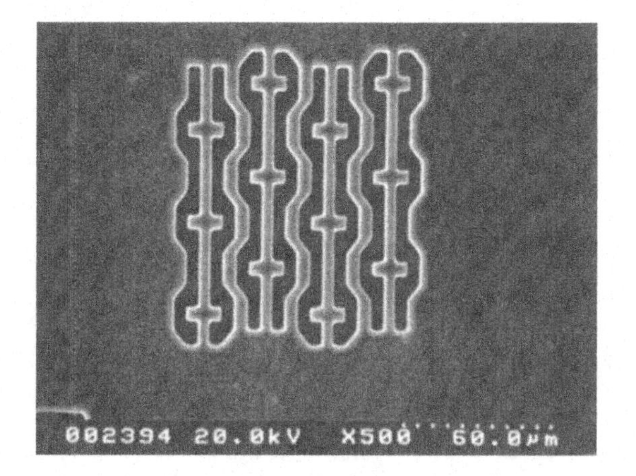

Figure 4.51 SEM photograph showing an example of a cell projection mask made of Si.

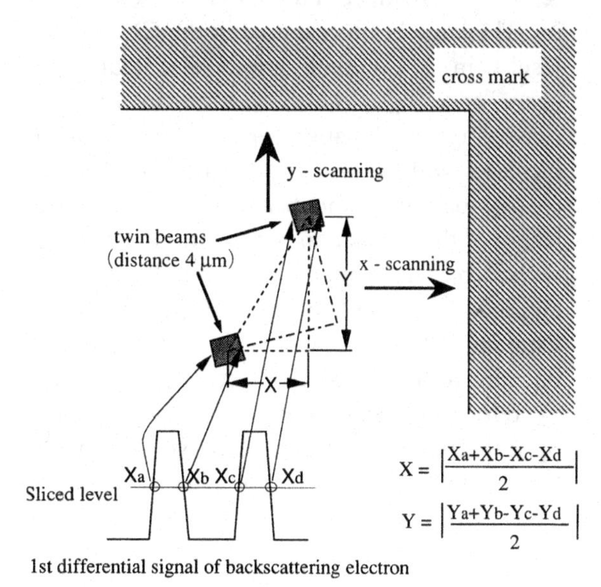

Figure 4.52 Measurement method for rotation and magnification.

edge positions X_a–X_d, Y_a–Y_d of the twin beams for both directions. The rotation error $\Delta\theta$ is calculated from the ratio of Y to X, and the magnification error ΔM is calculated from the distance between the twin beams, according to the equations shown in Figure 4.52. The CP optics is tuned until $\Delta\theta$ and ΔM become zero.

4.1.3.6 *Total system and its performance*

To achieve productive-level throughput and overlay accuracy, a lot of new technologies have been developed and are used in the HL-800D.[80] A cell projection method was combined with continuous writing technology in which the stage velocity changes variably depending on the pattern density; namely, the stage moves slowly for a high-density pattern area and vice versa.

The SEM photograph in Figure 4.53 shows that CP lithography can be used for sub-quarter-micron-level device fabrication. The silicon mask can make 0.2-μm cell patterns on a wafer. The repeated cell patterns are exposed using a CP beam and the peripheral patterns are exposed using a VS beam. The stitching error in the boundaries between a CP beam and a VS beam is less than 50 nm.

The throughput is an important criterion for judging practical use and the results of throughput analysis (assuming a resist sensitivity of $1\,\mu C/cm^2$) are shown in Figure 4.54. The writing time consists of shot-number-independent time and dependent time. The independent time consists of times for beam calibration, mark detection, wafer loading, and stage movement between stripes. It is an overhead time and is almost 170 s. The shot-number-dependent time consists of resist exposure time and electrical settling time. The sensitivity dependence of the throughput is shown in Figure 4.55 assuming a shot number of 3×10^8/wafer. This figure shows that more than 10 wafers/h can be obtained for a shot number smaller than 3×10^8/wafer and a resist sensitivity of $5\,\mu C/cm^2$.

4.1.3.7 *Future issues and possibilities*

The CP method will stimulate innovation for future ULSI production lines because of its high resolution and high throughput. However, reducing net writing time and overhead time are issues remaining to be solved. To reduce the net writing time, it is important to reduce shot number and increase resist sensitivity. The large beam size used in the CP method results in a Coulomb effect[81] that blurs the beam edge by electron–electron interaction. According to Equation (4.7), image blur δ increases with beam current I (the product of current density J and beam area l^2) and the length of the interaction path L, but it decreases with increasing beam energy V and beam semi-angle α. An example of the blurring due to the Coulomb effect is shown in Figure 4.56. Image blur due to the Coulomb effect may be reduced by refocusing. Some CP systems use a small optical column and an objective lens with large α and try to refocus for each shot.[75]

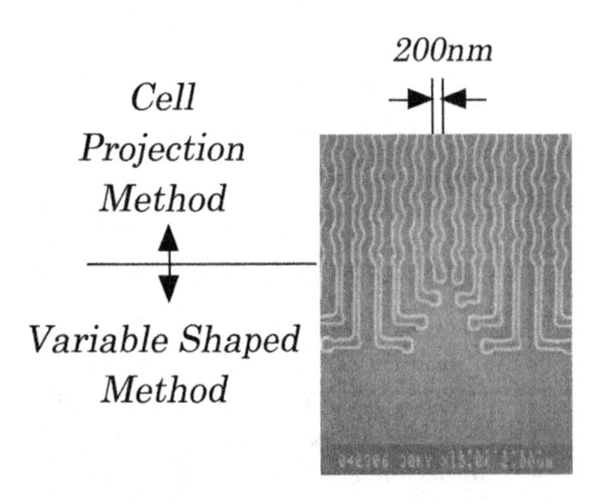

Figure 4.53 SEM photograph showing exposure of 0.2-μm pattern. Memory cell and peripherals are exposed by cell projection (CP) and variably shaped (VS) beam method.

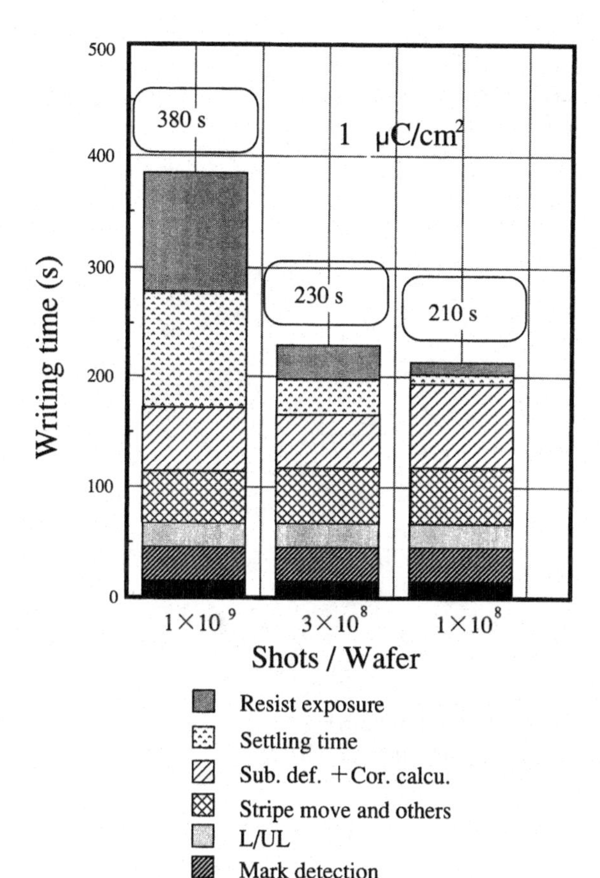

Figure 4.54 Throughput analysis for CP-plus-VSB method.

Some other ways of avoiding the blurring problem have also been proposed. Image blur is inversely proportional to image size. So attempts to reduce the interaction effects have led to the projection of macroscopic pattern beams. For example, projection exposure with variable-axis immersion lenses[82] and the scattering with angular limitation in projection electron-beam lithography (SCALPEL)[83] system have been developed. Another interesting example of obtaining high throughput is the array of micro-columns reported by T. H. P. Chang.[84] These new methods will require more technological progress but they seem to offer potential alternatives to optical lithography.

4.1.4 Multibeam lithography system

4.1.4.1 *Introduction*

With the conventional electron-beam systems such as a rectangular-shaped beam system or a Gaussian-beam system, electron beams are flashed sequentially in rectangular or circular units, or spots, to make exposure patterns. The finer an exposure pattern is, the smaller the exposure spots must be, and the greater the number of spots must be. But the use of more spots results in a lower throughput. The relationship between the exposure pattern size and throughput is shown in Figure 4.57 for several techniques used for electron-beam exposure.

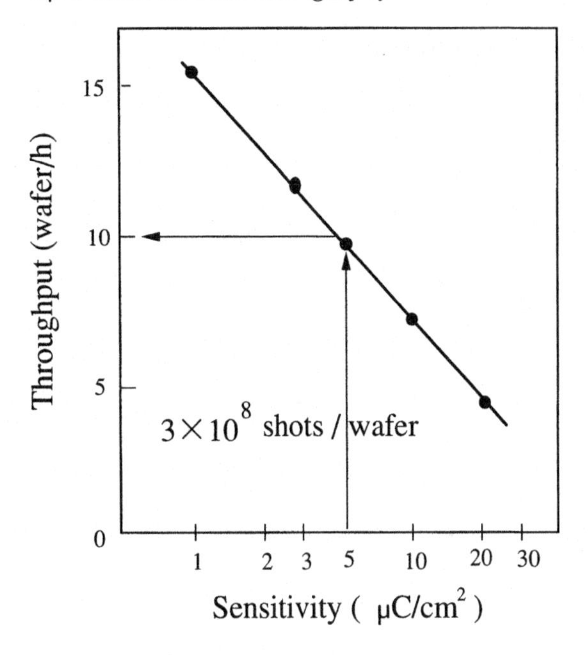

Figure 4.55 Sensitivity dependence of the throughput for CP-plus-VSB method.

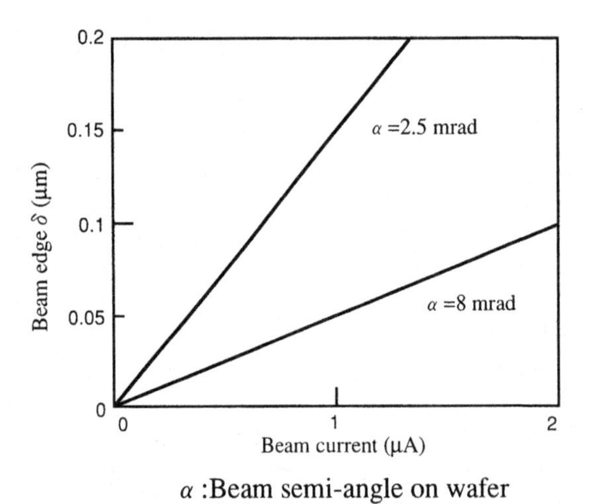

Figure 4.56 Beam blurring caused by the Coulomb effect.

The following example shows how the throughput of a conventional electron-beam exposure system is calculated. Let us assume that 0.1-μm patterns are formed on an 8-inch wafer with an exposure area ratio of 20%. The size of the area that must be exposed by electron beams is then about 6.3×10^3 mm^2 per wafer. To make the 0.1-μm pattern, it is necessary to divide the pattern into exposure spots of 0.1 μm in size. Let us divide the area to be exposed into 0.1-μm squares. The number of exposure spots on an 8-inch wafer will then be about

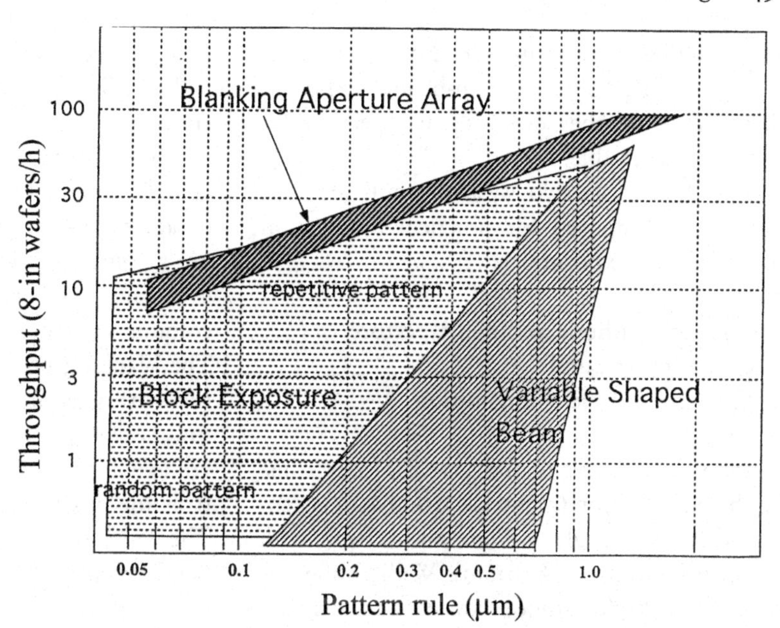

Figure 4.57 Exposure-pattern-size dependence of throughput for various electron-beam exposure techniques. Block exposure is a one-shot exposure technique using a stencil mask. Blanking aperture array is a multibeam exposure technique.

6.3×10^{11}. The time required for each spot, or the exposure cycle, is determined by the sensitivity of the resist, the current density obtained, and the settling time of the deflectors. With the best system currently available, the exposure time for each spot is about 100 ns. The required exposure time for an 8-inch wafer is therefore:

$$6.3 \times 10^{11} \text{spots} \times 100 \times 10^{-9}\ \text{s/spot} = 6.3 \times 10^{4}\ \text{s} = 17.5\ \text{h}. \quad (4.8)$$

We can thus conclude that the conventional electron-beam exposure techniques are not suitable for mass production.

As demonstrated in this example, the low throughput of conventional electron-beam systems results from the individual exposures of an enormous number of spots. This means that one-shot exposure of multiple spots may enable the electron-beam exposure technique to be used for direct wafer patterning on a mass production scale. One example of a system using this one-exposure, multiple-spot technique is the mask-pattern projection system described in Subsection 4.1.3. This technique is useful in the production of devices with repetitive patterns, such as memory cells and logic gates. It is not particularly useful, however, for random patterns such as the wiring patterns of logic circuits. This is because the number of patterns that can be prepared on a mask is limited, and most of the area of non-repetitive patterns must be exposed with rectangular beams.

With multibeam exposure, multiple spots can be exposed simultaneously, and high efficiency can be achieved even for random patterns. One multibeam exposure system, for example, has 1×10^{3} beams generated simultaneously and controlled separately to process 1×10^{3} spots on a wafer at the same time. The exposure time required to expose an entire wafer surface where there are 10^{3} spots in one shot can be estimated as shown below:

$$6.3 \times 10^{11} \times 5\ \text{spots} \times 10^{-3} \times 100 \times 10^{-9}\ \text{s/spot} = 315\ \text{s} = 5.25\ \text{min}. \quad (4.9)$$

This multibeam exposure technique can therefore increase the electron-beam exposure throughput to the level where direct wafer exposure is feasible.

The following two system configurations have been proposed for practical application of multibeam exposure:

(1) single-cathode and multiaperture configuration, and
(2) multicathode configuration.

Cathode here refers to the source of electrons. With configuration (1), an electron beam is generated from a single cathode and is split with apertures into multiple beams. With configuration (2), multiple cathodes are used to generate multiple electron beams. The beams are controlled with a single column or with multiple columns.

The comb probe printer described by Lischke *et al.*,[85] and the Blanking Aperture Array (BAA) exposure system described by Yasuda *et al.*[86] are systems based on configuration (1). Figure 4.58 shows schematically the electron optical column structure of the comb probe printer. The blanking plate has 1024 rectangular apertures arranged in 20-μm-pitch rows. Each aperture is coupled with a blanking electrode. The voltage applied to each blanking electrode controls the onset and offset of the electron beam. The width of the beam path, measured near the blanking plate, is about 20 mm. All electron beams passing through their respective aper-

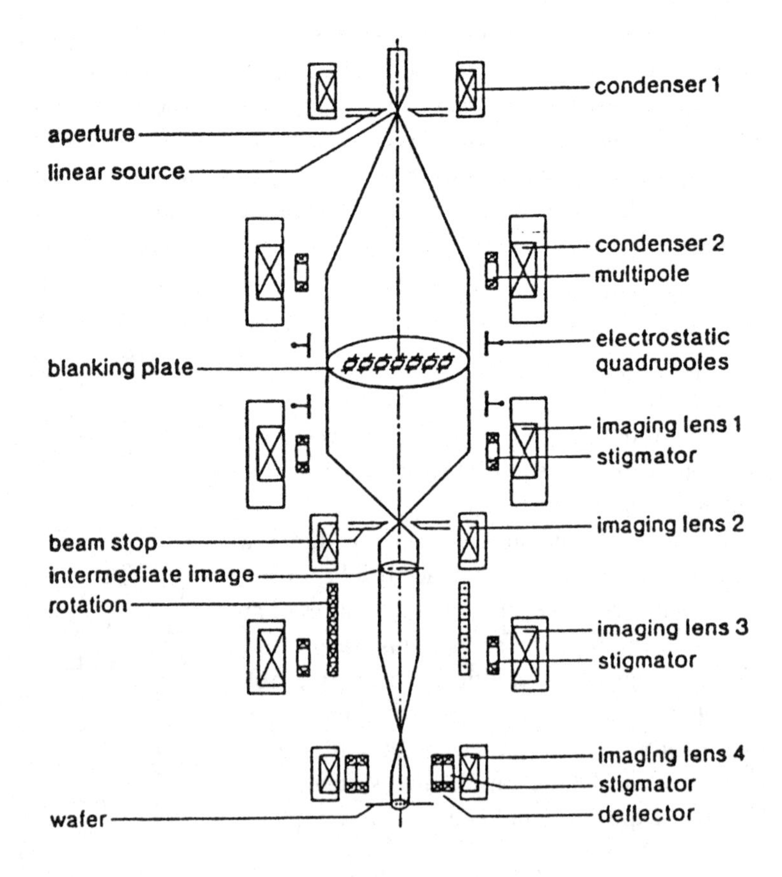

Figure 4.58 Comb probe printer column. The 1024 apertures on the blanking plate are arranged within a 20-mm length. The image reduction ratio from the blanking plate to the wafer is 312 : 1.

tures must form spots whose resolution, current density, and cross-section are uniform across the wafer surface. An exposure system based on configuration (1) has been developed. This system enables the column structure and the beam to be controlled in order to obtain such uniform multibeam conditions.

Examples of systems based on configuration (2) include the lithography-wand system described by Jones *et al.*[87] and the arrayed microcolumn described by Chang *et al.*[88,89] Figure 4.59 shows the structure of a field-emission gun and a beam extraction electrode in the lithography-wand system. By using a microfabrication technique, a large number of field-emission guns can be arranged close to each other on a Si wafer. The most important point of the wand system is how to control the electron guns so that uniform beam characteristics can be attained over the wafer. The following sections describe the latest multibeam exposure systems based on configurations (1) and (2).

4.1.4.2 *Single-cathode and multiaperture configuration (BAA exposure system)*

Figure 4.60 shows a BAA exposure system, and Figure 4.61 shows the layout of the aperture array. A total of 1024 apertures are arranged in 64 columns at 50-μm pitches in the X-direction (horizontal) and 16 rows at 75-μm pitches in the Y-direction (vertical). The even-numbered and odd-numbered rows are staggered by 25 μm (half the column pitch). Each aperture is 25-μm square, and the overall aperture array area is 3.2×1.2 mm^2. The maximum beam path width required near the blanking plate is about 1/6 of that required for the comb probe printer shown in Figure 4.58.

Gold plating forms a blanking electrode at each aperture edge. Each electrode is about 40-μm thick in the beam path. Wiring patterns connected to the electrodes are linked to the 1024 blanking amplifiers separately in order to control the blanking electrodes independently. A 10-V potential is applied for beam blanking.

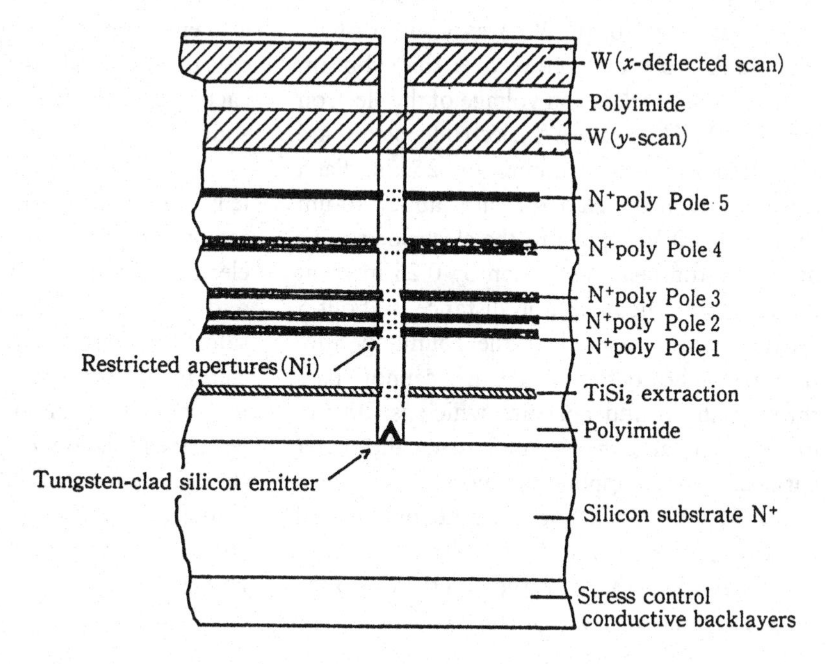

Figure 4.59 Cross-section of lithography wand and field-emission electron gun.

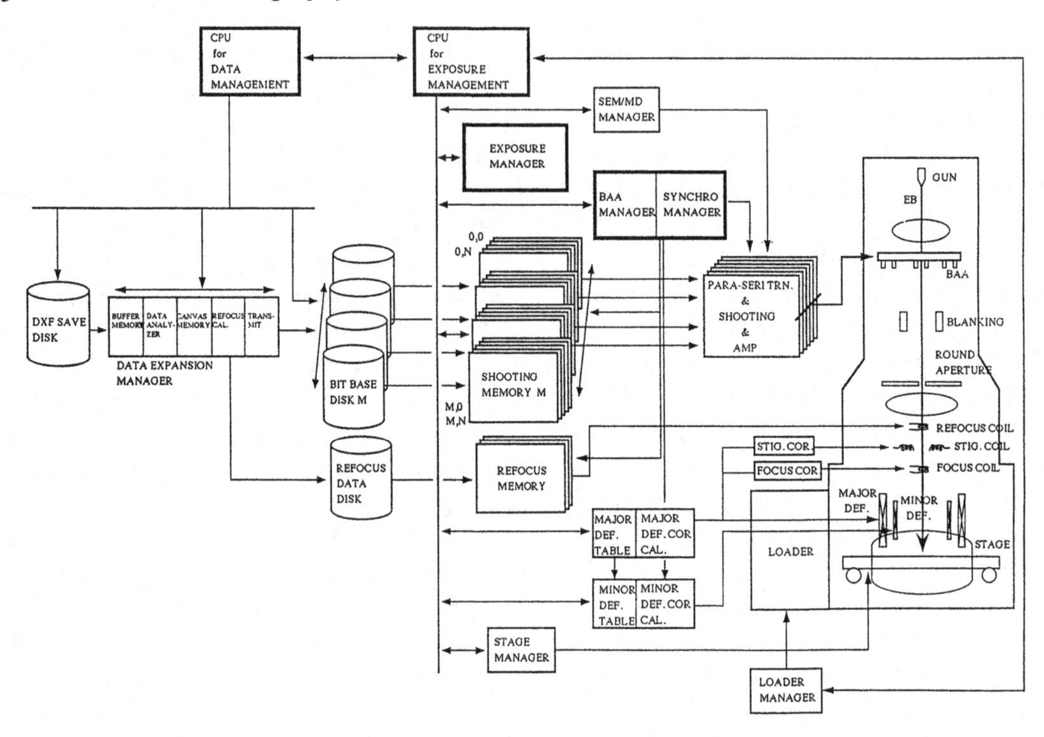

Figure 4.60 BAA (blanking aperture array) exposure system.

Figure 4.62 shows a cutaway diagram of the column structure of the BAA exposure system. It is similar to the structure of a shaped-beam column. The acceleration voltage of the electron gun is 30 kV. The BAA electrode plates are separated from one another by 25 µm. Each electrode is, in the electron path, about 40-µm thick. The 10 V applied to the electrode results in an electron-beam deflection by 0.25 mrad at the aperture. The deflection distance is about 5 40 µm when measured at the round-aperture position. The deflected beams cannot pass through the round aperture which is 200 µm in diameter, and they cannot reach the wafer surface. The corresponding spots are left in a beam-off state. Switching the potential applied to each aperture on and off enables the electron beams passing through the apertures to be turned on and off. The image reduction ratio from the aperture array to the wafer is 312 : 1. The entire aperture array exposed on the wafer covers an area of $10.24 \times 3.84\ \mu m^2$. The size of each aperture image (spot size) is then 0.08-µm square.

There are two kinds of deflectors in the final lens: an electromagnetic deflection coil (the major deflector) and an electrostatic deflection electrode (the minor deflector). A stigmator and a dynamic focus coil are used to compensate for aberrations of the deflected beams. The maximum beam current passing through the final lens is about 6.5 µA. The current varies depending on how many apertures are in the beam-on state. Differences in beam current change the imaging condition because of the Coulomb interaction of the electrons. This change is compensated for with a refocus coil. A current proportional to the number of

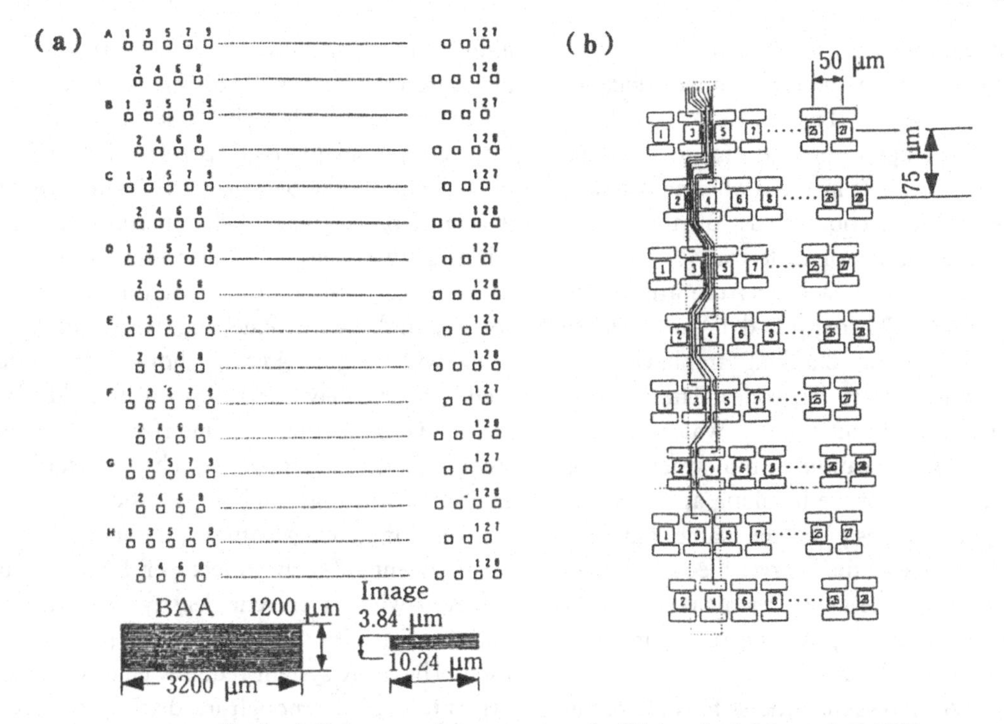

Figure 4.61 Layout of apertures on the BAA unit. (a) A total of 1024 apertures (each 25-μm square) are arranged in 64 columns in the X-direction and 16 rows in the Y-direction. (b) The aperture pitches in the X- and Y-directions are respectively 50 and 75 μm. (Wiring patterns are connected to electrodes in this figure.)

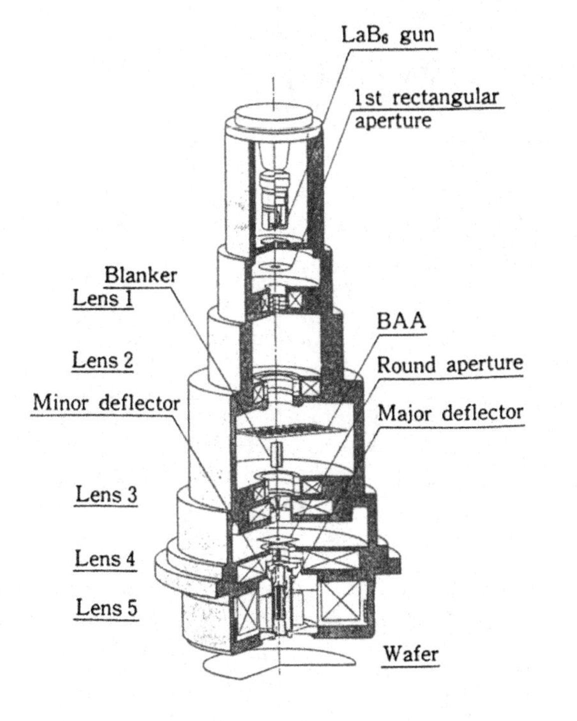

Figure 4.62 Structure of the BAA exposure column.

beam-on apertures is supplied to the refocus coil to fine-tune the overall convergence of the electron beams. Figure 4.63 is a SEM photograph of aperture array images exposed on a wafer. A similar image is obtained at each point in the deflection fields of the major deflector and the minor deflector.

Blanking data for each aperture are loaded into a register. The register data are used to switch on or off the blanking amplifiers connected to the aperture electrodes. The aperture array image is scanned on a wafer surface. During the image scanning on a wafer, the on state and off state of the blanking apertures are shifted synchronously in the same scanning direction as that of the arrays. Electron beams passing through the apertures arranged in the scanning direction expose the same predetermined points on the wafer.

Figure 4.64 shows the process for making the BAA structure. The substrate is a silicon plate with a boron-doped layer 20-μm thick. The surface of the substrate is thermally oxidized to a depth of 500 nm and then covered with a MoSi film 0.3 μm thick. SF_6 RIE etching makes wiring patterns, and beam apertures are made by Cl_2 RIE. After the electrode patterns are formed by gold plating, apertures are created by wet-etching the back surface. Figure 4.65(a) is a SEM photograph of the apertures and the wiring patterns on which electrodes are not yet formed. The size of each aperture is 25-μm square, and each connecting wire pattern is 2-μm wide. Figure 4.65(b) shows where electrode patterns have been formed by gold plating. To minimize interference from surrounding electrodes, each aperture is encircled with a grounding electrode.

Even for random patterns, a throughput of up to twenty 8-inch wafers per hour can be achieved when a current density of 50 A/cm^2 and a resist with a sensitivity of 2 μC/cm^2 are used. The BAA system requires eight exposure cycles to provide enough irradiation to form a pattern of the wafer. Accordingly, the blanking electrodes and amplifiers must be able to handle high-speed blanking signals (at 200 MHz or more).

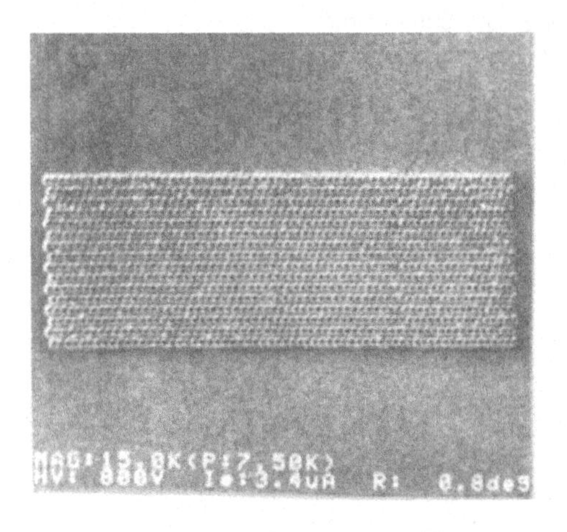

5 μm

Figure 4.63 SEM photograph of aperture array resist image on a wafer.

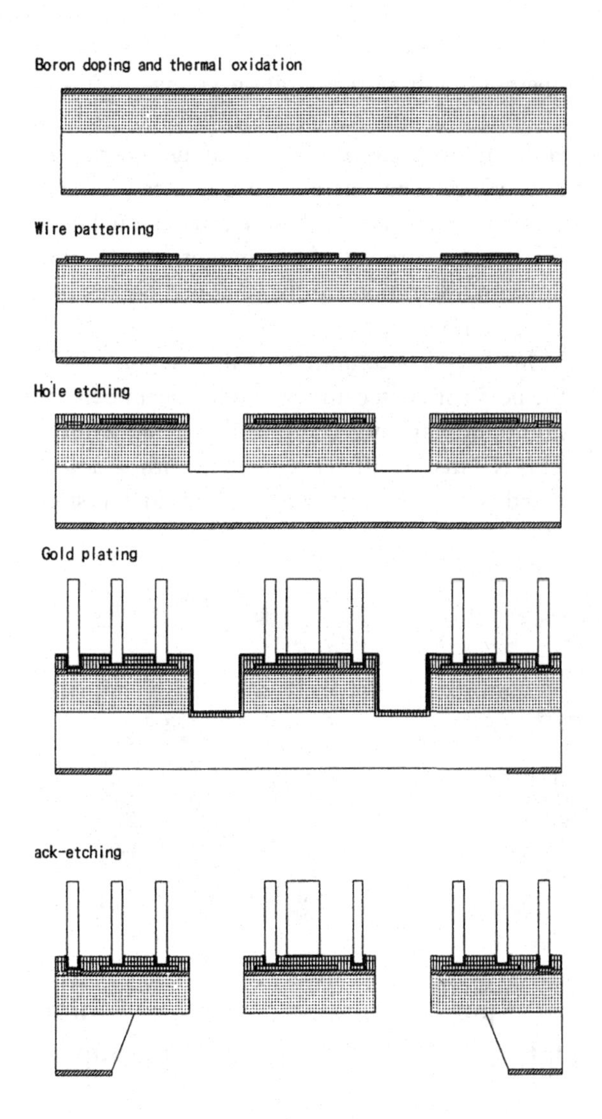

Figure 4.64 Fabrication process used to make the BAA (blanking aperture array).

4.1.4.3 *Multicathode configuration arrayed microcolumn*

The arrayed microcolumn is an electron-beam exposure system based on the multicathode and multilens configuration. As shown schematically in Figure 4.66, the arrayed microcolumn has a number of field-emission electron sources arranged in grids. The length and diameter of each column are 1–2 mm. In the example shown in Figure 4.66, there is only one column per chip on a wafer, but in an actual exposure system there may be more than one column for each chip. For a 20 × 20-mm chip, for example, there may be 1 to 10 columns per chip, or 60 to 600 columns on the entire surface of an 8-inch wafer. Each column exposes over an area 1× to 1/10× the chip size. The entire wafer surface can be exposed within the time required for one column to scan through the corresponding exposure region. The throughput obtained with an arrayed microcolumn is about 60 times faster than that of a conventional single-beam exposure system for one column/chip, or 600 times faster for 10 columns/chip. A throughput of more than ten 8-inch wafers per hour can be obtained with the latter 10-columns/chip system.

A microcolumn (Figure 4.67) consists of units containing a field-emission cathode and an anode plate (an electron gun), an electrostatic lens (an EINZEL lens), an electrostatic deflector, and a stigmator (OCTAPOLE). The units are fabricated using Si micromachining technology. The field-emission electron gun is operated with an acceleration voltage of 1 kV or less. The low acceleration voltage enables electron-beam exposure with minimum proximity effect. The microcolumn is a few orders of magnitude smaller than a conventional column, and a corresponding amount of reduction in beam aberration can be obtained because of the column geometry.

In the microcolumn shown in Figure 4.67, the field-emission cathode and the beam-extraction anode are not combined in a single unit. This is a major difference from the lithography-wand system of Jones *et al.*[87] The field-emission

Figure 4.65 SEM photographs of a BAA unit: (a) apertures and wiring patterns where blanking electrodes are not yet formed; (b) aperture arrays where blanking electrode patterns have been formed by gold plating.

5 μm

(a) before gold plating

5 μm

(b) after gold plating

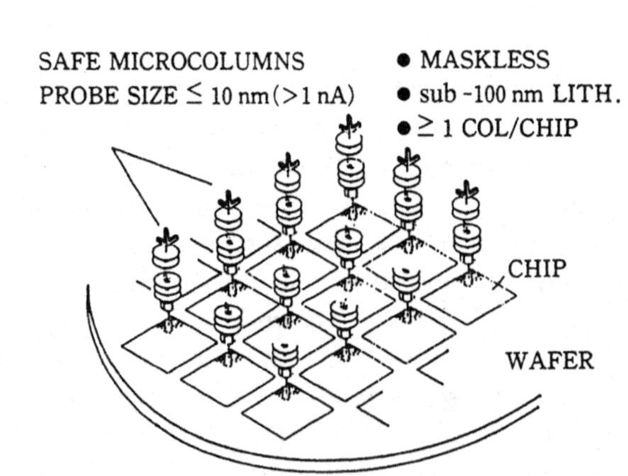

Figure 4.66 Direct wafer exposure by arrayed microcolumns. Field-emission columns 1–2-mm long are arranged in grids, one column per chip on the wafer. SAFE: self-aligned field emitter.

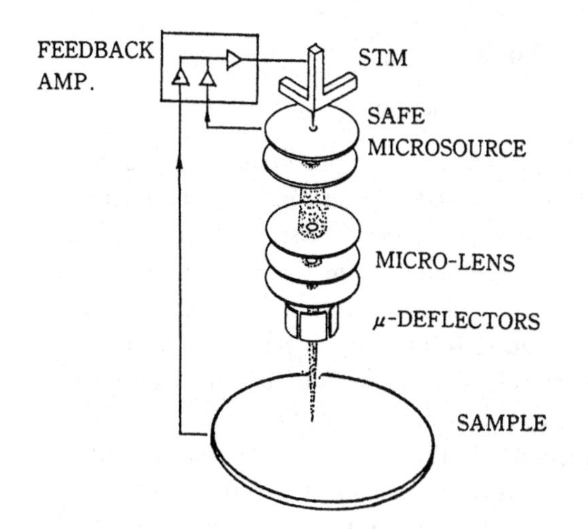

Figure 4.67 Structure of a microcolumn. The microcolumn consists of a field-emission cathode, a beam-extraction anode, an electrostatic lens (EINZEL lens), and an electrostatic deflector (OCTAPOLE). The field-emission cathode is mounted on a microactuator (STM) which is used to adjust the cathode positioning in the column.

cathode is mounted on a microactuator such as a scanning tunneling microscope (STM). The actuator is used to adjust the cathode position. The aperture of the plate closest to the cathode is about 1 μm in diameter. The beam characteristics of the electron gun vary with the position of the cathode surface, the distance between the cathode and the beam-extraction electrode, and the voltage applied to the cathode. To stabilize the irradiation of the beam from the electron gun, the STM is used to control the position and the direction of the cathode.

The electrostatic lens is an EINZEL lens consisting of three electrode plates. The upper and lower electrode plates are kept at the same potential (usually a ground potential) while the voltage applied to the middle electrode is varied to obtain the desired convergence. The deflector is an electrostatic deflector containing eight electrodes. The deflection voltage applied to the deflector is superimposed by using a compensating voltage for correcting astigmatic aberration.

Although the microcolumn is a very small structure, it still has all the units for an independent column: an electron source, an electrostatic lens, an electrostatic deflector, and an aberration correction unit. Each of these column units must be adequately controlled in order to expose fine patterns. Establishing the techniques needed to control the column units is a major target in the development of microcolumn systems. There should be no problem in increasing the number of microcolumns in the multibeam system.

4.1.4.4 *Summary and future challenges*

Multibeam exposure is useful in increasing the throughput of an electron-beam exposure system. Both (1) the single-cathode and multiaperture configuration and (2) the multicathode configuration are available. With either configuration, the number of beams needed to obtain a practical throughput is about 1×10^3. The BAA exposure system is based on configuration (1). In this system, arrayed apertures with blanking electrodes are used to make multiple beams irradiate a wafer. Geometrical aberrations such as astigmatism and field curvature are compensated so that each electron beam has a uniform profile. The arrayed microcolumn is based on configuration (2). Column units (electron source, deflector, and beam correction unit of each microcolumn) are controlled individually to stabilize each.

With the BAA system, a high-speed blanking (200 MHz or more) is needed in order to obtain a practical throughput. Column units in the arrayed microcolumn must be well-controlled so that the beam characteristics of each column are the same and the patterns exposed by one microcolumn match those exposed by the others. There are some technical difficulties concerning multibeam exposure systems. However, it is expected that the difficulties will be overcome and electron-multibeam exposure will be put to practical use.

4.2 Mask writing

4.2.1 Introduction

Electron-beam exposure systems (EB systems) have been used for mask writing on production lines since the EBES™ system was developed at AT&T Bell Labs.[90] The EBES™ system was the first machine using a round Gaussian beam, a continuously moving stage, and a raster scan with a small deflection width. Figure 4.68 shows the electron optics and the writing method of the EBES™. Electron beams are scanned perpendicularly to the continuously moving direction of the stage. The EBES™ sys-

(a)
Electron gun cathode
Grid
Anode
Centering coils
Blanking plates
Stop
High-frequency deflection coils
Final lens
Ion pump
Beam limit aperture
1st lens
2nd lens
ferrite
Silvered screening tube
Ion pumps
Mark detect current amplifier
Diffusion pumps

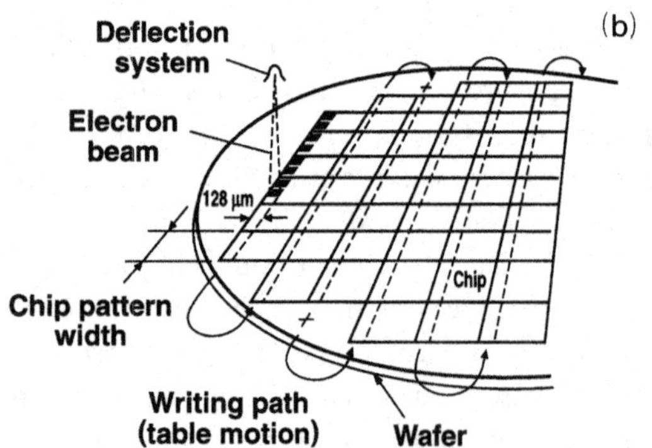

Figure 4.68 (a) Electron optical system of the EBES™. (b) Writing method of the EBES™

tem had a throughput much higher than that of the conventional systems with a step-and-repeat (S&R) stage. Stage stepping time is lost time in S&R systems, but the number of stage steppings is greatly reduced in a system with a continuously moving stage, and this reduces the amount of lost time. The continuously moving stage was a revolutionary concept used in the EB direct writing machine and subsequently was accepted by the semiconductor industry as a mask fabrication system. The MEBES™ system, which was a commercialized system based on the EBES™, is now widely used all over the world.[91]

EB systems before the EBES™ adopted the S&R method in which electron-beam writing was done on the stationary stage.[92] After an exposure is finished, the stage moves to the next position. To increase throughput, the deflection field has to be enlarged to reduce stepping time. Beam positioning accuracy and beam resolution, however, deteriorate drastically with increasing deflection field because of deflection aberrations.

To increase throughput, the variably shaped beam (VSB) method was developed.[45] Systems combining a VSB with a S&R stage have been used in mask fabrication.[93] However, the writing accuracy of VSB systems is not as good as that of Gaussian beam systems. This is because the VSB system has a complex electron optical column and it cannot accept the MEBES™ data format, which is the *de facto* standard, so it is difficult for VSB systems to win widespread acceptance. As the LSI design rule decreases, however, a system using a round Gaussian beam has to use a smaller beam size, and this results in a considerable reduction of throughput. The use of VSB systems is, therefore, becoming inevitable in advanced mask-making because it provides a high throughput.[94] It is essential to improve the writing accuracy of VSB systems and to develop data-conversion software from VSB data to MEBES™ data and from MEBES™ data to VSB data.

4.2.2 Mask trends

The miniaturization of LSI devices is expected to continue. LSI cost has been drastically reduced by reducing chip size and then by increasing wafer size. Computer downsizing has increased the demand for further reductions in chip cost. Soon, devices with a design rule of 0.18 μm will be fabricated by using a KrF exposure system together with a resolution-enhancement technology (RET) such as a phase-shifting mask (PSM) and/or off-axis illumination (OAI). And 0.13-μm patterning will be made possible by using an ArF exposure system and RET. For pattern sizes below 0.13 μm, it may be necessary to use non-optical lithography: EB aperture projection[95] or X-ray lithography.[96] Although non-optical lithography is potentially useful for increasing resolution, EB aperture projection faces throughput problems and X-ray lithography faces problems because the total overlay accuracy is low.

Mix-and-match between optical and EB lithography will be accepted where EB lithography assists optical lithography in patterning of some critical layers such as gates and contact holes.[97] Overlay accuracy in mix-and-match is not good because of distortion of the optical stepper lens. To improve overlay accuracy, stepper-lens distortion is measured and fed back as mask-pattern-position data. Lens distortion is corrected by changing the mask-pattern position. An EB aperture projection system also requires an aperture mask.[98]

Device fabrication using X-ray lithography has been tried.[99] The greatest difficulty in realizing X-ray lithography is to improve mask-pattern-placement accuracy, which is mainly caused by a lack of writing accuracy of an EB mask-writing system. If an ultra-accurate X-ray mask-pattern writer can be realized, X-ray lithography will be brought closer to application on production lines.[100]

Some non-optical lithographies as well as optical lithography, require accurate masks. The key to making an accurate mask is to develop an accurate mask-writing system, and an ultra-high-accuracy system is particularly strongly required for X-ray lithography.

Figure 4.69 shows the relationship between depth of focus (DOF) and line-and-space (L&S) pattern size. Resolution is clearly improved by using a halftone (HT) mask together with off-axis (annular) illumination. The HT mask has been used in production-line exposure systems with g-line, i-line, and KrF sources. For ArF wavelengths, HT masks are under development. A Levenson-type PSM is the most effective with regard to improving resolution, but its theoretical advantages are very difficult to realize in practice due to the complexity of LSI patterns. To fabricate a Levenson-type PSM, a phase-shifter pattern has to be written on the LSI pattern. Fabrication of a Levenson-type PSM therefore requires an alignment function for a mask-pattern writer.

Resolution is improved by using RET but mask critical dimension (CD) control is becoming increasingly severe with decreasing design rule. Figure 4.70 shows the relationship between mask CD error on the wafer and DOF, which is obtained by simulation calculation.[101] If a DOF of 0.9 μm is required for the model pattern of 256-Mb DRAM (Figure 4.70), an acceptable mask CD uniformity for the mask with a magnification of four (4× mask) is ±16 nm. This mask CD uniformity is more severe than the wafer CD uniformity of 27.5 nm (10% of 0.257 μm). It is difficult to achieve this level of CD uniformity because the CD uniformity obtainable with the most advanced mask writer is about ±30 nm. If one wishes to relax the required mask CD uniformity of ±16 nm, the DOF has to be less than 0.9 μm or/and the exposure tolerance has to be severe. If DOF and exposure latitude cannot be decreased, the required mask CD uniformity becomes severe.

Pattern fidelity deteriorates if pattern size is reduced to less than the exposure wavelength. Figure 4.71(b) contains a SEM photograph of

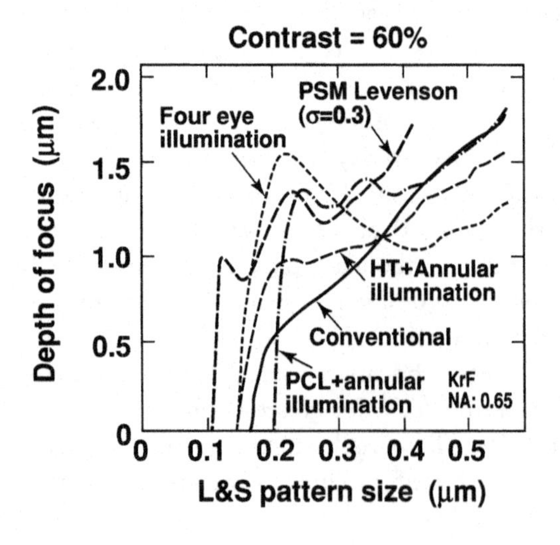

Figure 4.69 Dependence of depth of focus (DOF) on pattern size for several kinds of exposure conditions. A Levenson-type phase-shifting mask (PSM) is the most effective, but it is difficult to use in real LSI patterning. Halftone (HT) masks combined with annular illumination have already been used in production lines. PCL: phase contrast lithography, in which a phase-shift filter is inserted at pupil or optics in stepper.

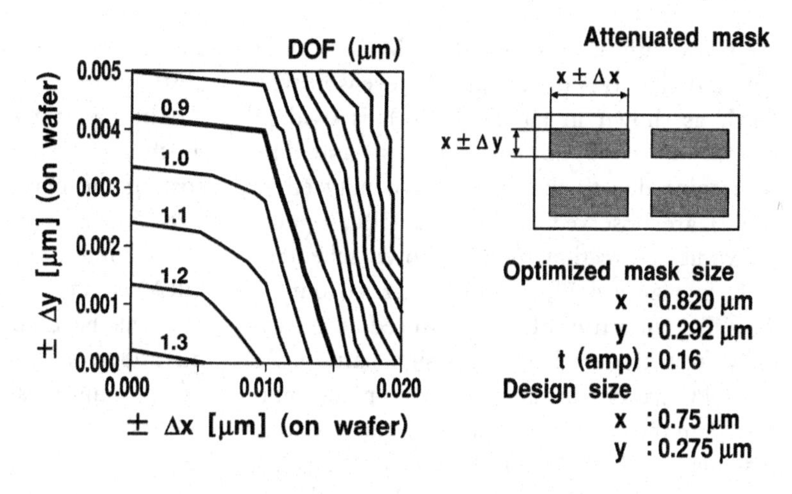

Figure 4.70 Dependence of mask CD error (on wafer) on DOF. Mask CD error greatly reduces DOF. Exposure wavelength: 248 nm (KrF); exposure latitude: 15%; mask bias: 0.07 μm in x-direction and 0.017 μm in y-direction; transmittance of HT mask: 0.16 in amplitude (2.56% in transmittance); NA: 0.47; and coherent factor: 0.71 were assumed.[13]

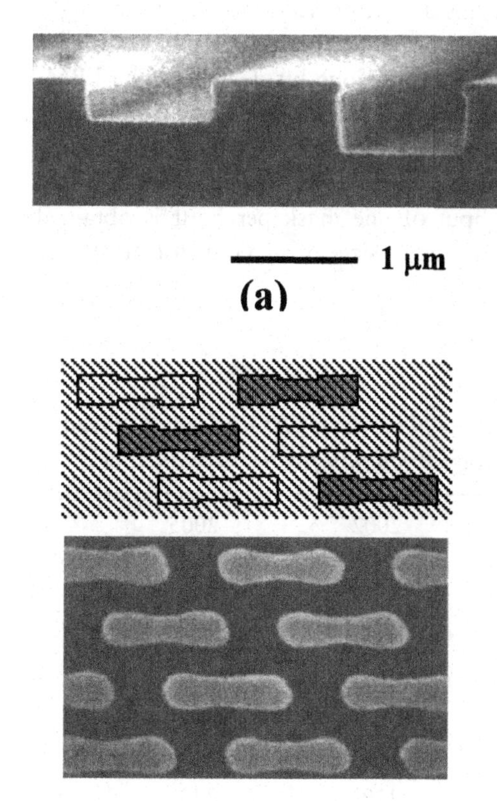

Figure 4.71 SEM photographs: (a) Structure of a dual-trench Levenson mask. Shallow- and deep-etched areas respectively correspond to phase angles of 0 and π.[14] (b) *Upper*: 1-Gb-DRAM test pattern. White and dotted patterns represent shallow (0) and deep (π) etched areas. *Lower*: A resist pattern produced by using the mask in the upper part. Corners are rounded and pattern length is considerably shortened.

the resist pattern of a 1-Gb-DRAM test device with a design rule of 0.18 μm. A Levenson-type PSM, as shown in Figure 4.71(a), is used in producing it.[102] Corner rounding and line shortening due to the optical proximity effect (OPE) are observed. To reduce OPE, optical proximity correction (OPC), shown schematically in Figure 4.72, has to be carried out.[103] An OPC pattern has to be produced automatically from the original LSI design pattern. As the OPC pattern includes a much smaller pattern than the original LSI pattern, mask-writing systems with high resolution and small address size are required for producing OPC masks.

Target mask specifications are listed in Table 4.3. The mask size of the stepper is limited by the exposure field size and nowadays a size of 6 inches (4×) is used. The throughput of optical exposure systems is improved with increasing reticle size. The step-and-scan exposure system has recently become widely accepted as the main tool for use below the 0.25-μm design rule. This system can use a larger size mask than a conventional optical stepper. The size standard of the next-generation mask will change from 6 inches (150 mm) to 9 inches (230 mm).

Although throughput of exposure systems is much improved by enlarging the mask size, the cost of the mask is increased with increasing size. A 9-inch mask size has cost advantages in volume production of LSIs such as DRAMs. On the other hand, a 6-inch mask size will be used to produce devices in small quantities, such as logic devices. So, selection of optimum mask size will improve cost effectiveness.

4.2.3 Mask-writing system

EB writing systems using a round Gaussian beam need to use smaller beams as the LSI design rule decreases. Table 4.4 shows throughput of EBM-130/40 (Toshiba Machine Co.) for DRAMs with various bit densities. The EBM-130/40 adopts a round Gaussian beam, a continuously moving stage, and a raster scanning method as shown in Figure 4.73.[94,104] A throughput of one mask per hour is obtained using a beam size of 0.5 μm for a 4-Mb-

Table 4.3 *Target mask (4×) specification requirements. Image placement, CD uniformity, mean to target and data volume are from the 'National Technology Roadmap for Semiconductors'.*[115]

Year	1997	1999	2002	2005
Design rule (μm)	0.25	0.18	0.13	0.10
Image placement (nm)	52	36	28	20
CD uniformity (nm)	26–36	18–26	13–18	9–14
Mean to target (nm)	20	12	8	6
Data volume (gigabytes)	8	32	128	512
Mask size (mm)	152	230	230	230

Note: Range of values reflects different target depending on pattern-types.

Figure 4.72 Mask pattern without OPC (*left*) and with OPC (*right*).

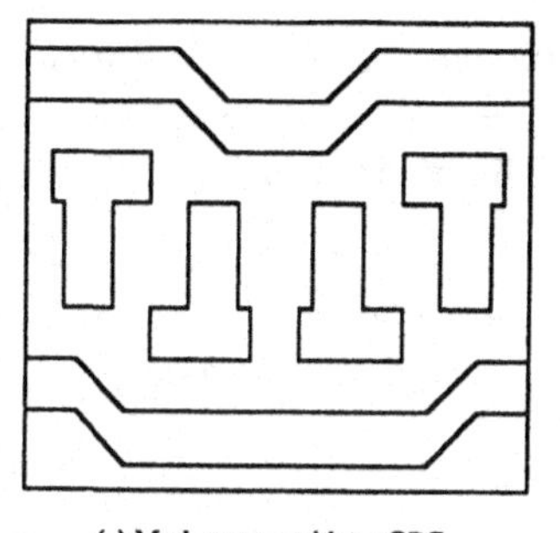

(a) Mask pattern without OPC.

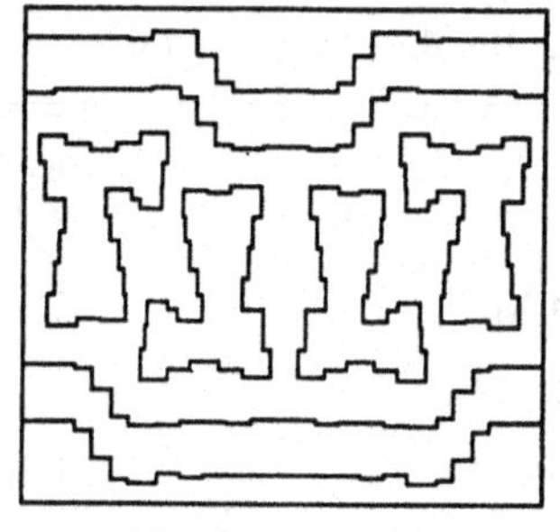

(a) Mask pattern with OPC.

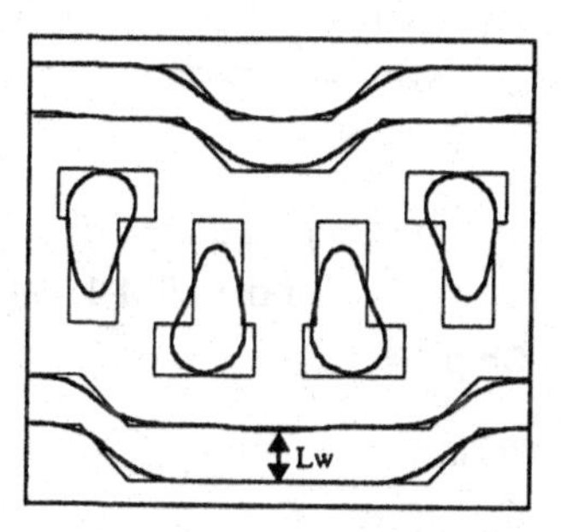

(b) Printed resist image with nominal dose and just focus.

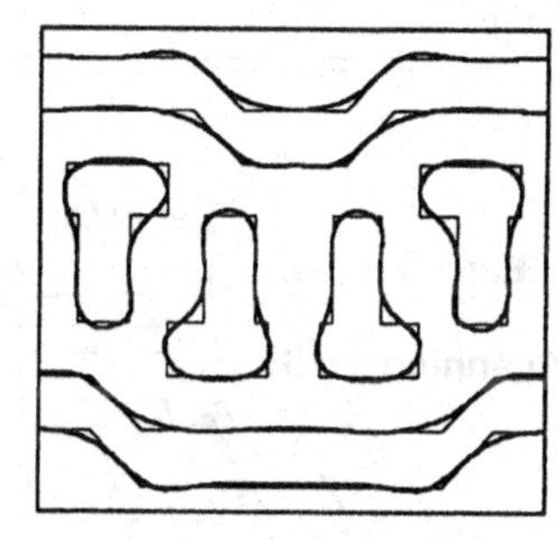

(b) Printed resist image with nominal dose and just focus.

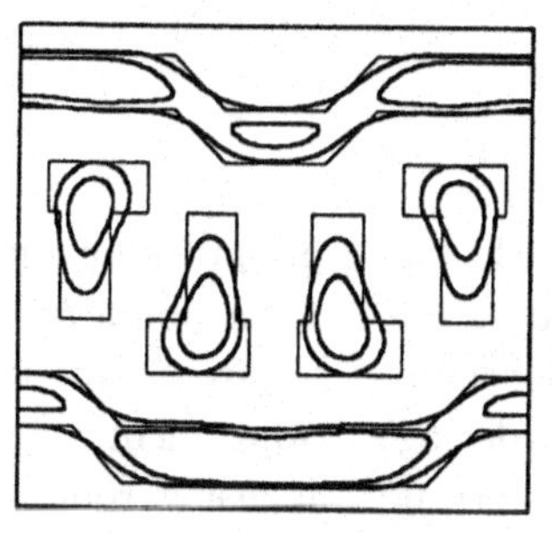

(c) The largest and smallest printed resist images within process window of ±5% exposure latitude and ±0.75μm DOF.

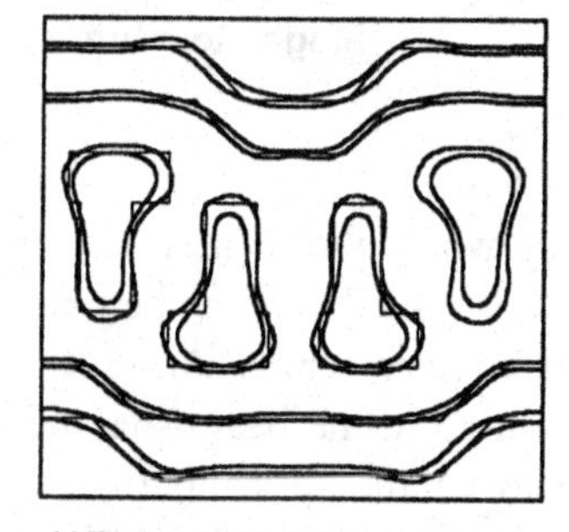

(c) The largest and smallest printed resist images within process window of ±5% exposure latitude and ±0.75μm DOF.

DRAM-class pattern. For a 64-Mb-DRAM-class pattern, however, throughput is decreased remarkably (to 0.05/h) because beam diameter has to be reduced to 0.01 μm.

To increase throughput, MEBES IV adopts a Zr–W thermal-field-emission electron gun. The current density of a beam with a diameter of 0.1 μm is 400 A/cm^2. Figure 4.74 is a schematic view of the electron optical column of the MEBES IV. The data transfer rate is increased from the 80 MHz of MEBES III to 160 MHz.[105]

Another way to increase throughput is to develop systems that use a VSB. Figure 4.75 shows an example of such a system, the EX-8, which uses a VSB, a continuously moving stage, and vector scanning methods.[94] It can generate triangular beams as well as rectangular beams by using a key-holder-shaped aperture, as

Table 4.4 *Throughput of the round Gaussian system EBM-130/40.*[6,16]

DRAM capacity	Reticle pattern size (μm)	Beam diameter (μm)	Throughput (h^{-1})
4 Mb	3.5 (5×)	0.5	1.3
16 Mb	2.5 (5×)	0.25	0.3
64 Mb	1.5 (5×)	0.1	0.05
256 Mb	1.0 (4×)	0.05	0.0125
1 Gb	0.6 (4×)	0.05	0.0125

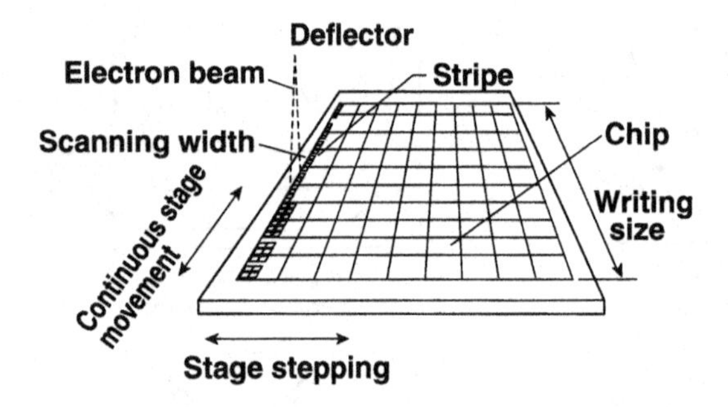

Figure 4.73 Writing method of the EBM-130/40.

shown. Oblique lines with angles of 45 °, 135 ° 225 ° and 315 ° can be written at high speed by triangular beams. An in-lens octopole deflector is used in the beam-shaping system in order to shorten the optical path, and the shorter optical column reduces beam blur arising from the space-charge effect. A single-crystal LaB_6 cathode with a flat surface having a diameter of 70 μm was developed.[106] The axial orientation of the cathode is ⟨100⟩. Very homogeneous beams are obtained from the flat-surface LaB_6 cathode and illuminate the 1st beam-shaping aperture. Homogeneous aperture illumination is very important for generating shaped beams with a high uniformity in intensity profile. The cathode service life is about 5000 hours. The beam is accurately positioned by main deflectors and sub-deflectors composed of electrostatic octopoles. The field size of the main deflector is about 1-mm square and that of the sub-deflector is about 64-μm square.

An automatic column program controls lens adjustment for illumination, magnification and focusing, rotation adjustment of the 1st and 2nd shaping apertures, alignments between the lens center and the optical axis, and alignments between the shaping and limiting aperture centers and optical axis. A beam-calibration program carries out beam-size calibration, beam-rotation calibration, astigmatism correction, and distortion correction. The mask height is monitored during pattern writing and its data are transferred to the deflection-control unit for deflection-width correction and to the lens-control unit for dynamic focusing.

The mechanical structure of the EX-8 is illustrated in Figure 4.76. Mask blanks with sizes of 5 and 6 inches are stored in rotary cassette

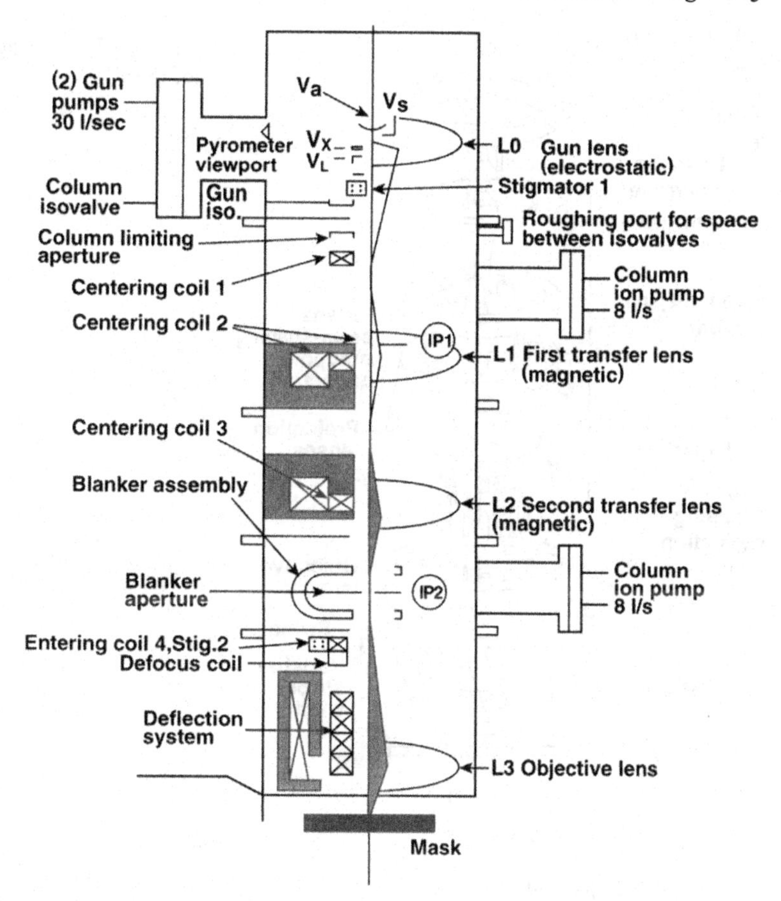

Figure 4.74 Schematic view of electron optical column of the MEBES IV.

magazines. A robot takes the mask blank of the desired size and carries it into the input/output (I/O) chamber. After the I/O chamber is evacuated, the blank is brought into the loading chamber and loaded onto the shuttle. The shuttle is then carried into the writing chamber. A delineated mask on another shuttle is unloaded and comes back to the delineated-mask stocking position. Loading and unloading motion is done at the same time to reduce overhead time. Throughput is about 1 mask/h for a 64-Mb-DRAM-class pattern and the EX-8 can produce about 500 six-inch masks per month.

Data conversion from CAD/LSI data to EB data is important, and Figure 4.77 shows an example of the data flow of a mask-making system composed of the data-conversion system, the EB-writing system, and the database-inspection system. For a 64-Mb-DRAM-class pattern, average time needed to convert from GSD II data to EX-8 data is about 10 min. Although difficult layers require much more time for conversion, the speed of this conversion system is high enough for 64-Mb-DRAM-class devices.

As mask pattern size decreases, proximity effect correction is becoming essential for improving CD linearity with decreasing mask-pattern size. Figure 4.78 shows mask-pattern error versus pattern size for isolated lines and isolated windows. Patterns down to a size of 4 μm are obtained (within an error of 10% of

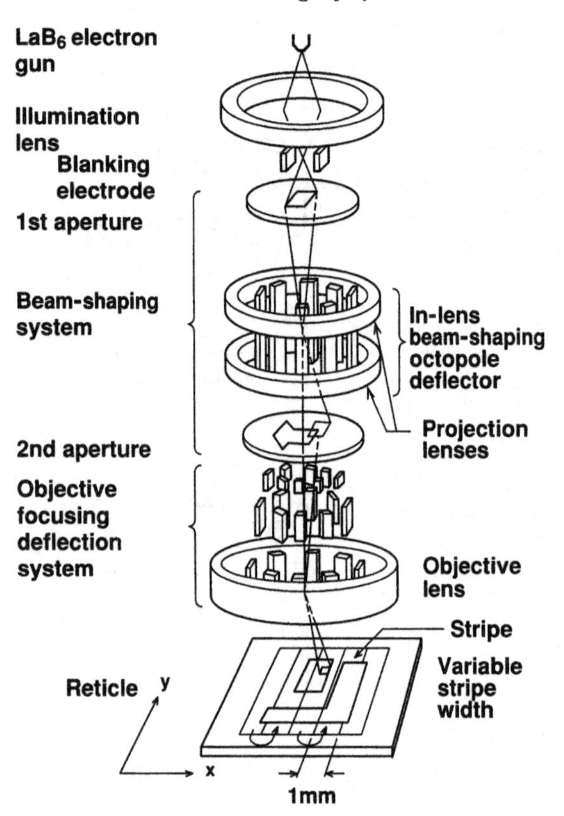

Figure 4.75 Writing method of the EX-8.

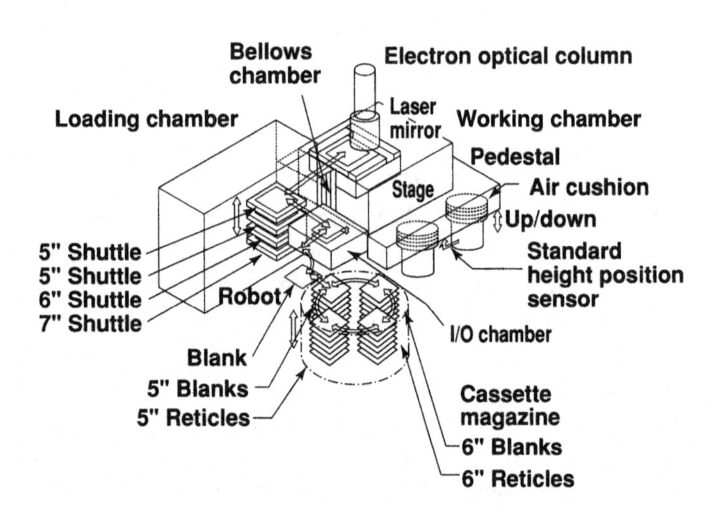

Figure 4.76 Mechanical structure of the EX-8.

pattern size) when the acceleration voltage is 20 kV. A pattern with a size of 2 μm, corresponding to a 16-Mb-DRAM-class mask (mask magnification of four), was formed with a pattern-size error of 10% at an acceleration voltage of 15 kV.[94] A small pattern is accurately formed with decreasing acceleration voltage. The information in Table 4.5 indicates the effect

Table 4.5 *Acceleration-voltage dependence of forward-scattered and back-scattered electron ranges.*[19]

Acceleration voltage (kV)	10	12	15	50
Forward-scattered electron range (μm)				
Resist thickness : 1.0 μm	1.08	0.77	0.51	0.066
Resist thickness : 0.5 μm	0.30	0.22	0.15	0.024
Back-scattered electron range (μm)	0.63	0.85	1.25	9.66

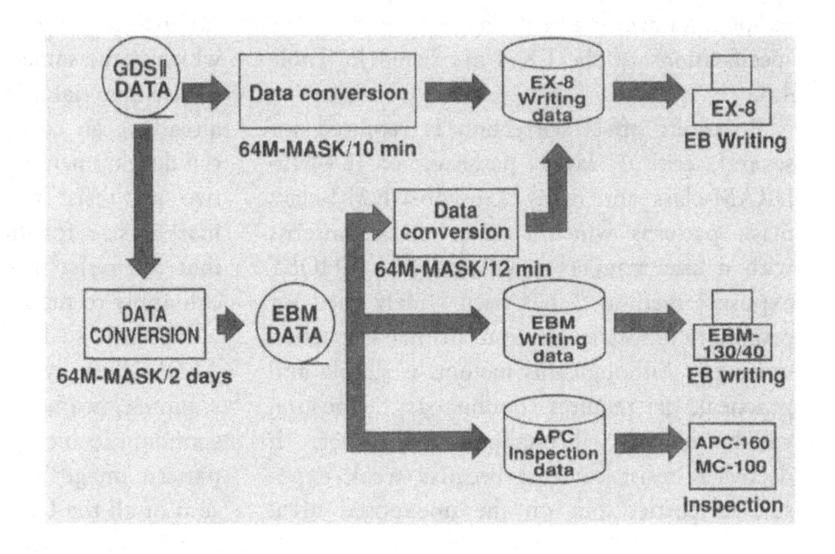

Figure 4.77 Data flow of a mask-making system composed of a data-conversion system, an EB-writing system of EX-8 and EBM-130/40, and a database-inspection system of APC-160 and MC-100.

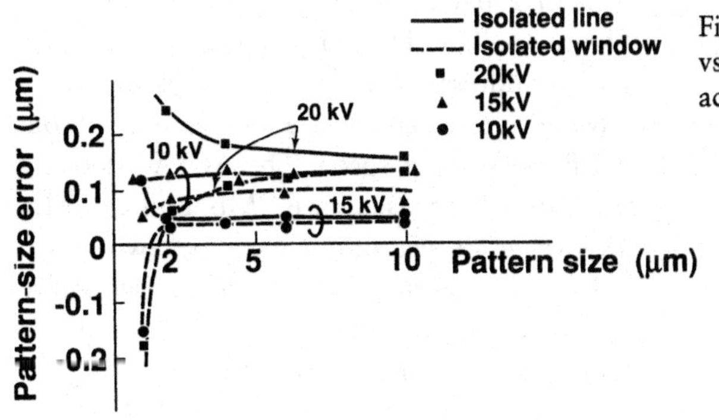

Figure 4.78 Pattern-size error vs pattern size for various acceleration voltages.

of acceleration voltage on forward-scattered and back-scattered electron ranges. The back-scattered electron range reduces with decreasing acceleration voltage, and it is less than 1.25 μm at an acceleration voltage of 15 kV.[107] On the other hand, the forward-scattered range increases with decreasing acceleration voltage. The forward-scattered range is 0.15 μm under the conditions of an acceleration voltage of 15 kV and a resist thickness of 0.5 μm. The EX-8 was designed to write a mask with a minimum feature size pattern of 1.5 μm without correcting the proximity effect. Therefore, the EX-8 adopted an acceleration voltage of 15 kV. The specifications of the EX-8 are listed in Table 4.6.

Proximity effect correction is required for several critical layer patterns of 64-Mb-DRAM-class and many 256-Mb-DRAM-class mask patterns which contain mask patterns with a size from 1 to 0.8 μm. The GHOST exposure method[108] has been widely used for proximity effect correction in mask pattern writing.[98] Although this method is simple and practical, it requires double-pass exposure, which reduces throughput. Moreover, it decreases resist contrast because weak exposure is carried out on the unexposed area.

Table 4.6 *Specifications of the EX-8.*[6]

Purpose	64-Mb-DRAM-class
Reticle size (inch)	5, 6 and 7
Minimum pattern size (μm)	1.5
Placement accuracy (nm)	50
CD uniformity (nm)	40
Stitching accuracy (nm)	35
Acceleration voltage (kV)	15
Maximum beam size (μm)	2.55
Address size (nm)	10

Therefore, a new proximity effect correction is required.

Laser-beam writing has recently become a widely accepted technique. Figure 4.79 shows the optical architecture of a laser-beam writing system, the ALTA-3000,[109] which uses 32 parallel writing beams, a continuously moving stage, and raster scanning. Beams are modulated with a 32-channel acousto-optic modulator with a 50-MHz channel modulation rate. Beam size is 0.35 μm. A 1-watt Ar-ion CW laser operating at a wavelength of 363.8 nm is used as the exposure source. The ALTA-3000 can use an i-line (wavelength 365 nm) resist which is the same as that of the wafer process, whose material and process developments have already been completed. Since EB-resist-material development for mask making is not attractive for resist makers because of the small market size for these resists, it is important that the resist material for the wafer process can apply to mask making.

The gray-scale concept was introduced in the ALTA-3000 system. In conventional raster scanning, beams are switched on or off corresponding to bit pattern data of 1 or 0. Since a pattern image is determined by the intensity sum of all the Gaussian beams exposed for the pattern, pattern edges can be positioned only in increments of address size. The gray-scale exposure, which dynamically changes laser-beam intensity during pattern writing, allows placement of a pattern edge between address units (Figure 4.80). The gray-scale concept can reduce address unit size from a real size of 266 nm to an effective size of 8.33 nm without sacrificing writing speed. The ALTA-3000 can therefore write masks for 64-Mb- and 256-Mb-DRAMs at a speed of one mask per two hours.

To reduce writing errors, the offset multi-pass exposure concept is used in the ALTA-3000. As shown in Figure 4.81, reprinting the

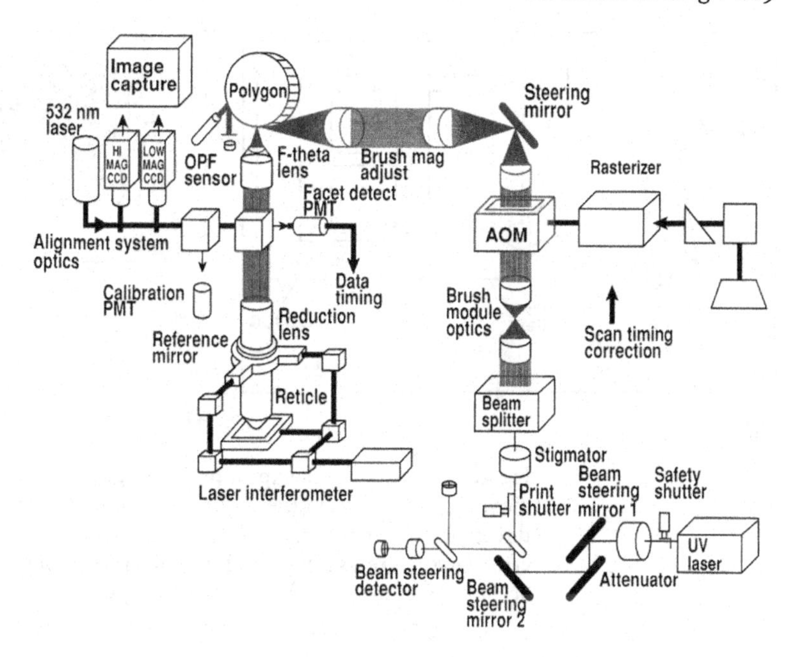

Figure 4.79 Optical architecture of the laser-beam writing system ALTA-3000.

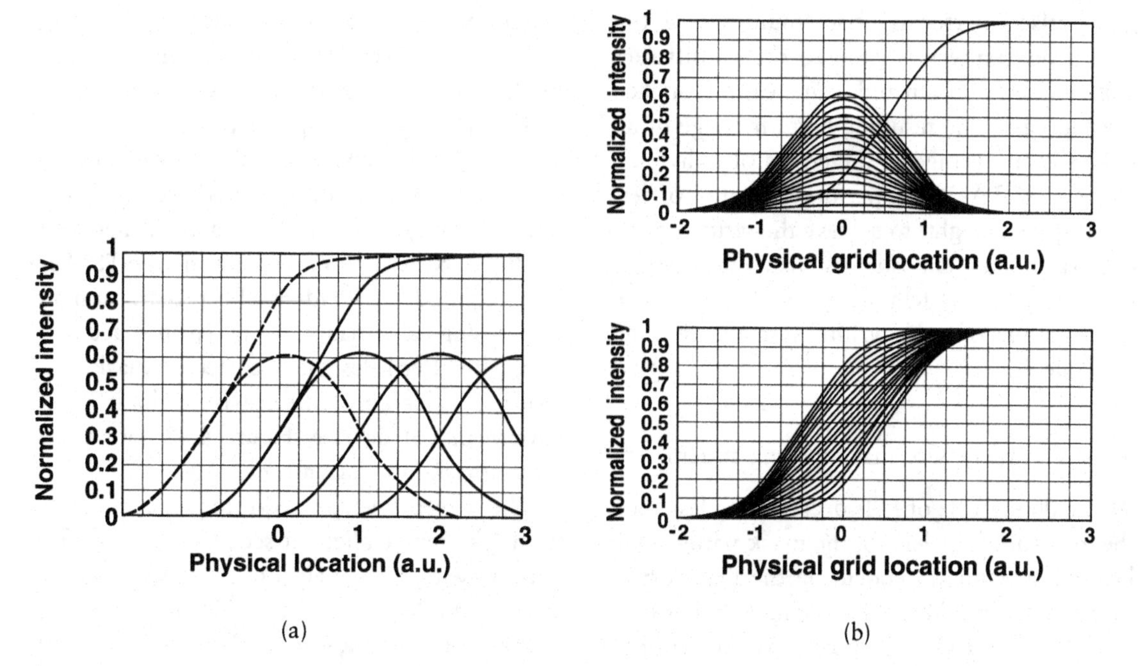

Figure 4.80 (a) Conventional writing method of a raster scan. The Gaussian beam is positioned at the grid and beam intensity is always constant. (b) Gray-scale concepts used. Laser-beam intensity can be changed in 16 steps, and changing the beam intensity can move beam edge position.

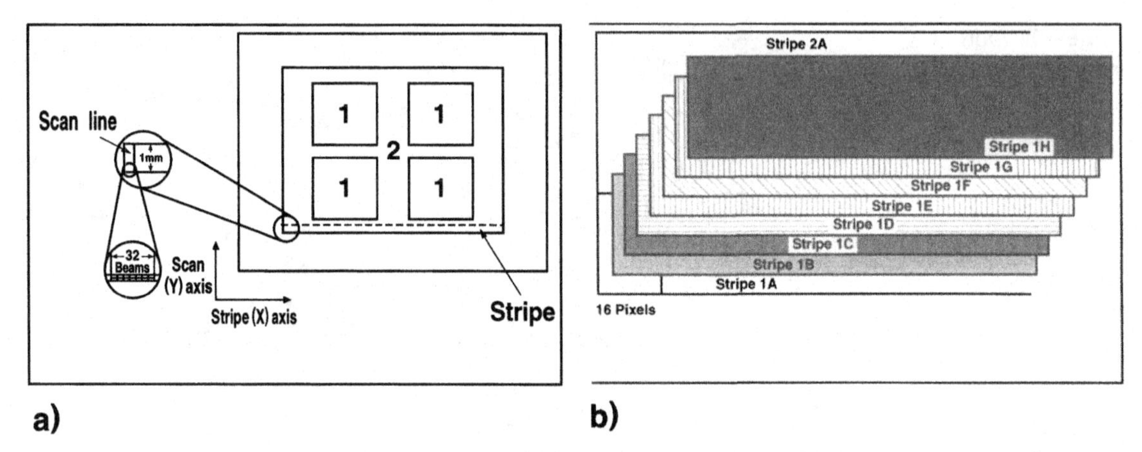

Figure 4.81 Multi-pass exposure. Writing method of the ALTA-3000. (a) The stage moves continuously in the X-direction and 32 laser beams are scanned in the Y-direction. (b) Stripes from 1A to 1H are written by slightly shifting the stripes. All error factors are averaged and placement accuracy, CD uniformity, and stitching accuracy are improved by multi-pass exposure.

pattern data with different scan and stripe origins (offset) performs offset multi-pass exposure. Every pattern is exposed with a different combination of beam and scan location for each writing pass. The resulting pattern image of all passes is an average of the errors of individual passes. ALTA-3000 performs eight offsets (multi-pass of eight) to achieve the writing quality. Multi-pass exposure is also effective for improving EB system accuracy.[110] The main specifications of mask-writing systems are listed in Table 4.7.

4.2.4 Future view of mask-writing system

As the design rule of LSIs decreases to less than the exposure wavelength, the mask pattern will become strongly dependent on the LSI fabrication process of lithography, etching, and design. Mask bias and OPC patterns have to be optimized according to allowable mask CD error, allowable corner rounding, required DOF, allowable exposure latitude and so on, and these quantities differ from layer to layer. For each layer, simulation calculations can also determine the optimum mask type from among a conventional mask, an HT-type PSM, and a Levenson-type PSM. Layout design of LSI devices is strongly dependent on mask type (PSM or conventional mask). Layout design has to be especially modified for using a Levenson-type PSM. Automatic generation of shifter patterns for a Levenson-type PSM is also required.[111] Loading effect correction in dry-etching also has to be taken into account in mask-pattern design. A corrected mask pattern is considerably different from the original pattern of LSI devices. Figure 4.82 shows the future mask-fabrication flow.[112]

The Levenson-type PSM requires a mask-writing system to align the second layer of shifter patterns with the first layer of LSI patterns. Since the second layer of the shifter pattern of a Levenson-type mask is formed on the chromium island pattern on the quartz plate, a charging problem occurs when an EB is used for second-layer pattern writing. In pattern writing on insulators, laser-beam writing has an advantage over

Table 4.7 *Specifications of mask-writing systems.*

System name	EX-8	ALTA-3000	HL-800M	JBX-7000 II	MEBES-4500
Manufacturer	Toshiba	ETEC[a]	Hitachi	JEOL[b]	ETEC[a]
Target DRAM capacity (Mb)	64	64/256	256	64/256	64/256
Beam type	VSB	Laser	VSB	VSB	Round
Minimum pattern size on wafer (μm)	0.35	0.35/0.25	0.25	0.35/0.25	0.35/0.25
Image placement (nm)	50	40	40	40/50	40
CD uniformity (nm)	40	30	30	30/40	50
Mask size					
(inches)	6	9	6	6	6
(mm)	150	230	150	150	150
Writing speed (min/mask)	45	120	60	–	–

[a] ETEC Systems Inc. (USA)
[b] JEOL Ltd (Japan)

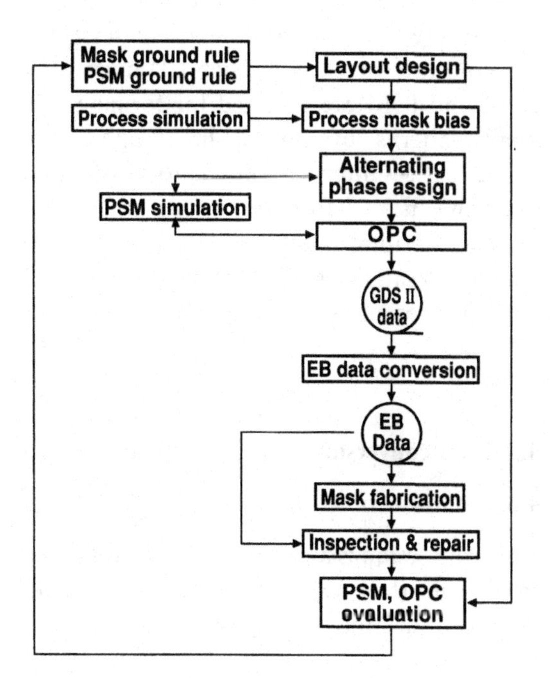

Figure 4.82 Future mask-fabrication flow.

EB writing in terms of the charging problem. The ALTA-3000 is attractive for second-layer pattern writing of a Levenson-type PSM.

For writing small patterns by EB, proximity effect correction is essential. The GHOST exposure method is popular for mask writing, but it has the disadvantage of narrow process latitude. Proximity effect correction based on dose modulation does not decrease resist contrast, which leads to good CD uniformity due to wide resist process latitude. The CD uniformity requirements in mask making are becoming increasingly severe. Proximity effect correction based on dose modulation must therefore be developed.

The amount of pattern data in proximity effect correction for mask making is 16 times greater than that for direct writing because the area of a 4× mask is 16 times that of a 1× chip. The speed of proximity effect correction for a mask pattern should be therefore much faster than that of direct writing.

Controlling the acceleration voltage is important for proximity effect correction. When the range of back-scattered electrons is much larger than the pattern size, the back-scattered electron density is uniform in a small area. Proximity effect correction becomes simple in this situation, when a high acceleration voltage is used, and when the effect of forward-scattered electrons is negligibly small.[113] However, impinged electron beams, especially when the acceleration voltage is high, generate heat in a glass plate. This heating leads to the resist-heating problem. The resist-heating problem in EB mask writing is much more serious than that in EB direct writing and in X-ray mask writing.[114] Power dissipation is proportional to the square of acceleration voltage, so a highly sensitive resist is required for mask-pattern writing in order to make the resist-heating problem less severe.

The key to X-ray lithography will be an ultra-accurate mask-writing system. Proximity effect correction for pattern writing on a heavy metal film used for an X-ray mask will be made easy by using 100-kV acceleration voltage. Various key technologies will be developed for EB writing of the optical masks needed to make 1-Gb DRAMs and beyond. These technologies will also be useful for developing a practical X-ray mask-writing system.

4.2.5 Summary

In mask fabrication, EB lithography systems with a round Gaussian beam, a continuously moving stage, and raster scanning, such as the MEBES™ system, have been widely used.

As the design rule of LSI devices decreases, high-throughput systems with high resolution are urgently required. Progress has been made on three fronts: development of VSB systems, improvement of the MEBES™ system, and development of the laser-beam systems.

Mask requirements are becoming increasingly severe as the design rule decreases below the exposure wavelength. High CD uniformity, OPC, and PSM are required for forming such small patterns by optical lithography. Any future mask-writing system has to satisfy these requirements.[115] Moreover, highly accurate X-ray mask-writing systems will also be required. The key to satisfying all these requirements will be to develop ultra-accurate technology.

4.3 **Data conversion**

4.3.1 Introduction

Data conversion is a process for generating electron-beam (EB) exposure data from layout designs of a large-scale integrated device (LSI) and is one of the most critical processes in pattern fabrication using EB lithography. Advanced LSIs have a huge number of shapes in a single chip; for example, the most critical layers such as poly-Si wiring of a 64-Mb dynamic random access memory (DRAM) have more than 1×10^8 shapes. Moreover, improving critical-dimension (CD) accuracy and achieving wide process latitude require that complicated shape operations are performed. Therefore, a high-speed computing and a large data-storage are required. Furthermore, this time and volume have increased very rapidly because the packing density of LSIs has been increased by a factor of four and the allowable CD error has been decreased by about 30% per three years. Accordingly, it will continue to be necessary to increase the throughput of the data conversion and to reduce the data volume in EB lithography. In addition, the shape operations for achieving a higher CD accuracy have become crucial.

This section describes the data-conversion flow and data-conversion technologies that have been developed for sub-half-micron devices. Particular attention is given to the details of data-conversion methods used to increase throughput and to eliminate the narrow shapes that degrade CD accuracy.

4.3.2 Data-conversion flow

Figure 4.83 illustrates the data-conversion flows for two common EB writing schemes: variably shaped EB writing and raster-scan EB writing using a Gaussian round beam. In data conversion for variably shaped EB writing, the shape operations for overlap removal and tone reversal are necessary. Overlap removal is essential for avoiding excessive EB exposure in the areas where shapes overlap. Tone reversal is the shape operation for generating shapes over the area where no shapes exist. It is performed when fabricating patterns for wiring processes by using a positive-type EB resist. Since positive-type EB resists have been commonly used when making masks because of their ability to provide high resolution, tone reversal is an important shape operation.

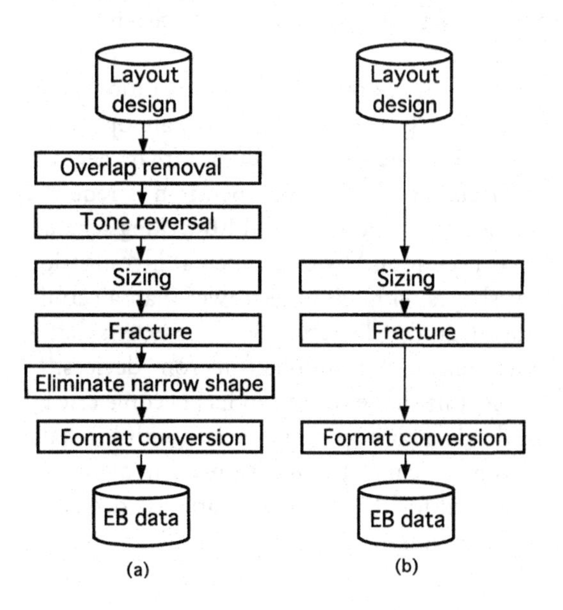

Figure 4.83 Data-conversion flow. (a) Variably shaped EB writing scheme. (b) Raster-scan EB writing scheme.

Algorithms for overlap removal and tone reversal become simple when hierarchical data structures including many array references are converted to a flat structure. It has been found, however, that when this is done the data-conversion time and the data volume of a 16-Mb DRAM or a more densely packed device exceeds levels that are practical.[116] To solve this problem, several researchers have developed data-conversion algorithms based on hierarchical structures.[116–120] The details of this approach will be explained in Subsection 4.3.3.

In data conversion for the raster-scan EB writing scheme, on the other hand, the shape operations of overlap removal and tone reversal are not required. EB exposure shapes are converted into bit-maps where each pixel value is 1 or 0; this operation is performed in the EB writing system just before EB exposure and the shape overlaps are removed by this bit-map conversion. Tone reversal is also performed quite easily by reversing each pixel value from 1 to 0 or from 0 to 1.

'Sizing' is the shape operation used to swell or reduce the shape of original layout designs. It has been found to be very beneficial for widening the process latitude and getting steeper resist walls;[121,122] therefore, it has become an indispensable shape operation for fabricating LSI patterns with sub-half-micron and smaller dimensions. Since this shape operation is efficient in EB lithography, it is performed for both the variably shaped EB and the raster-scan EB writing schemes.

Although sizing is a useful shape operation, it requires the use of an intricate algorithm. To perform sizing accurately for complicated LSI patterns, the following five processes are required: (1) overlap removal of input shapes; (2) tracing of all the vertices that can be connected and construction of polygons; (3) recognition of windows inside the polygons; (4) calculation of new vertex positions with consideration of the inside windows; and (5) overlap removal for all the polygons resulting from the new vertex positions. Given the complexity of the sizing algorithm, it is clear that the computing time required for sizing is much greater than that required for overlap removal or tone reversal. The sizing throughput has therefore become one of the greatest problems in data conversion. Taking advantage of the method using hierarchical structures, which was developed for overlap removal and tone reversal, has increased this throughput. Using the distributed processing method developed by Otto *et al.* has further increased it.[123] The details of these methods will be described in Subsection 4.3.3.

'Fracturing' is the shape operation for partitioning polygons into the following primitive shapes: (1) rectangles; (2) trapezoids that have two horizontal edges or two vertical edges; and (3) triangles which have at least one horizontal or one vertical edge. This shape operation is required for both the variably shaped EB and the raster-scan EB writing schemes because EB writing systems can handle only the primitive shapes.

There is another shape operation that improves CD accuracy; that is, an operation eliminating the narrow shapes that degrade CD accuracy. This shape operation is required only for the variably shaped EB writing scheme. It has been found by experimental investigation that shapes with width narrower than a certain value degrade CD accuracy to the extent that is unacceptable for sub-half-micron devices.[124] Hence, this shape operation has become crucial for the data conversion of devices with sub-half-micron or smaller dimensions. An algorithm developed for this shape operation is described in Subsection 4.3.4.

4.3.3 High-throughput data conversion

Layout designs of LSIs have hierarchical data structures using many array references of cells. In other words, there are many repetitive shapes in layout designs. If the repetitive shapes are expanded, the total number of shapes that need to be processed becomes huge. The best way to ensure a high throughput is thus to perform the shape operations without expanding the repetitive shapes.

Several approaches to doing this have been developed.[116–120] In the technique developed by Moriizumi *et al.*,[116] layout designs are searched for large-scale two-dimensional repetitive shapes, and the interactions between the two-dimensional repetitive shapes and the other peripheral shapes are investigated. Then, a unique hierarchical structure with two levels is constructed as shown in Figure 4.84. The distinctive features of this structure are that a set of several shapes, which is called a 'unit', is array-referenced in the areas where there are no shapes except the 'units' and all the shapes in a 'unit' must be entirely within the area of the 'unit'. The size of a 'unit' is defined as the pitch of the array reference. This reference method is called an 'area-unit reference'. When the unique hierarchical structure is constructed, the shape operations of overlap removal and tone reversal inside the areas prescribed in 'area-unit reference' are performed only on the shapes in the 'unit'. This technique is especially effective for memory devices, which have many large-scale two-dimensional repetitive shapes.

This technique can also be used in the sizing operation. To maintain the shape continuity between adjacent 'units', sizing is performed on 3 × 3 array 'units' as shown in Figure 4.85. After sizing, the center part of the 3 × 3 array 'unit' is clipped as a resultant 'unit'. Likewise, for maintaining the shape continuity between the outermost 'units' and the peripheral shapes located just outside the reference area, the outermost 'units' along the border of the reference area must be expanded. As a

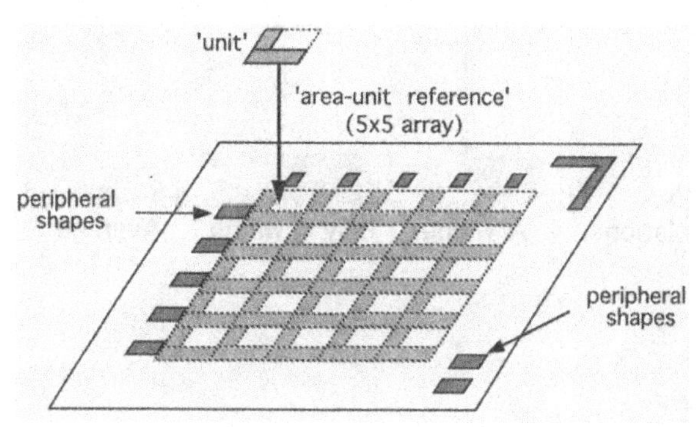

Figure 4.84 Unique hierarchical structure. 'Units' are referenced in the area where there is no shape except the 'units'.

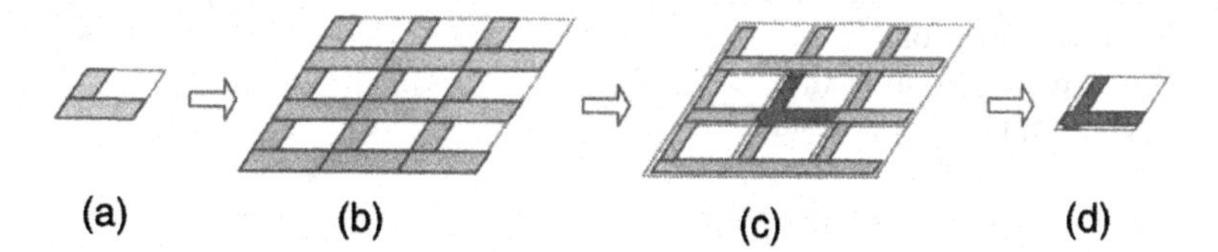

Figure 4.85 'Sizing' algorithm using the unique hierarchical structure: (a) Shapes in original 'unit'. (b) 'Unit' is expanded to a 3 × 3 array. (c) Sizing is performed on the 'unit' array. (d) Center part is clipped as a resultant 'unit'.

result, the reference area is reduced by the size of the outermost 'units'. However, most of the reference area survives. Throughput is, therefore, much higher than when using the conventional method based on a flat structure. The efficacy of applying the technique to sizing has been evaluated for a 64-Mb DRAM. As shown in Figure 4.86, the throughput of sizing has been increased about five-fold.

The technique described above is very effective for memory LSIs which have large-scale two-dimensional repetitive shapes, but it does not work for peripheral shapes of memory LSIs and logic LSIs. A distributed processing method using a number of computers has therefore been developed. Figure 4.87 shows a distributed processing method. The whole area of a LSI chip is divided into small regions with a redundant frame, which is required for maintaining the continuity of shapes lying across a border of the small regions. The efficacy of the distributed processing system using ten computers connected with networks has been investigated for a 64-Mb DRAM. As shown in Figure 4.88, sizing throughput has been increased roughly seven-fold. By using both these methods, the throughput of sizing has been increased dramatically; that is, by a factor of 35.

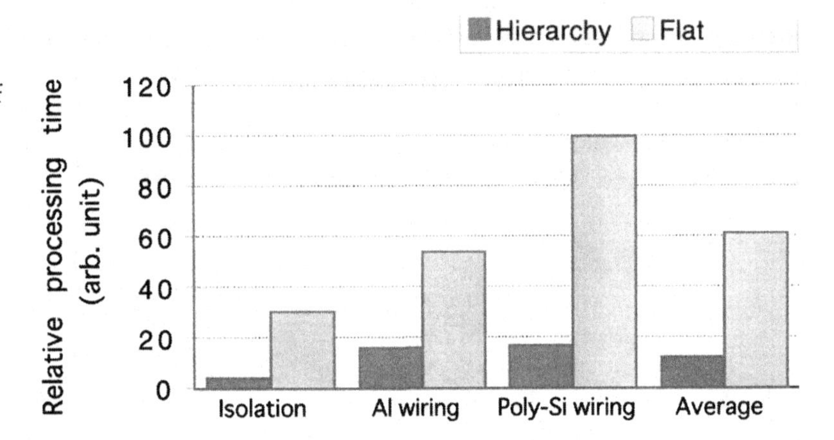

Figure 4.86 Throughput of sizing using hierarchical structures compared to that of flat structure. A 64-Mb DRAM was used for the evaluation.

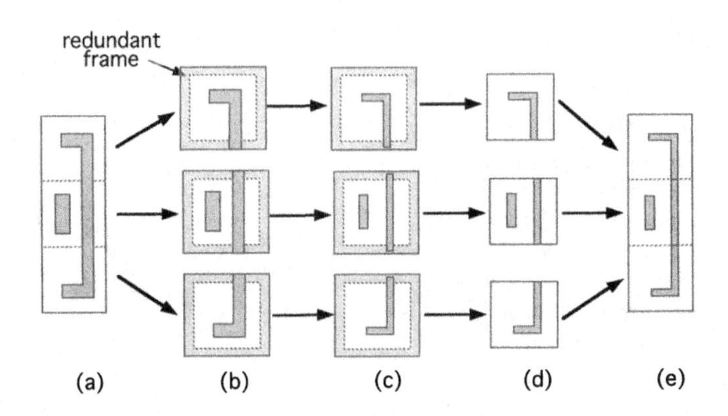

Figure 4.87 Distributed processing method: (a) layout design; (b) the layout design is divided into small regions with redundant frames; (c) a shape operation such as 'sizing' is performed on shapes in the small regions concurrently; (d) after the operation, the shapes inside the redundant frames are clipped; and (e) all shapes are merged.

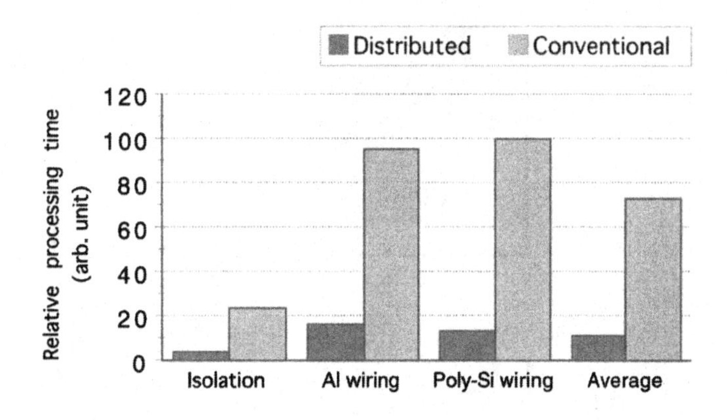

Figure 4.88 Throughput of 'sizing' using the distributed processing method compared with the conventional method. A 64-Mb DRAM was used for the evaluation, and distributed processing was performed with ten computers.

4.3.4 Narrow-shape elimination

In the variably shaped EB writing scheme, the cross-sectional area of a shaped EB varies according to the input shapes. Thus, the current of the shaped EB changes. This variation of the EB current brings about a variation of the space-charge effect, which causes a blur of the EB. It is well-known that EB blur strongly affects not only the resolution limit but also CD accuracy.[125] Figure 4.89 shows that the size variation of EB exposure shapes degrades CD accuracy. The original layout design for the EB exposure shapes in Figure 4.89(a) and (c) are the same. However, the different size of EB exposure shapes can be generated because of different fracture directions. The EB blur at the left edge for the case of Figure 4.89(a) becomes much smaller than that for the case of Figure 4.89(c), as shown in Figure 4.89(b) and (d). Therefore, it is difficult to achieve the same size of resist patterns in spite of the same layout design.

The degradation of CD accuracy due to the size variation of EB exposure shapes has been investigated experimentally; Figure 4.90(a) shows the test structure. The designed narrow-shape width W_{NS} was varied from 0 to 1.3 μm. The designed large-shape width W_{LS} which includes the designed narrow shape and the designed space width W_{CD} were fixed at 4 and 1.5 μm, respectively, in order to keep the influences of the EB proximity effect constant. Patterns were fabricated on Cr masks using a 20-keV variably shaped electron beam, a positive-type resist, and a dry-etching process. Then the pattern width fabricated at W_{CD} which is directly influenced by the variation of the narrow-shape width W_{NS}, was measured. The results are shown in Figure 4.90(b). The pattern width changes drastically when the narrow-shape width W_{NS} becomes less than 0.5 μm. The difference between the measured width W_{CD} in the case where W_{NS} is 0 μm and that in the case where W_{NS} is 0.05 was about 0.06 μm. This difference is not acceptable for 5× reticles of sub-half-micron devices. These unacceptable EB exposure shapes are called 'narrow shapes' here.

The best solution to this problem is to develop a data-conversion method which either does not generate or else eliminates the narrow shapes. An algorithm for eliminating the narrow shapes has been developed, and its process flow is shown in Figure 4.91. First, the EB exposure data are searched for narrow shapes. Then, the narrow shapes and their adjacent shapes are merged into polygons. Finally, the polygons are partitioned into the primitive shapes by using a sophisticated fracture algorithm. This algorithm

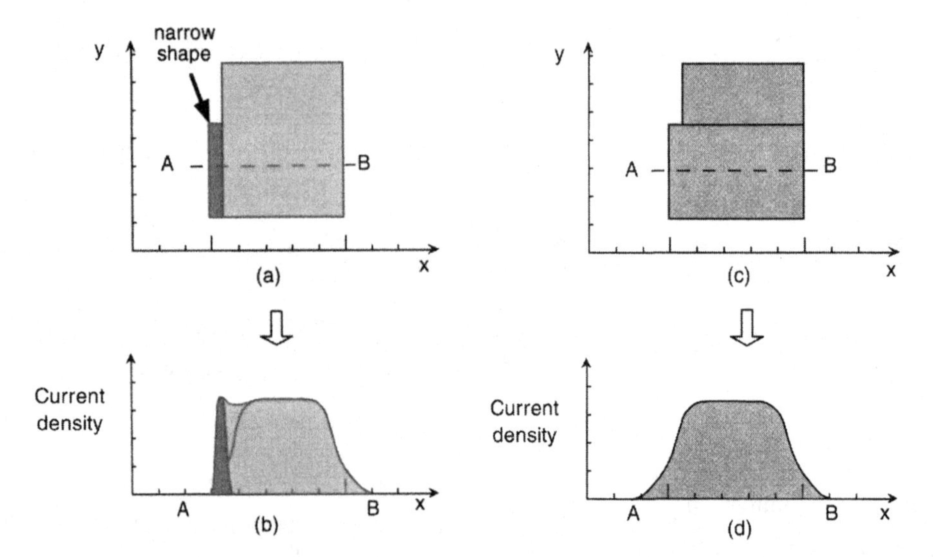

Figure 4.89 Influence of a narrow shape on EB blur of shaped EB. (a) Exposure shapes with a narrow shape; (c) exposure shapes without a narrow shape; (b) and (d) current-density distributions for exposure shapes (a) and (c). The original layout designs of (a) and (c) are the same. The EB blurs at the left edge are not the same because of the narrow shape.

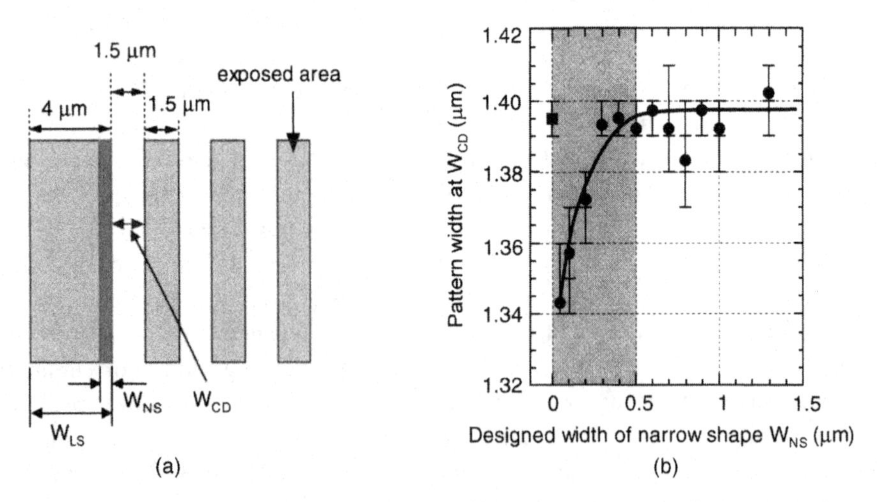

Figure 4.90 Influence of a narrow shape on pattern dimensions. (a) Test structure, and (b) relationship between the designed width of narrow shape W_{NS} and the width of the pattern fabricated at W_{CD}. The patterns were fabricated on a Cr mask using 20-keV variably shaped EB, a positive-type resist, and a dry-etching process.

delivers the best set of primitive shapes from a given polygon and is a key technology for eliminating narrow shapes.

The aim of the fracture algorithm developed by Nakao *et al.*[124] is to find a set of partitioning lines to fracture a polygon perfectly into primitive shapes. This set minimizes the generation of the narrow shapes. Partitioning lines are selected from all the possible partitioning lines in the following manner. First, partitioning lines which themselves generate no narrow shapes and also suppress narrow-shape generation by other partitioning lines are selected. Second, partitioning lines are selected which generate narrow shapes but suppress the generation by other partitioning lines of narrow shapes longer than those generated by themselves. Third, partitioning lines that generate no narrow shapes are selected. The selection of partitioning lines is completed when a polygon is perfectly fractured into primitive shapes.

Figure 4.92 is an example of what happens when this fracture algorithm is used. First, partitioning line *la* is selected because it results in no narrow shapes and it suppresses the generation of the narrow shape which would be made by partitioning line *lf*. Then partitioning line *lf* is also selected because it suppresses the genera-

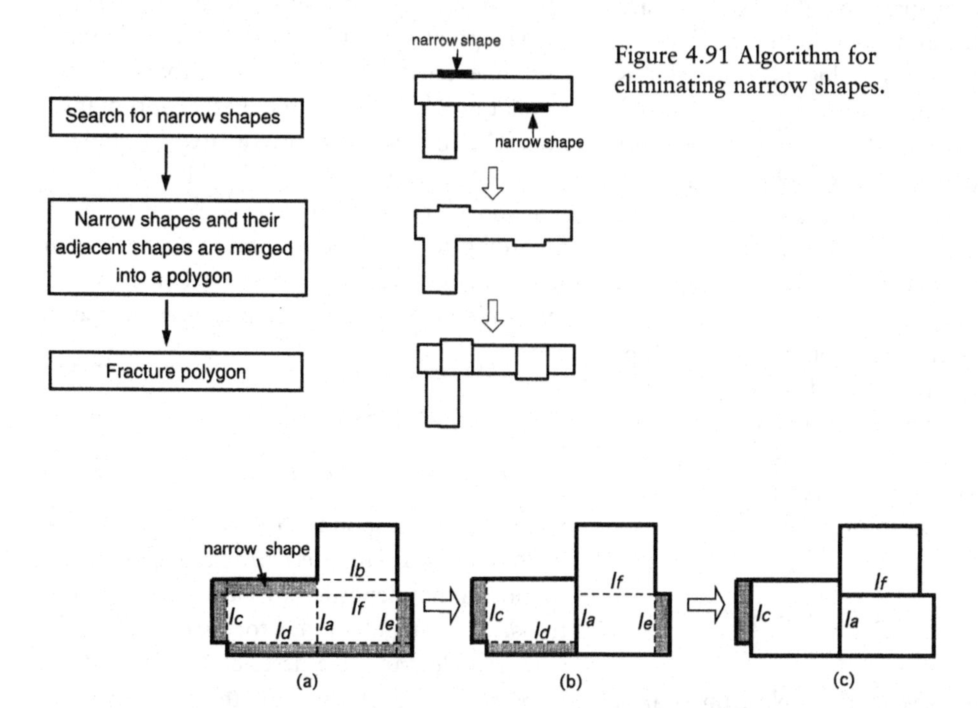

Figure 4.91 Algorithm for eliminating narrow shapes.

Figure 4.92 Fracture algorithm for minimizing narrow-shape generation. (a) All the possible partitioning lines are nominated. Shaded shapes are narrow shapes that would be generated by the nominated partitioning lines. (b) Partitioning line *la* is selected because it generates no narrow shape and it suppresses the narrow-shape generation that would be made by partitioning line *lf*. Partitioning line *lf* is selected because it also suppresses the narrow-shape generation that would be made by partitioning line *le*. (c) Partitioning line *lc* is selected because the length of the narrow shape made by itself is shorter than that made by partitioning line *ld*.

tion of the narrow shape that would be made by partitioning line *le*. Finally, partitioning line *lc* is selected because it generates a narrow shape but suppresses the generation of a narrow shape longer than that made by itself which would be made by partitioning line *ld*. The generation of narrow shapes is thus minimized by using the algorithm.

4.4 Proximity effect correction

4.4.1 Introduction

When an electron beam with uniform dose is used to make a pattern, the pattern size obtained after resist development deviates from the designed size. This phenomenon is called the proximity effect.[126] Pattern-size deviation depends on the size of the pattern itself and on that of the peripheral patterns. The size deviation reaches 0.2–0.3 μm, which is a crucially large error for making a sub-micron pattern. Figure 4.93(a)–(c) shows an example.[127]

Proximity effect correction is used to suppress the size deviation caused by the proximity effect and thus to obtain the correct pattern size. Figure 4.94(a)–(c) shows a resist pattern for which the proximity effect has been corrected using the dose modulation method.[121]

4.4.2 Proximity effect and point-spread function

When a pattern is written by electron beams, the injected electrons directly expose the resist on the substrate, and some electrons are back-scattered in the substrate and expose the resist at unexpected positions. This exposure by back-scattered electrons is an unwanted background part of the energy deposited in the resist. Moreover, the background part deviates (see Figure 4.98), depending on the size of the pattern itself and on that of the peripheral pattern, to cause the proximity effect.

The proximity effect can be quantitatively described by the point-spread function.[128-130] The point-spread function $g(x)$ expresses the energy deposited by electrons when the electrons are injected at the position $x = 0$. By using the point-spread function, we can express the deposited energy for a pattern as:

$$E(x) = \int g(x - x')D(x')\mathrm{d}x' \qquad (4.10)$$

where $D(x')$ is the dose at position x'. When proximity effect correction is not adopted, dose $D(x')$ is set at a uniform value at any position x'. The point-spread function, $g(x)$, is obtained experimentally[128,131] or by Monte Carlo simulation[129,132] using the Bethe approximation. The function is approximated by analytic functions, such as a double Gaussian function[129] or a triple Gaussian function,[130] according to the experimental conditions such as the material of the substrate. For example, the double Gaussian function is expressed as:

$$g(x) = C[g_\mathrm{f}(x) + \eta_E g_\mathrm{b}(x)], \qquad (4.11)$$

$$g_\mathrm{f}(x) = (1/\pi\sigma_\mathrm{f}^2)\exp(-x^2/\sigma_\mathrm{f}^2), \qquad (4.12)$$

$$g_\mathrm{b}(x) = (1/\pi\sigma_\mathrm{b}^2)\exp(-x^2/\sigma_\mathrm{b}^2). \qquad (4.13)$$

Here $g_\mathrm{f}(x)$ corresponds to the energy deposited by forward-scattered electrons, and $g_\mathrm{b}(x)$ corresponds to that deposited by back-scattered electrons. The ranges of the forward- and back-scattered electrons are described by σ_f and σ_b, respectively. The term η_E is the ratio between the total amount of energy deposited by back-scattered electrons and that deposited by forward-scattered electrons, and C is a constant. The values of these parameters depend on the acceleration voltage, substrate, resist thickness, and so on. Figures 4.95 and 4.96 show some examples of the dependences.[132,133]

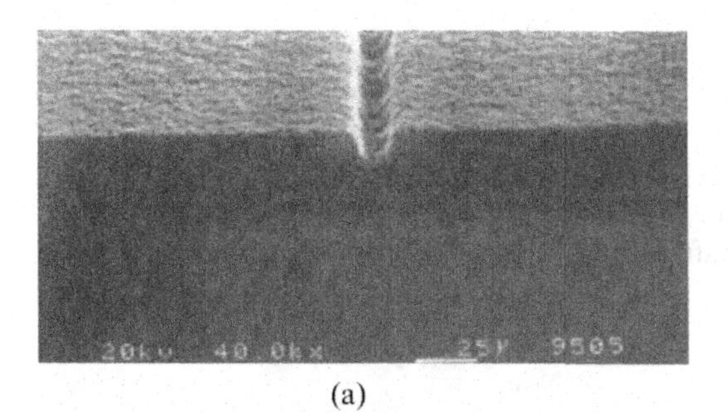

(a)

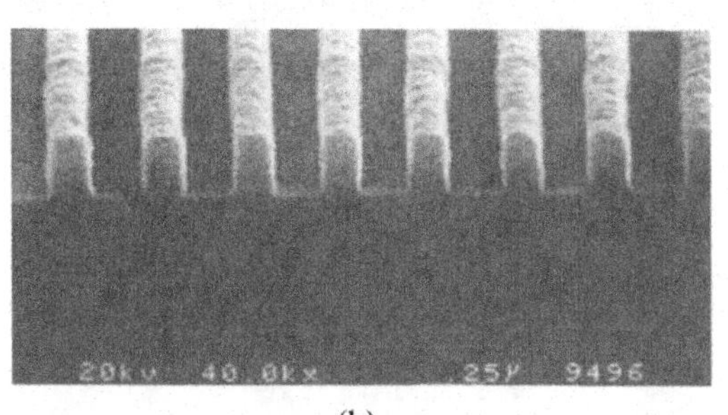

(b)

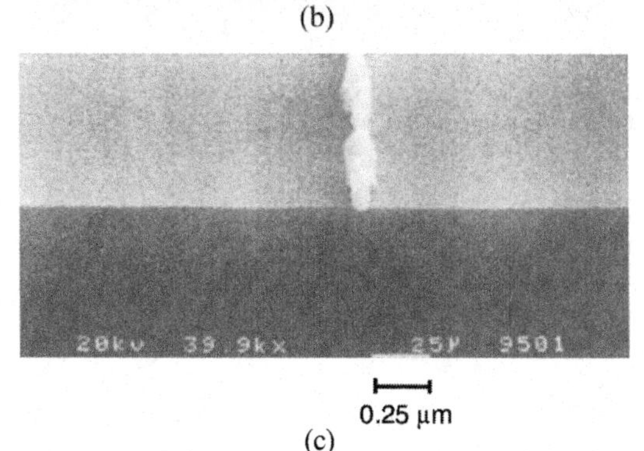

(c)

Figure 4.93 Degradation of resist image caused by proximity effect. Design rule 0.2 µm. When 1 : 1 line-and-space pattern (b) is made with the designed size, the isolated-space pattern (a) is not resolved, and the isolated-line pattern (c) is overdeveloped.

4.4.3 Correction methods

In correcting the proximity effect, it is assumed that the pattern size is determined only by the deposited energy and the threshold value, (T), as shown in Figure 4.97 (threshold model). Under this assumption, the proximity effect can be corrected by equalizing the energy deposited at the edges of all patterns.

Many types of correction methods have been proposed and examined. These are: (1) dose modulation;[134–137] (2) subsidiary exposure for correction, such as GHOST;[58,108,138,139] (3) pattern modulation method;[140] (4) a combination of (1) and (2); (5) usage of an intermediate layer;[141,142] and (6) usage of a low-acceleration-voltage electron beam.[143] To apply methods (1), (3), and (4), the optimum dose or optimum pattern size or both must be calculated before exposure. Subsections 4.4.5 and 4.4.6, respectively, explain the widely used GHOST and dose-modulation methods.

4.4.4 High acceleration voltage and proximity effect correction

The acceleration voltage is the most important parameter, determining the back-scattering and forward-scattering ranges (see Figure 4.96).

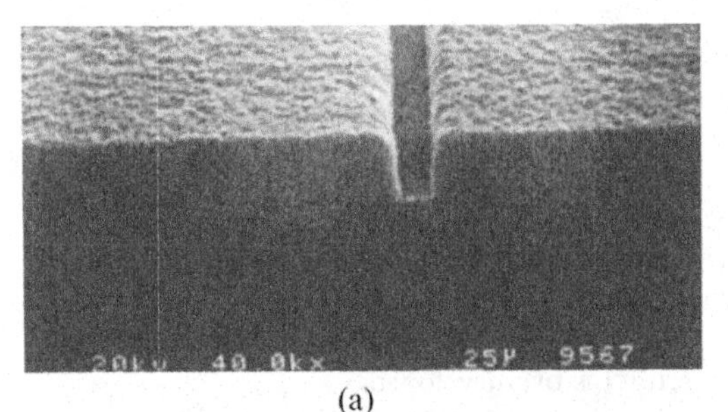

(a)

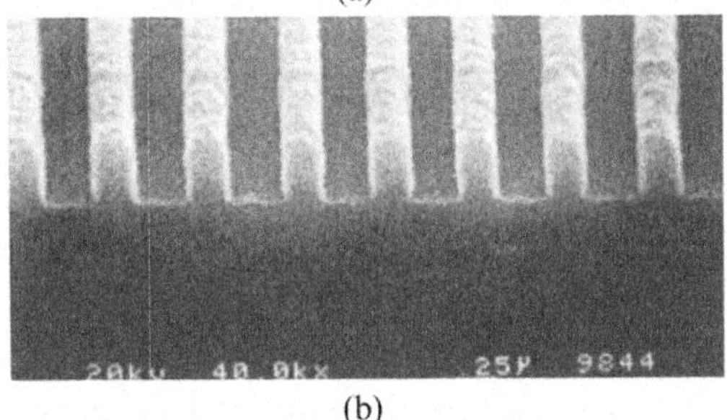

(b)

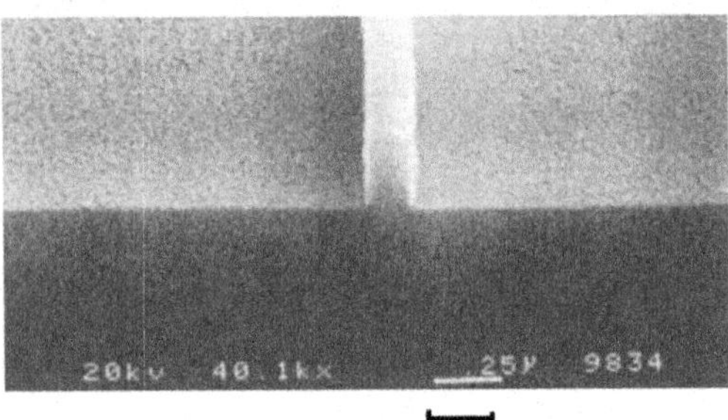

(c)

Figure 4.94 Example of proximity effect correction. Design rule 0.2 μm.
(a) Isolated-space pattern,
(b) 1 : 1 line-and-space pattern, and
(c) isolated-line pattern.

Recently a high acceleration voltage (such as 50 kV) has often been used because it reduces both the space-charge effect and the forward-scattering range of electrons in the resist. It is thus advantageous for making sub-micron patterns.

A high acceleration voltage is also advantageous for correcting the proximity effect. The advantages are listed here. The practical proximity effect corrections are based on these advantages.

- The forward-scattering range of electrons becomes considerably smaller than the minimum feature size. Thus it is sufficient to correct only for the effect of back-scattered electrons. This simple method is effective for both rapidly and accurately correcting the proximity effect.[135,136]
- The minimum feature size becomes sufficiently smaller than the back-scattering range. Then the concept of coarse graining (which is explained in Subsection 4.4.8) becomes effective, and high-speed and highly accurate correction can be carried out.

In the following subsections, we consider only the case where a high-voltage electron beam is used.

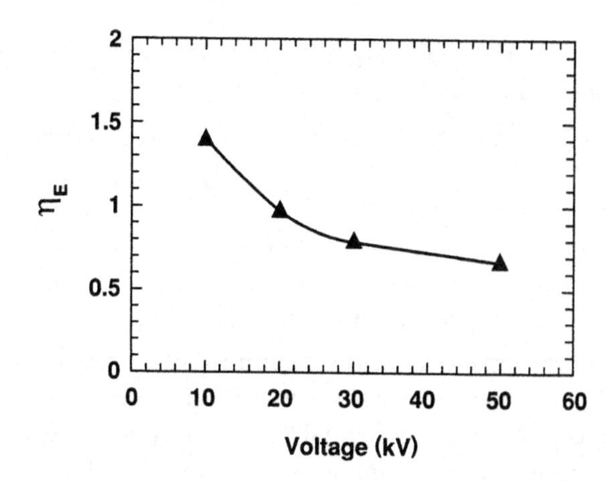

Figure 4.95 Acceleration voltage dependence of parameter η_E.

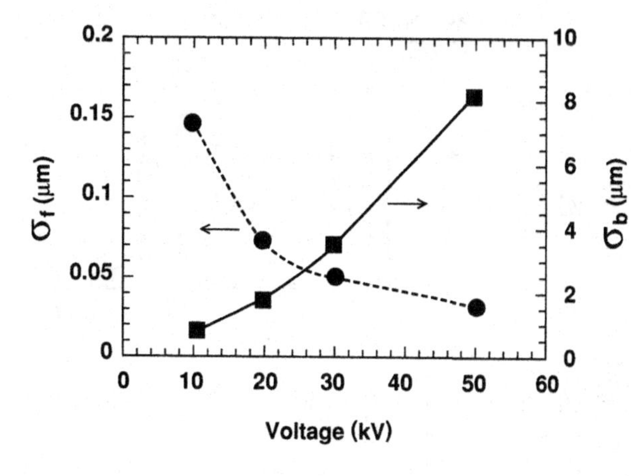

Figure 4.96 Acceleration voltage dependence of the scattering ranges σ_f and σ_b.

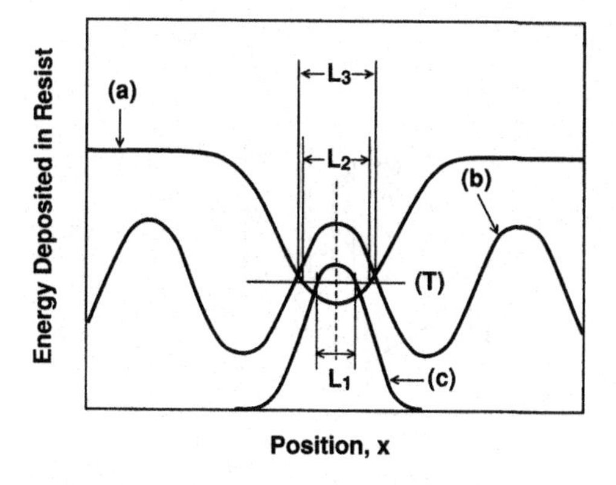

Figure 4.97 Threshold model; (T) shows the threshold value of deposited energy. Deposited energy of (a) isolated line; (b) 1 : 1 line-and-space; and (c) isolated space.

4.4.5 GHOST method

Figure 4.98 explains the sequence and the correction mechanism of GHOST.[108] When the pattern is written with a focused beam and a uniform dose D_0, the background part of the deposited energy deviates. When the hatched inverted pattern is exposed during subsidiary exposure with the blurred beam, a higher subsidiary dose is added for the sparse pattern than that added for the dense pattern. Thus, the combined exposures make the background part of the deposited energy almost homogeneous.

For the subsidiary exposure, optimum dose D_{sub} and optimum blurred beam radius R_{sub} are given:[108]

$$D_{sub} = [\eta_E/(1+\eta_E)]D_0, \tag{4.14}$$

$$R_{sub} = \sigma_b/(1+\eta_E)^{1/4} \tag{4.15}$$

One key factor which determines the quality of the resist pattern is the contrast of the deposited energy between the pattern region and non-pattern region. When GHOST is used and the subsidiary exposure is applied to the whole

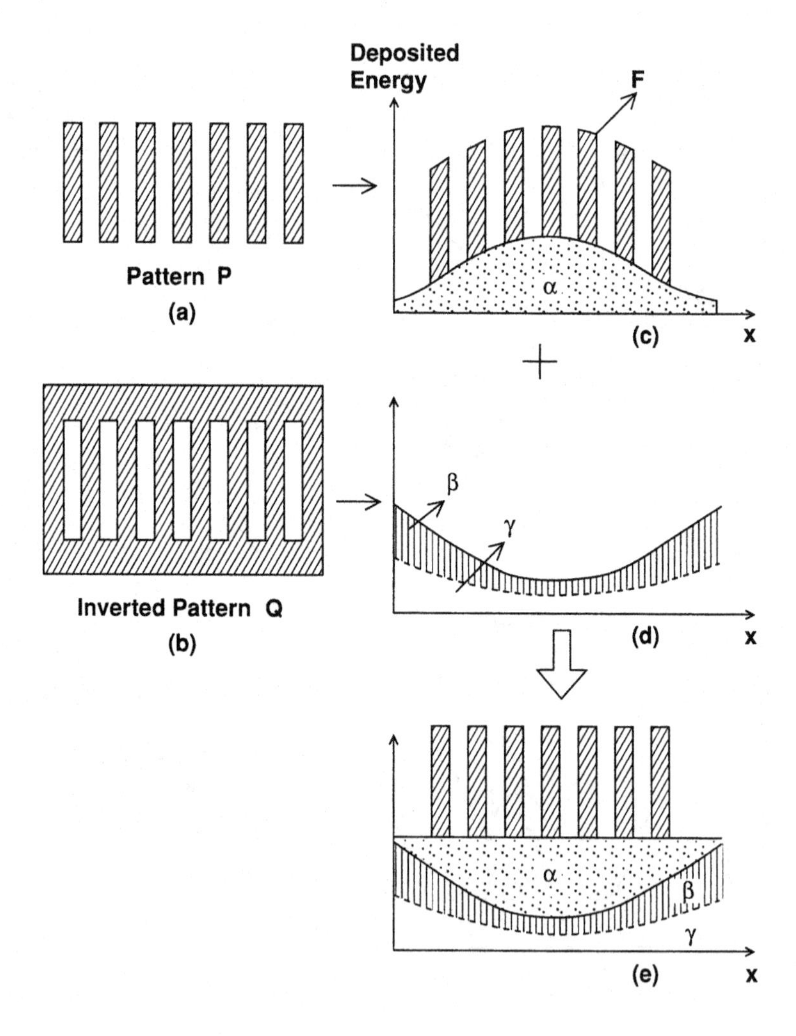

Figure 4.98 GHOST method. (a) Pattern P is exposed with a focused beam and a uniform dose (pattern exposure); (c) energy deposited by patterning exposure (F: energy deposited by forward-scattering electrons; α: by back-scattering electrons). (b) Hatched inverted pattern Q is exposed with a defocused beam to obtain correction (subsidiary exposure); (d) energy deposited by subsidiary exposure (β: energy deposited by forward-scattering electrons; γ: by back-scattering electrons); and (e) total deposited energy.

region of the LSI pattern, the contrast for the whole region is set at that of the worst case. This contrast degradation is a disadvantage of GHOST. Another disadvantage is that the electron optical system (EOS) must be controlled in order to obtain a focused and a defocused beam. This control can degrade the throughput and the accuracy of the EOS.

On the other hand, GHOST does not require the optimum dose be calculated. This is one of its advantages not shared with the dose-modulation method. Another advantage of GHOST is that the exposure system does not need a function for modulating the dose. For example, GHOST is a simple and effective correction tool for the SCALPEL system,[138] with which it is difficult to modulate dose according to position.

GHOST has been used for making X-ray masks with a 100-kV electron beam and has yielded a correction accuracy of ±7 nm when the pattern size is 0.15 μm.[144]

4.4.6 Dose modulation

4.4.6.1 *Equation for proximity effect correction and matrix method*

Figure 4.99 shows the dose-modulation method. This method equalizes the deposited energy at all regions of the pattern, by changing the exposure dose shot-by-shot[134] to correct the proximity effect. Compared with GHOST, the dose-modulation method has the following advantages: (1) Degradation of contrast between the deposited energy in the pattern region and that in the non-pattern region occurs only for a dense pattern; (2) control of the EOS is not needed to make a defocused beam; (3)

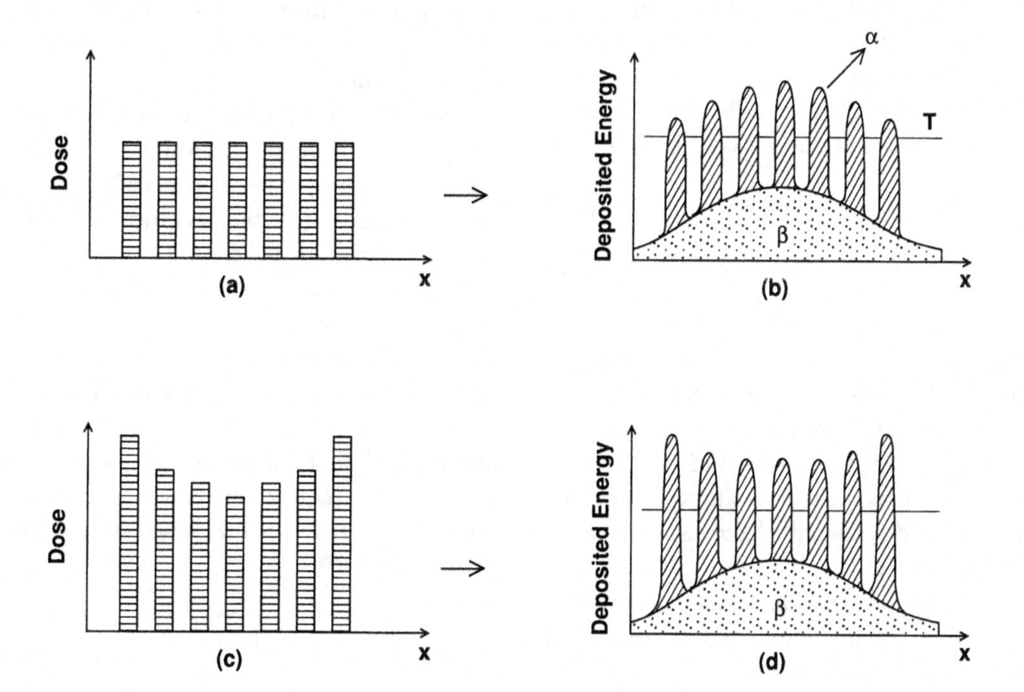

Figure 4.99 Dose-modulation method. (a) Dose without proximity effect correction; (b) corresponding deposited energy. (c) Dose with correction; and (d) corresponding deposited energy.

because the subsidiary exposure is not needed, the total exposure time is less than it would be when using GHOST. These are the advantages of dose modulation.

For dose modulation, the optimum correction dose must be obtained before exposure by using a suitable algorithm. In principle, the correction dose at position x can be obtained by solving the two-dimensional integral equation,[135]

$$D(x)/2 + \eta_E \int D(x') g_b(x - x')\, dx' = 1, \quad (4.16)$$

under the following conditions:

$$D(x) > 0 \text{ (pattern; exposed area)} \quad (4.17)$$

$$D(x) = 0 \text{ (no pattern).} \quad (4.18)$$

The integral region of $|x - x'| < 3\sigma_b$ is sufficient for practical use. The factor (1/2) in the first term in Equation (4.16) is assigned to equalize the deposited energy of the pattern edge for all patterns.[145]

It must be noted that the following condition cannot be imposed instead of condition (4.18):

$$D(x'')/2 + \eta_E \int D(x')\, g_b(x'' - x')\, dx' = 0, \quad (4.19)$$

where x'' indicates any position in the non-pattern region. The reason for this is that when condition (4.19) is used, a correction dose with negative value is obtained for the non-pattern region; that is, a non-valid solution.[146] For the same reason, the proximity effect due to the forward-scattered electrons cannot be completely corrected.

When using dose modulation, optimum dose is assigned to small figures. These are called elements in the present section. To obtain accurate correction, (1) the size of an element must be sufficiently smaller than the back-scattering range, at least for the region where the pattern density changes sharply, or (2) some kind of interpolation must be used when the size of an element is comparable with the back-scattering range. For simplicity, it is assumed in the present section that condition (1) holds throughout the whole region of the LSI pattern.

Because the size of an element is sufficiently small, the integral equation is modified to the matrix equation:

$$E = D(x_i) + \eta_E \Sigma D(x_i)\, S(x_j) \times \exp\left[-(x_i - x_j)^2/\sigma_b\right], \quad (4.20)$$

where i, j are elements, and the center and the area of element j are denoted x_j and S_j, respectively.

Although the optimum dose can be obtained by solving the matrix equation (matrix method),[134] the method requires a long calculation time because the number of calculations N_c is proportional to the 6th power of the product of the back-scattering range σ_b and the density of an element ρ.

To obtain the optimum correction dose with a high calculation speed, several algorithms for the calculation have been proposed. These algorithms include (1) the iteration method,[143] (2) an approximate formula method[135–137] (below), (3) Fourier transformation,[146,147] and (4) neural network.[148]

The following subsections explain the iteration method and the approximate formula method. These methods are faster than the matrix method because the number of calculations they require increases with the 4th power rather than the 6th power of the element density ρ.

4.4.6.2 *Iteration method*

The iteration method uses a recursion formula for calculating the correction dose, and the convergent value of the correction value is used as the optimum dose. An example of the recursion formula[149] is:

$$D_{k+1}(x_i) = D_k(x_i)/[E_k(x_i)], \qquad (4.21)$$

where D_{k+1} and D_k are, respectively, the correction doses obtained by the $(k+1)$th and the kth recursions. The deposited energy $E_k(x_i)$ at the position x_i is calculated by using dose $D_k(x)$ in Equation (4.10). Figure 4.100 shows the effectiveness of using this correction method when making X-ray masks.[117]

The calculation time is reported to be nine minutes[117] when a 0.3-μm-rule LSI device and a hierarchical structure are used, the number of figures (elements) is 3.4×10^6, and the calculation speed of the computer is 200 MIPS (millions of instructions per second).

4.4.6.3 *Approximate formula*

Instead of calculating the exact optimum correction dose, the approximate formula is used for calculating correction dose. Although the calculated correction dose is not exact, the value of the error is permissible. Equations (4.22) and (4.23) express the approximate formula:[135,146]

$$D(x) = 1/[1/2 + \eta_E U(x)], \qquad (4.22)$$

where the energy $U(x)$, deposited by the back-scattering electrons, is:

$$U(x) = \int g_b(x - x')\,dx'. \qquad (4.23)$$

Almost all the calculation time is used for evaluating the function $U(x)$. Furthermore, this evaluation can be done rapidly.[136] Thus, high-speed correction calculation is obtained using the approximate formula method.

The formula (4.22) is exact for the region where the pattern size is sufficiently smaller than the back-scattering range and the pattern density is homogeneous. The correction error appears at the position where the pattern density changes sharply, although the correction error is, at most, about 3–5%.[135] It is reported

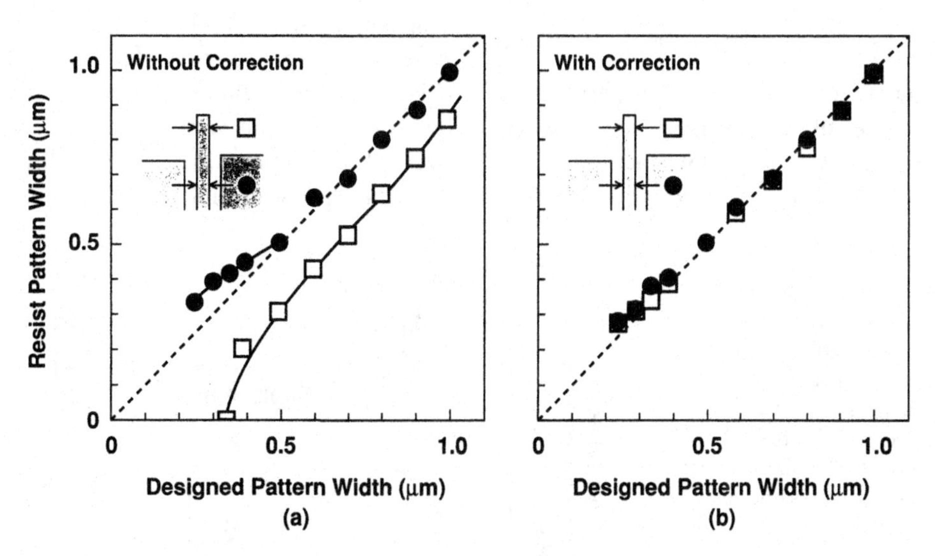

Figure 4.100 Correction accuracy obtained when using the iteration method[117] in X-ray mask fabrication. The acceleration voltage is 20 kV and the resist is 0.5 μm thick.

that a correction accuracy of ±10 nm is obtained for a minimum feature pattern of 0.3 μm, when this method and the direct-writing system with a 50-kV acceleration voltage are used.[150] When using a direct-writing system to make a 256-Mb-DRAM-class device and when using a computer with a calculation speed of 200 MIPS, the calculation time is about 15 minutes.[149]

4.4.7 Maximum size of pattern

The optimum correction dose is assigned to the element (or the region) with a finite size of $\Delta_r \times \Delta_r$. Because the element is exposed with a uniform dose, the deposited energy deviates from the uniform value within the element. The deviation is a sort of correction error (called edge error in this section). The value of the error becomes worst at the edge of a large pattern. By calculating the inclination of the deposited energy E at the edge, for a given permissible error $\pm\Delta E$, the maximum size of an element Δ_r can be roughly evaluated as:[139]

$$\Delta_r = \Delta E / \{\eta_E / [(2 + \eta_E)\sigma_b \sqrt{\pi}]\}. \quad (4.24)$$

This guideline also gives the permissible maximum size of a character beam when using the character projection method.[133,151] Equation (4.24) also gives a rough estimation of the permissible alignment error of the subsidiary exposure when using GHOST for correcting the proximity effect.

4.4.8 Coarse graining; a fast method for proximity effect correction

The time needed to calculate the optimum correction dose increases rapidly with increasing pattern density of the LSI; that is, the 'generation' of LSI. Similarly, when using GHOST, the time for subsidiary exposure also increases with each generation of LSI. Many types of methods for fast proximity effect correction have been proposed. These include coarse graining,[58,139,145,152–155] such as the representative figure method and the pattern area density map method; dedicated hardware;[145,153] a reference table;[136] parallel processing;[156] and hierarchical data access of LSI patterns.[117,136]

The concept of coarse graining is effective for many aspects of correcting the proximity effect. Figure 4.101 shows schematically the basic idea of coarse graining. When the back-scattering range is 10 μm, the energy deposited by the back-scattering electrons for two lines of 0.1-μm width is almost the same as that for one line of 0.2-μm width. This is because the sizes of the figures are considerably smaller than the back-scattering range. It is found that

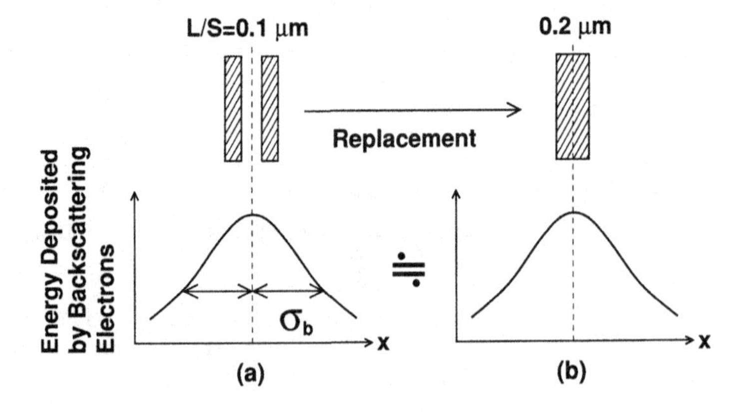

Figure 4.101 Coarse graining – case 1. Because the pattern size is sufficiently smaller than the back-scattering range σ_b, the energy deposited by back-scattering electrons for the two figures with the size of 0.1 μm is almost the same as that for one figure with a size of 0.2 μm

the detailed information of the fine pattern is irrelevant in correcting the proximity effect; only the coarse information of a fine pattern, such as the area and center of gravity, is relevant. Therefore, the concept of coarse graining holds, that is, the coarse information of a pattern is sufficient for accurately correcting the proximity effect. By using the concept of coarse graining, high-speed correction can be provided without sacrificing correction accuracy. Figure 4.102 shows schematically that total area and center of gravity can be used as coarse information.

The coarse information must be calculated in individual small regions after dividing the whole region of a LSI into small regions. The maximum size of the small region depends on what types of coarse information are used.

CASE 1 (see Figure 4.101): When only the area is used as coarse information, the maximum size of the small region must be much smaller than the back-scattering range σ_b. This is because the error caused by using the coarse information depends on the small region size $\Delta_r \times \Delta_r$ as (Δ_r/σ_b).

CASE 2: When both area and center of gravity are used as coarse information, the maximum size of the small region becomes comparable with the back-scattering range σ_b. This is because the error behaves as $(\Delta_r/\sigma_b)^2$.

In both cases, the error can be made negligibly small by controlling the small-region size.[133,147] When the concept of coarse graining is used in the dose-modulation method, the procedure for obtaining the optimum dose is as follows.

STEP 1: Divide the LSI pattern into small regions whose size is comparable to or smaller than the back-scattering range, and obtain the coarse information for each individual small region.

STEP 2: Calculate the correction dose using a suitable algorithm, such as the approximate formula shown in Subsection 4.4.6.3, using the coarse information in each small region instead of the original LSI pattern.

For example, in each small region, the original features are replaced by simple rectangles whose coarse information is the same as that of the original figures. When the approximate formula is used for obtaining the correction dose, the energy $U(x)$ deposited by the back-scattering electrons can be evaluated using the

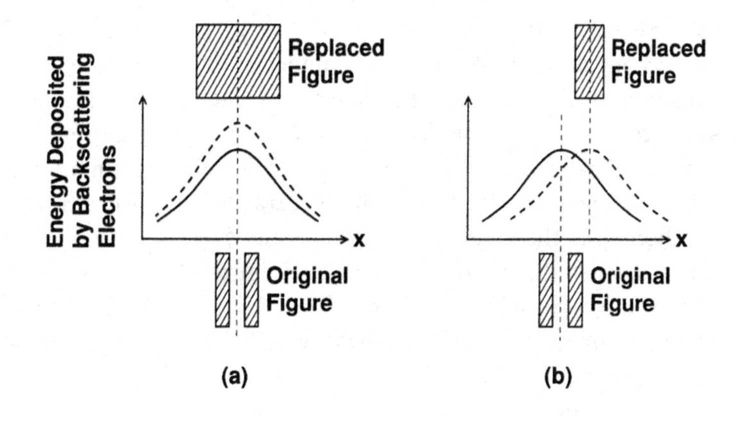

Figure 4.102 There are two conditions for replacing figure to obtain the right correction: both area and center of gravity of the replaced figure must be the same as that of the original figure. If either condition is not satisfied, the energy deposited by the back-scattering electrons for the replaced figure (dashed line) disagrees with that for the original figure (solid line), as shown in (a) and (b), respectively.

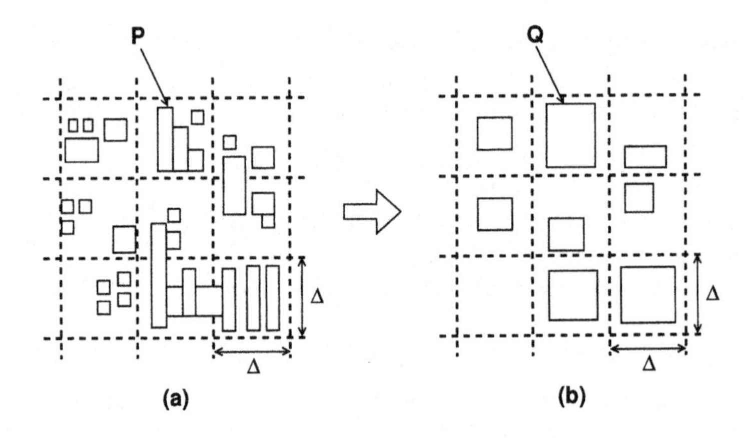

Figure 4.103 Simplification of figures (representative figure method). For proximity-effect correction, the complicated and fine figures of an LSI pattern (a) can be replaced by simple figures (b) whose coarse information is the same as that of original figures in individual small regions. For the dose-modulation method, the optimum dose can be calculated by using the replaced figures. For GHOST, simplification of figures is applied to the inverted pattern of (a), and one shot (Q) replaces many shots (P).

simple replaced rectangles (Figure 4.103) instead of complicated original figures as:

$$U(x_i) = \Sigma S(x_j)\exp[-(x_i - x_j)^2/\sigma_b]. \quad (4.25)$$

When the concept of the coarse graining is not used, almost all of the time needed to calculate the optimum correction dose is used to obtain the energy deposited by the back-scattering electrons. The calculation time increases with pattern density ρ at least as ρ^2.

On the other hand, when coarse graining is used, the calculation of the deposited energy is carried out as in STEP 2, and the number of the calculations depends on the density of each small region, not on the density of the original pattern. This means that the time for STEP 2 is independent of the LSI 'generation'. Furthermore, the number of small regions is much smaller than that of the original figures. For example, the former is about 1/10–1/100 of the latter, under the condition that the acceleration voltage is 50 kV, the back-scattering range is 10 μm, and the minimum feature size is 0.2 μm. The calculation time is therefore much less than it would be without using coarse graining.

It is evident from the above discussion that the larger the ratio between the back-scattering range and the minimum feature size, the more effective is coarse graining. Because the back-scattering range increases with acceleration voltage, it is more effective to use coarse graining at higher voltages.

Real-time correction systems that use the dose-modulation method have been developed. One system is for a direct-pattern-writing system applicable to devices with a minimum feature size of 0.3 μm. The other system can be used to fabricate reticles whose minimum feature size is 0.5 μm. The concept of coarse graining is one of the key technologies in these real-time correction systems.

In applying coarse graining to GHOST, the hatched inverted pattern (see Figure 4.98) in each individual small region is replaced by a simple rectangle whose coarse information is the same as that of the inverted pattern. By using the simplified rectangles for the subsidiary exposure, the number of shots, and therefore the time for the subsidiary exposure, is reduced.[58]

The concept of coarse graining is also effective for reducing the amount of the data for the optimum dose.[151] Coarse graining is already an effective and powerful tool for correcting the proximity effect.

4.4.9 Applications and future problems

Proximity effect correction is used in many applications of electron-beam lithography: direct writing using a variably-shaped beam[134,135] or character projection,[133,151] reticle fabrication,[154] X-ray-mask fabrication,[133,144,157] and the SCALPEL system.[138]

In the case of direct writing, a pattern that was made in the previous processes remains in the lower layer on the substrate. The proximity effect includes the effect of the electrons that are scattered by these patterns. Methods to correct this effect have been proposed.[145] Furthermore, methods to correct part of the effect due to the forward-scattered electrons have also been proposed.[137]

With regard to the character-projection method, it is reported that the dose-modulation method can correct the pattern-size deviation caused by the proximity effect as well as that caused by the space-charge effect.[151] The pattern-size deviation caused by the space-charge effect has been reported to be suppressed to less than ± 5 nm.

With the increase in the number of devices contained within a LSI, the minimum feature size of LSI patterns will continue to decrease and more accurate proximity effect correction will be required. When the minimum feature size becomes less than 0.2 μm, correction accuracy in the order of nanometers must be obtained reliably. To obtain such a high accuracy, it will be necessary to study many aspects of the proximity effect. For example, it has been reported that the triple Gaussian function must be used.[130] It has also been reported that the fast secondary electrons play an important role in the proximity effect.[141] To make deep-submicron patterns, we will need more effective concepts, technology, and knowledge about the proximity effect.

4.5 Summary

Electron-beam (EB) lithography is the only technology available commercially for writing patterns with submicrometer dimensions. Therefore, EB systems have been used to fabricate the minute devices used in research and development and whatever masks are used for optical lithography in mass production. There is a strong demand for higher-performance and higher-density Si LSIs with ever-smaller design rules. In the research and development of such devices and in mask fabrication, high resolution and high accuracy to the order of a nanometer are required. Therefore, EB is now expected to become one of the next-generation lithography tools for design rules of 0.1 μm or less. Because of this, the throughput of electron-beam lithography needs to be improved.

To achieve high throughput, several types of apparatus have been proposed, such as EB projection lithography (cell, block or character projection), blanking aperture array, multibeam lithography and SCALPEL. However, there are still some problems in obtaining high throughput and high accuracy simultaneously with these methods. Much effort has, thus, been spent on solving these problems, and we expect that some advanced type of EB apparatus will soon be used in the mass-production line.

4.6 References

1. J. Kelly, T. Groves and H. P. Kuo, *J. Vac. Sci. Technol.*, **19**, 936 (1981)
2. D. Stephani, E. Kratschmer and H. Beneking, *J. Vac. Sci. Technol.*, **B1**, 101 (1983)
3. M. A. Gesley, F. J. Hohn, R. g. Viswanathan and A. D. Wilson, *J. Vac. Sci. Technol.*, **B6**, 2014 (1988)
4. T. Yoshimura *et al.*, *Jpn. J. Appl. Phys.*, **35**, 6421 (1996)
5. H. Gamo, K. Yamashita, F. Emoto and S. Namba, *J. Vac. Sci. Technol.*, **B3**, 117 (1985)
6. H. Nakazawa *et al.*, *J. Vac. Sci. Technol*, **B6**, 2019 (1988)
7. Y. Ochiai, M. Baba, H. Watanabe and S. Matsui, *Jpn. J. Appl. Phys.*, **30**, 3266 (1991)
8. A. N. Broers, W. W. Molzen, J. J. Cuomo and N. D. Wittels, *Appl. Phys. Lett.*, **29**, 596 (1976)
9. Z. W. Chen, G. A. C. Jones and H. Ahmed, *J. Vac. Sci. Technol.*, **B6**, 2009 (1988)
10. M. A. McCord *et al.*, *J. Vac. Sci. Technol.*, **B10**, 2764 (1992)
11. M. Gesley, *J. Vac. Sci. Technol.*, **B10**, 2451 (1992)
12. A. Claßen, S. Kuhn, J. Straka and A. Forchel, *Microelectronic Eng.*, **17**, 21 (1992)
13. R. Broßardt, F. Otz and B. Rapp, *Microelectronic Eng.*, **27**, 139 (1995)
14. G. A. C. Jones, S. Blythe and H. Ahmed, *J. Vac. Sci. Technol.*, **B5**, 120 (1987)
15. H. Hiroshima *et al.*, *J. Vac. Sci. Technol.*, **B13**, 2514 (1995)
16. H. G. Craighead, *J. Appl. Phys.*, **55**, 4430 (1984)
17. S. P. Beaumont, P. G. Bower, T. Tamamura and C. D. W. Wilkinson, *Appl. Phys. Lett.*, **38**, 436 (1981)
18. W. Chen and H. Ahmed, *Appl. Phys. Lett.*, **62**, 1499 (1993)
19. R. E. Howard *et al.*, *J. Vac. Sci. Technol.*, **B1**, (No. 4), 1101 (1983)
20. K. Kurihara *et al.*, *Jpn. J. Appl. Phys.*, **34**, 6940 (1995)
21. W. Langheinrich, A. Vescan, B. Spangeberg and H. Beneking, *Microelectronic Eng.*, **17**, 287 (1992)
22. W. Langheinrich and H. Beneking, *Jpn. J. Appl. Phys.*, **32**, 6218 (1993)
23. J. Fujita *et al.*, *J. Vac. Sci. Technol.*, **B13**, 2757 (1995)
24. C. D. Gutsche, *Calixarenes*, Royal Soc. Chem., Cambridge, 1989
25. Y. Ohnishi, J. Fujita, Y. Ochiai and S. Matsui, *Microelectronic Eng.*, **35**, 117 (1997)
26. N. Mita, US Patent No. 5,143,784 (1992)
27. J. Fujita *et al.*, *J. Vac. Sci. Technol.*, **14**, 4272 (1996)
28. J. Fujita, Y. Ohnishi, Y. Ochiai and S. Matsui, *Appl. Phys. Lett.*, **68**, 1297 (1995)
29. L. D. Jackal *et al.*, *Appl. Phys. Lett.*, **45**, 698 (1984)
30. Y. W. Yau, R. F. W. Pease, A. A. Iranmanesh and K. J. Polasko, *J. Vac. Sci. Technol.*, **19**, 1048 (1981)
31. A. Sugita and T. Tamamura, *J. Electrochem. Soc.*, **135**, 1741 (1988)
32. P. A. Peterson, Z. J. Radzimski, S. A. Schwalm and P. E. Russell, *J. Vac. Sci. Technol.*, **B10**, 3088 (1992)
33. M. A. McCord and T. H. Newman, *J. Vac. Sci. Technol.*, **B10**, 3083 (1992)
34. E. Munro, J. Orloff, R. Rutherford and J. Wallmark, *J. Vac. Sci. Technol.*, **B6**, 1971 (1988)
35. K. Ishii and T. Matsuda, *JJAP Series 4, Proc. 1990 Intl. MicroProcess Conf.*, pp. 48–51 (1990)
36. W. Brunger, *Microelectronic Eng.*, **27**, 135 (1995)
37. Y. Ochiai, S. Manako, S. Samukawa, K. Takeuchi and T. Yamamoto, *Microelectronic Eng.*, **30**, 415 (1996)
38. S. Manako *et al.*, *Jpn. J. Appl. Phys.*, **33**, 6993–7 (1994)
39. T. Azuma *et al.*, *Jpn. J. Appl. Phys.*, **30**, 3138–41 (1991)
40. M. Ono *et al.*, *ibid.*, 119–22 (1991)
41. H. Kawaura *et al.*, Extended abst. *1996 Intl. Conf. on Solid State Devices and Materials, Yokohama*, p. 22 (1996)
42. H. Kawaura *et al.*, *55th Ann. Device Res. Conf.*, p. 14 (1997)
43. H. Kawaura *et al.*, Extended abst. *1997 Intl. Conf. On Solid State Devices and Materials, Hamamatsu*, p. 572 (1997)
44. T. Fujino *et al.*, *Jpn. J. Appl. Phys.*, **31**, 4262 (1992)
45. H. C. Pfeiffer, *J. Vac. Sci. Technol.*, **15**(3), 887 (1978)
46. M. Fujinama *et al.*, *J. Vac. Sci. Technol.*, **19**, 941 (1981)
47. R. Yoshikawa *et al.*, *J. Vac. Sci. Technol.*, **B5**, 70 (1987)
48. Y. Sakitani *et al.*, *J. Vac. Sci. Technol.*, **B10**, 2579 (1992)

49. H. Yasuda, K. Sakamoto, A. Yamada and K. Kawashima, *Jpn. J. Appl. Phys.*, **30**, 3098 (1991)
50. H. C. Pfeiffer, *IEEE Trans. Electron Devices,* **ED-26**, 663 (1979)
51. M. H. Shearer *et al.*, *J. Vac. Sci. Technol.*, **B4**, 64 (1986)
52. M. Fujinami, N. Shimazu, T. Hosokawa and A. Shibayama, *J. Vac. Sci. Technol.*, **B5**, 61 (1987)
53. D. R. Herriot, R. J. Collier, D. S. Alles and J. W. Stafford, *IEEE Trans. Electron Devices,* **ED-22**, 385 (1975)
54. E. V. Weber, *Optical Eng.*, **22**, 190 (1983)
55. T. Takigawa, K. Kawabuchi, M. Yoshimi and Y. Kato, *Microelectronic Eng.*, **1**, 121 (1983)
56. Y. Takahashi *et al.*, *J. Vac. Sci. Technol.*, **B10**, 2794 (1992)
57. T. Abe and T. Takigawa, *J. Appl. Phys.*, **65**, 4428 (1989)
58. T. Abe, S. Yamasaki, R. Yoshikawa and T. Takigawa, *Jpn. J. Appl. Phys.*, **30**, L528 (1991)
59. Y. Kato *et al.*, *Proc. SPIE*, **393**, 62 (1983)
60. A. Sugita and T. Tamamura, *J. Electrochem. Soc.*, **135**, 1741 (1988)
61. K. Ishii and T. Matsuda, *JJAP Series 4, Proc. 1990 Intl. MicroProcess Conf. Chiba, Japan*, p. 48 (1990)
62. K. Hattori *et al.*, *J. Vac. Sci. Technol.*, **B11**, 2346 (1993)
63. M. Nakasuji and H. Wada, Abst. *Microcircuit Engineering Conf.*, Cambridge (1978)
64. K. Hattori *et al.*, Extended abst. *19th Conf. on Solid State Devices and Materials, Tokyo* (1987)
65. K. Hattori *et al.*, *JJAP Series 3, Proc. 1989 Intl. MicroProcess Conf. Kobe, Japan*, p. 59, (1989)
66. H. Anze *et al.*, *Jpn. J. Appl. Phys.*, **31**, 4248 (1992)
67. K. Takamoto, T. Matsuda, F. Omata and T. Okubo, *J. Vac. Sci. Technol.*, **B5**, 561 (1987)
68. B. H. Keok, T. Chisholm, J. Romijn and A. J. V. Run, *Jpn. J. Appl. Phys.*, **33**, 6971 (1994)
69. H. Ohta *et al.*, *Jpn. J. Appl. Phys.*, **32**, 6044 (1993)
70. K. Hattori *et al.*, *Jpn. J. Appl. Phys.*, **33**, 6966 (1994)
71. M. Ogasawara, K. Ohtoshi and K. Sugihara, *Jpn. J. Appl. Phys.*, **34**, 6655 (1995)
72. T. H. Newman *et al.*, High-resolution shaped electron beam lithography, *Micro Lithography World* **16**, March/April (1992)
73. L. K. Hanes and A. Morris, Advanced direct write electron beam lithography for GaAs monolithic microwave integrated circuit production. *J. Vac. Sci. Technol.*, **B7**(6), 1426 (1989)
74. Y. Sakitani *et al.*, Electron-beam cell-projection lithography system, *J. Vac. Sci. Technol.*, **B10**(6), 2759 (1992)
75. N. Yasutake *et al.*, *Jpn. J. Appl. Phys.*, **31**, 4241 (1992)
76. H. Elsner *et al.*, *Microelectronic Eng.*, **23**, 85 (1994)
77. E. Goto, T. Soma and M. Idesawa, *J. Vac. Sci. Technol.*, **15**(3), 883 (1978)
78. H. Satoh, Y. Nkayama, N. Saitou and T. Kagami, *Proc. SPIE,* **2254**, *Photomask and X-ray Mask Technology*, pp. 122–32 (1994)
79. Y. Someda *et al.*, *J. Vac. Sci. Technol.*, **B12**(6), 3399 (1994)
80. N. Saitou and Y. Sakitani, *Proc. SPIE,* **2194**, 11 (1994)
81. H. G. Jansen, *Coulomb Interactions in Particle Beams*, Academic, Boston (1990)
82. H. C. Pfeiffer, *Jpn. J. Appl. Phys.*, **34**, 6658 (1995)
83. J. A. Liddle *et al.*, *Jpn. J. Appl. Phys.*, **34**, 6663 (1995)
84. T. H. P. Chang, D. P. Kern and M. A. McCord, *J. Vac. Sci. Technol.*, **B7**, 1855 (1989)
85. B. Lischke *et al.*, *JJAP Series 3, Proc. 1989 Intl. MicroProcess Conf.*, p. 47, (1989)
86. H. Yasuda *et al.*, *Jpn. J. Appl. Phys.*, **32**, 6012 (1993)
87. G. W. Jones, S. K. Jones, M. Walters and B. Dudley, *J. Vac. Sci. Technol.*, **B6**, 2023 (1988)
88. T. H. P. Chang, D. P. Kern and L. P. Muray, *J. Vac. Sci. Technol.*, **B10**, 2743 (1992)
89. L. P. Muray, U. Staufer, D. P. Kern and T. H. P. Chang, *J. Vac. Sci. Technol.*, **B20**, 2749 (1992)
90. D. R. Herriot, R. J. Collier, D. S. Alles and J. W. Stafford, *IEEE Trans. Electron Devices*, **ED-22**, 385 (1975)
91. J. L. Freyer and K. P. Standiford, *Solid State Technol.*, **26**, 165 (1983)
92. T. H. P. Chang, A. D. Wilson and H. Ting, *Proc. 7th Intl. Conf. on Electron and Ion Beam Sci. Technol.* ed. R. Bakish, p. 392, Electrochem. Soc. (1976).

93. L. C. Hsia and E. V. Weber, *J. Vac. Sci. Technol.,* **B3**, 128 (1985)
94. T. Takigawa *et al., J. Vac. Sci. Technol.,* **B8**, 1877 (1990)
95. H. C. Pfeiffer and G. O. Langner, Extended abst., *8th Intl. Conf. on Electron and Ion Beam Sci. Technol. (Seattle, WA)* p. 833, Electrochem. Soc. Princeton, N.J. (1978)
96. D. L. Spears and H. I. Smith, *Electronics Lett.,* **8**, 102 (1972)
97. M. Horiguchi *et al., 1995 IEEE Intl. Solid-State Circuits Conf., Tech. Digest,* 252 (1995)
98. Y. Nakayama, Y. Sohada, N. Saitou and H. Itoh, *Proc. SPIE* **1924**, 183 (1993)
99. Y. Nishioka *et al., 1995 IEEE Intl. Electron Devices Mtg., Tech. Digest,* 903 (1995)
100. S. C. Nash *et al., Jpn. J. Appl. Phys.,* **33**, 6678 (1994)
101. S. Inoue *et al., Proc. SPIE,* **2197**, 99 (1994)
102. H. Kanai *et al., Proc. SPIE,* **2793**, 165 (1996)
103. K. Tsukuda *et al., Jpn. J. Appl. Phys.,* **34**, 6652 (1995)
104. T. Takigawa, K. Shimazaki and N. Kusui, *Proc. SPIE,* **632**, 173 (1986)
105. M. Gesley, *J. Vac. Sci. Technol.,* **B9**, 2949 (1991)
106. T. Takigawa and I. Sasaki, *Proc. 10th Intl. Conf. on Electron and Ion Beam Sci. Technol. (Montreal),* 135 (1982)
107. T. Takigawa *et al., J. Vac. Sci. Technol.,* **B9**, 2981 (1991)
108. G. Owen and P. Rissman, *J. Appl. Phys.,* **54**(6), 3573 (1983)
109. B. J. Grenon *et al., Proc. SPIE,* **2322**, 50 (1994)
110. S. Ohki, T. Matsuda and H. Yoshihara, *Jpn. J. Appl. Phys.,* **32**, 5933 (1993)
111. K. Ooi, K. Koyama and M. Kiryu, *Jpn. J. Appl. Phys.,* **33**, 6774 (1994)
112. T. Takigawa *et al., Proc. SPIE,* **2512**, 180 (1995)
113. T. Abe, *J. Vac. Sci. Technol.,* **B14**, 2474 (1996)
114. E. Kratschmer and T. R. Groves, *J. Vac. Sci. Technol.,* **B8**, 1898 (1990)
115. *The National Technology Roadmap for Semiconductors,* 1998 update, Semiconductor Industry Association, 1998, p. 28
116. K. Moriizumi *et al., Proc. SPIE,* **1089**, 93–102 (1989)
117. K. Moriizumi *et al., Jpn. J. Appl. Phys.,* **29**, 2584 (1990)
118. K. Koyama *et al., J. Vac. Sci. Technol.,* **B6**, 2061 (1988)
119. K. Koyama *et al., Jpn. J. Appl. Phys.,* **28**, 2329 (1989)
120. M. Hintermaier *et al., J. Vac. Sci. Technol.,* **B9**, 3043 (1991)
121. A. Muray and R. L. Lozes, *J. Vac. Sci. Technol.,* **B8**, 1775 (1990)
122. M. G. Rosenfield, A. R. Neureuther and C. H. Ting, *J. Vac. Sci. Technol.,* **19**, 1242 (1981)
123. O. W. Otto and A. K. Griffith, *J. Vac. Sci. Technol.,* **B6**, 2048–52 (1988)
124. H. Nakao *et al., Proc. SPIE,* **2793**, 398–409 (1996)
125. M. G. Rosenfield *et al., J. Vac. Sci. Technol.,* **B5**, 114–19 (1987)
126. J. S. Greenich and T. Van Duzer, *J. Vac. Sci. Technol.,* **10**, 1056 (1973)
127. T. Nakasugi, private communication.
128. J. S. Greenich, *J. Vac. Sci. Technol.,* **16**, 1749 (1979)
129. T. H. P. Chan, *J. Vac. Sci. Technol.,* **6**, 1271 (1975)
130. (For an example) S. J. Wind *et al., J. Vac. Sci. Technol.,* **B7**, 1507 (1989)
131. (For an example) S. Uchiyama, S. Ohki and T. Matsuda, *Jpn. J. Appl. Phys.,* **32B**, 6028 (1993) and references cited therein.
132. (For an example) A. Moniwa, H. Ymaguchi and S. Okazaki, *J. Vac. Sci. Technol.,* **B10**, 2771 (1992) and references cited therein.
133. T. Fujino *et al., Jpn. J. Appl. Phys.,* **33B**, 6946 (1994)
134. M. Parikh, *J. Vac. Sci. Technol.,* **15**, 931 (1978)
135. J. M. Pavkovich, *J. Vac. Sci. Technol.,* **B4**, 154 (1986)
136. T. Abe *et al., J. Vac. Sci. Technol.,* **B7**, 1524 (1989)
137. T. R. Groves, *J. Vac. Sci. Technol.,* **B11**, 2746 (1993)
138. J. A. Liddle, G. P. Watson, S. D. Berger and P. D. Miller, *Jpn. J. Appl. Phys.,* **34B**, 6672 (1995)
139. S. Yamasaki *et al., Jpn. J. Appl. Phys.,* **30B**, 3103 (1991).
140. N. Sugiyama and K. Saitoh, *Trans. IECE Japan,* **E63**, 198 (1980)
141. E. A. Dobisz, C. R. K. Marrian, L. M. Shirey and

M. Ancona, *J. Vac. Sci. Technol.*, **B10**, 3067 (1992)
142. D. F. Kyser and M. Parikh, *IBM Tech. Disclosure Bull.*, **21**, 2496 (1978)
143. D. F. Kyser and C. H. Ting, *J. Vac. Sci. Technol.*, **16**, 1759 (1979)
144. M. A. Gesley and M. A. McCord, *J. Vac. Sci. Technol.*, **B12**, 3478 (1994)
145. F. Murai *et al.*, *J. Vac. Sci. Technol.*, **B10**, 3072 (1992)
146. D. P. Kern, *Proc. 9th Intl. Conf. Electron and Ion Beam Sci. Technol.*, Electrochem. Soc., Princeton, New Jersey, p. 326 (1980)
147. M. E. Haslam *et al.*, *J. Vac. Sci. Technol.*, **B3**, 165 (1985)
148. R. C. Frye, *J. Vac. Sci. Technol.*, **B9**, 3054 (1991)
149. M. Hashimoto *et al.*, *Jpn. J. Appl. Phys.*, **30B**, 3058 (1991)
150. T. Nakasugi, T. Abe and S. Yamasaki, *Jpn. J. Appl. Phys.*, **34B**, 6644 (1995)
151. H. Yamashita, T. Tamura, E. Nomura and H. Nozue, *Jpn. J. Appl. Phys.*, **34B**, 6684 (1995)
152. T. Abe, S. Yamasaki, R. Yoshikawa and T. Takigawa, *J. Vac. Sci. Technol.*, **B9**, 3059 (1991)
153. T. Abe *et al.*, *Jpn. J. Appl. Phys.*, **30A**, 2965 (1991)
154. K. Kawasaki *et al.*, *Proc. XEL 95*, 14 (1995)
155. H. Eisenmann, T. Waas and H. Hartmann, *J. Vac. Sci. Technol.*, **B11**, 2741 (1993)
156. K. Moriizumi *et al.*, Extended abst., *38th Spring Mtg., Jpn. Soc. Appl. Phys., Tokyo,* (1991), p. 559 [in Japanese]
157. Y. Kuriyama, S. Moriya, S. Uchiyama and N. Shimazu, *Jpn. Appl. Phys.*, **33B**, 6983 (1994)

Ion-beam lithography 5

Masanori Komuro and Shinji Matsui

5.1 Principles

Lithographic performance depends on the interaction of radiation with a resist material. Fabrication speed is limited by the sensitivity of the resist which, in turn, is related to the energy deposited in the resist and to the overall efficiency with which this energy is converted to chemical changes. The resolution depends on the lateral spread of the radiation and the contrast of the resist.

Chemical changes in resist materials can be induced by ion-beam exposure. The kinetic energy of incident ions is dissipated through collisions with electrons and nuclei. In evaluating the chemical changes in resist materials, one must therefore take into account the contributions of energy deposition due to both collision processes. As shown in Figure 5.1, where depth–dose functions of electrons and several ion species in polymethylmethacrylate (PMMA) are indicated,[1] the amount of energy deposited per unit path length is much greater for an ion beam than for an electron beam. Sensitivity measured in terms of incident charge per unit are is thus clearly expected to be higher for an ion beam than for an electron beam. With regard to electronic collisions, the number of chemical events per unit absorbed energy for ion beams is almost equal to that for electron beams, which means that the main contribution to the chemical changes is from secondary electrons generated by the primary beam.[1] With regard to nuclear collisions, on the other hand, the number of chemical events per unit deposited energy is smaller because part of the energy absorbed in nuclear collisions is dissipated in displacing atoms.

Since the electronic collisions dominate in the case of light (i.e., low-mass) ionic projectiles, such as protons, the resist sensitivity for a light-ion beam is proportional to that for an electron beam (Figure 5.2).[2] In the case of high-energy beams, say those of more than a few mega-electron-volts ion incidence, however, the large amount of energy transferred along the track of the incoming particles gives rise to nonlinear excitations that modify the relation between sensitivity and transferred energy.[3]

High resolution can also be expected when an ion beam is used because the short penetration depth of ions makes lateral spread in the resist film much less for an ion beam than for an electron beam. Figure 5.3 shows computer-simulated trajectories for a 20-keV electron beam and a 60-keV proton beam in PMMA resist on a Si substrate.[4] It is clear that the lateral spread of ions is less than that of electrons. Ion beams might thus provide high resolution without the proximity effect which limits the resolution of electron-beam (EB) lithography.

In addition to studies of polymer-resist exposures, there have been studies made of other fabrication processes that utilize the effects of ion-beam bombardment, such as enhanced chemical etching due to the creation of atomic defects,[5,6] atomic mixing at a heterogeneous interface,[7] and surface modification induced by ion-implantation effects.[8,9] Ion-beam lithography can be categorized into two types: focused-ion-beam (FIB) lithography, where the ion beam is focused down to a diameter less than 100 nm,[10] and masked-ion-beam (MIB) lithography where the mask pattern is replicated onto the resist film either through a

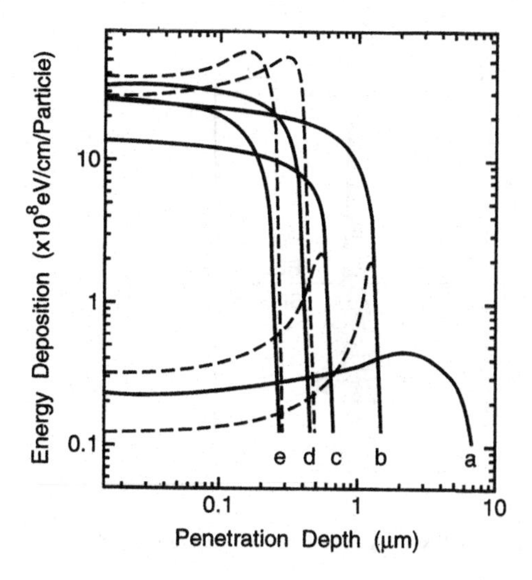

Figure 5.1 Calculated curves of energy deposition in PMMA resist as a function of the penetration depth.
a: 20-keV electron; b: 200-keV He^+; c: 60-keV He^+; d: 250-keV Ar^+; and e: 150-keV Ar^+, where solid and broken lines are of electronic and nuclear collision loss, respectively.

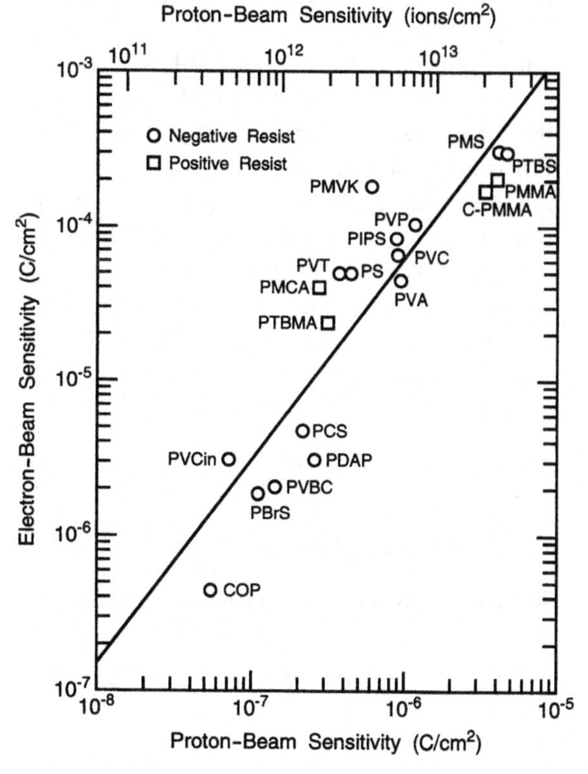

Figure 5.2 Sensitivities of various resist materials: 100-keV proton-beam vs 20-keV electron-beam exposure.

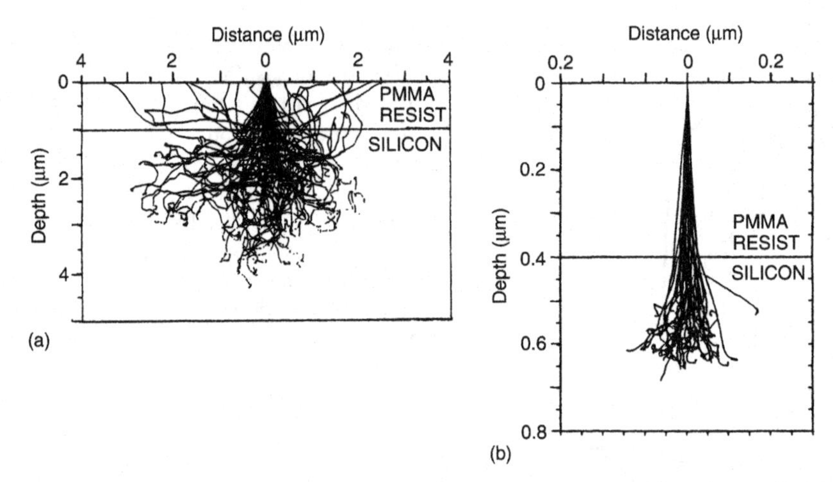

Figure 5.3 Computer-simulated trajectories in PMMA resist on Si: (a) 20-keV electron beam and (b) 60-keV proton beam.

gap of 10–30 μm between the mask and the wafer (proximity ion-beam lithography) or through demagnified optics (projection ion-beam lithography).[11] A bright ion source, called a liquid-metal ion source (LMIS), has often been used in FIB systems. A beam of ions such as Ga^+, Si^{++} or Be^{++} is extracted from this kind of source and is focused down to a diameter less than 100nm,[12] reportedly to a diameter of only 8 nm.[13]

A finely focused ion beam smaller in diameter than that obtained by using LMIS may be formed by using a gas-filled ion source of H^+ or He^+, because the brightness of the gas-filled source is expected to be 100× that of a LMIS.[14,15] On the other hand, light ions are commonly used in MIB lithography in order to get deeper penetration into the resist film even when rather low accelerating voltages are used. The resolution of the replicated pattern is limited mainly by the angular divergence of ions after they have passed through the mask. To date, a pattern width as narrow as 20 nm has been obtained using a hyperthin membrane mask.[16,17]

5.2 **Focused-ion-beam lithography**

A FIB can be utilized for maskless ion-implantation, maskless ion-etching, and high-resolution lithographic processes.[10,18,19,20,21,22,23,24,25] The sensitivity of a resist used for FIB lithography is two or more orders of magnitude higher than that of a resist used for EB lithography.[26] And, FIB is advantageous in the fabrication of submicron devices, because ion scattering in the resist is negligible and there is very little backscattering from the substrate.[27] Kubena and coworkers reported that they were able to produce patterns with features as small as 6–8 nm when using a 50-keV Ga^+ two-lens microprobe system.[13]

5.2.1 High-resolution patterns

FIB exposures were carried out with the JEOL JIBL-150 system.[28] The maximum accelerating voltage was 150 kV and the minimum beam diameter was 0.1 μm.

5.2.1.1 *Positive resist patterns*

PMMA positive resists were used to demonstrate high-resolution FIB lithography. Pre-baking was carried out in a N_2 atmosphere at 170 °C for 30 min. After the PMMA-coated wafers were exposed to ion beams, they were developed by dipping for 1 min into a solution at 25 °C containing one part methylisobutyl ketone (MIBK) per three parts isopropyl alcohol (IPA). They were then rinsed by dipping them in IPA for 1 min. The resists were exposed using various doses of Si and Be ions at various acceleration voltages. Figure 5.4 shows the resultant sensitivity curves. Although the sensitivity of PMMA resist depends on the ion species, its sensitivity to the 200-keV Be^{++} ions is an order of magnitude higher than its sensitivity to a 20-keV electron beam. The ion-energy dependence of the developed depth for the resist is shown in Figure 5.5. It is clear that Be ions are useful when a thick resist is used. Figure 5.6 shows a 0.1-μm-linewidth pattern in a 0.5-μm-thick PMMA resist fabricated using a 260-keV Be^{++} FIB at 1.5×10^{13} ions/cm^2 dose.

5.2.1.2 *Negative resist patterns*

A novolac-based negative resist (Shipley Co. SAL601-ER7) was used to demonstrate the high-resolution capability of FIB lithography.[6] After pre-baking for 30 min at 80 °C in a N_2 atmosphere, the resist on the wafer was exposed to 260-keV Be^{++} and 260-keV Si^{++} FIB. Post-exposure baking was carried out for 7 min at 105 °C in a N_2 atmosphere. The wafers were then developed for 3 min using SAL-MF6522 developer at 25 °C, and were rinsed with deionized water. The SAL601-ER7 resist sensitivity for 260-keV FIB exposure was measured and Figure 5.7 shows that the sensitivity obtained with post-exposure baking (PEB) is about 100× that obtained without it. A 0.1-μm-linewidth pattern in a SAL601-ER7 resist with 0.6-μm

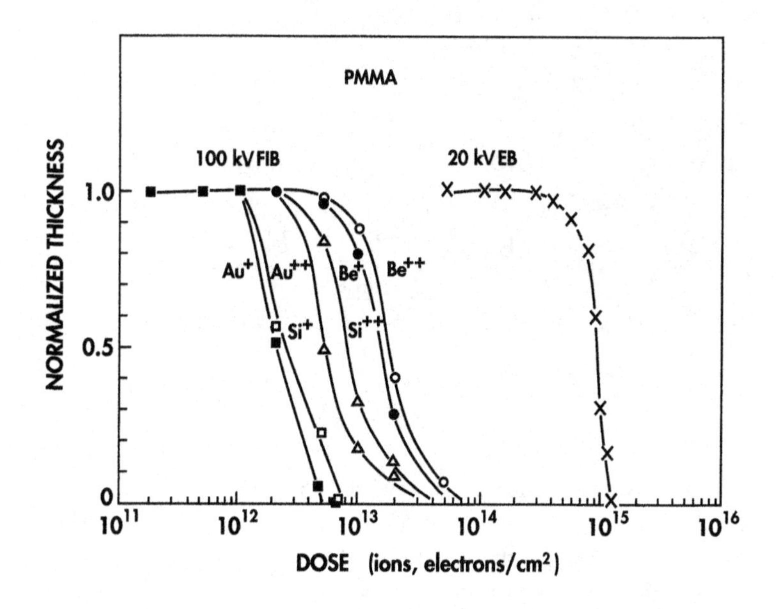

Figure 5.4 Resist sensitivity and ion ranges of PMMA resist for 100-kV FIB and 20-kV EB exposures.

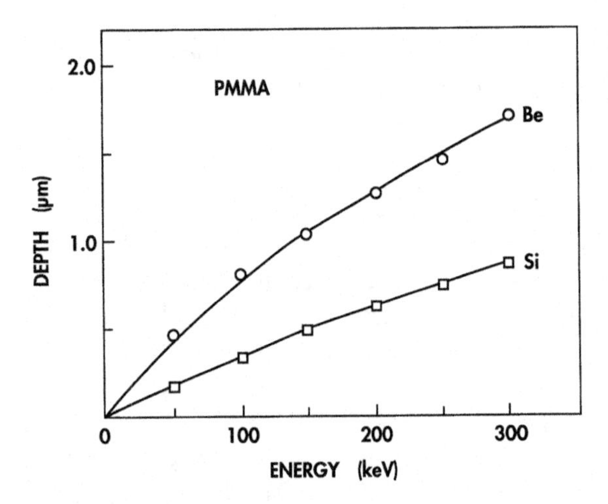

Figure 5.5 Ion-energy dependence for developed resist depth. Si-ion dose: 3.0×10^{12} ions/cm^2; Be-ion dose: 7.0×10^{13} ions/cm^2.

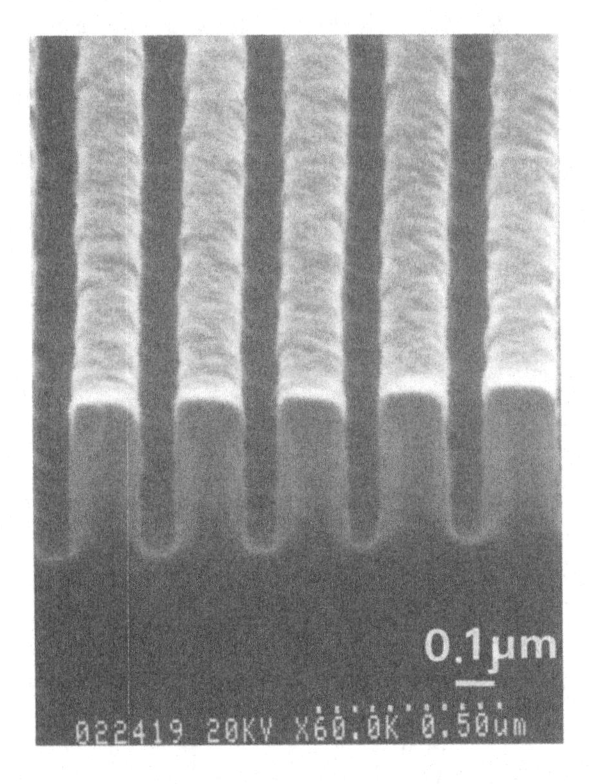

Figure 5.6 SEM photograph of 0.1-μm-linewidth PMMA resist patterns fabricated by 260-keV Be^{++} FIB at 1.5×10^{13} ions/cm^2 dose. Resist thickness is 0.5 μm.

thickness was fabricated using a 260-keV Be^{++} FIB at 2×10^{12} ions/cm^2 dose. As shown in Figure 5.8, this pattern had an excellent profile with vertical side walls.

5.2.2 Mark detection and overlay accuracy

The mark-detection procedure used in FIB direct writing is different from that used in electron-beam writing. When a FIB is scanned

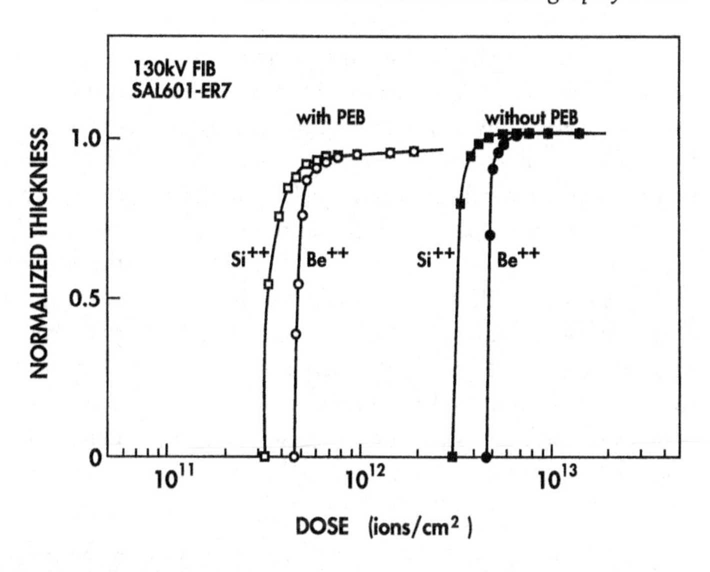

Figure 5.7 Comparison of SAL601-ER7 resist sensitivities with and without post-exposure baking (PEB). Exposure with 260-keV Be^{++} and Si^{++} FIB, respectively.

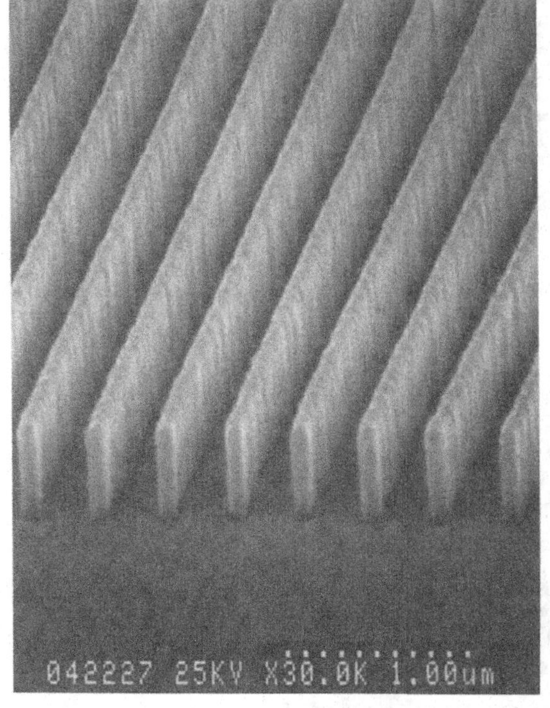

Figure 5.8 SEM photograph of 0.1-μm-linewidth SAL601-ER7 resist patterns fabricated at a dose of 2.0×10^{12} ions/cm^2 by 260-keV Be^{++} FIB. Resist thickness is 0.6 μm.

across the mark edge, secondary electrons are emitted from the target surface. Because these secondary electrons have an energy of only a few electron-volts, their escape depth is limited to a few angstroms. The secondary electron (SE) signal therefore depends on the surface morphology.

An example of a SE signal which was obtained by scanning a FIB across a mark edge is shown in Figure 5.9. The cathode-ray-tube (CRT) screen used to obtain these images is mounted on the system control panel. Signals obtained when the FIB was scanned in the x- and y-directions are indicated in (a) and (b), respectively. In each image, the upper signal shows the change in the SE signal; the middle signal shows the gate signal which was generated by an electronic detection circuit from the detected SE signal; and the lower signal shows the wave for determining the position which was generated from the gate signal. The SE signal was detected by a circular microchannel plate with a pinhole, through which the FIB was targeted. The mark was concave, with a depth of 1.7 μm, and the PMMA resist thickness was 0.5 μm. The FIB consisted of 260-

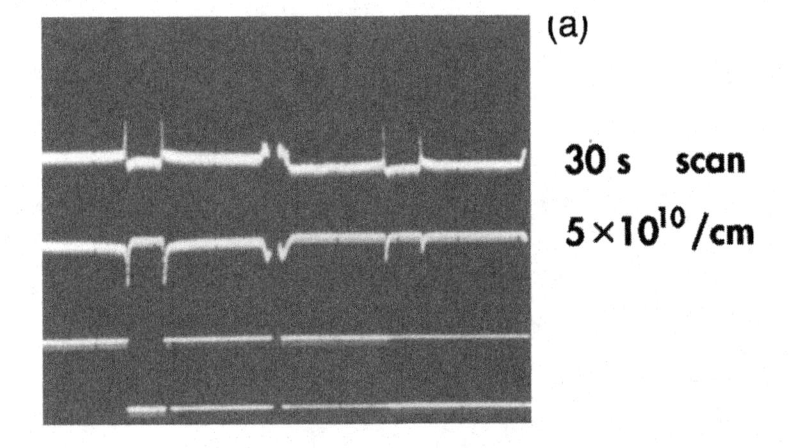

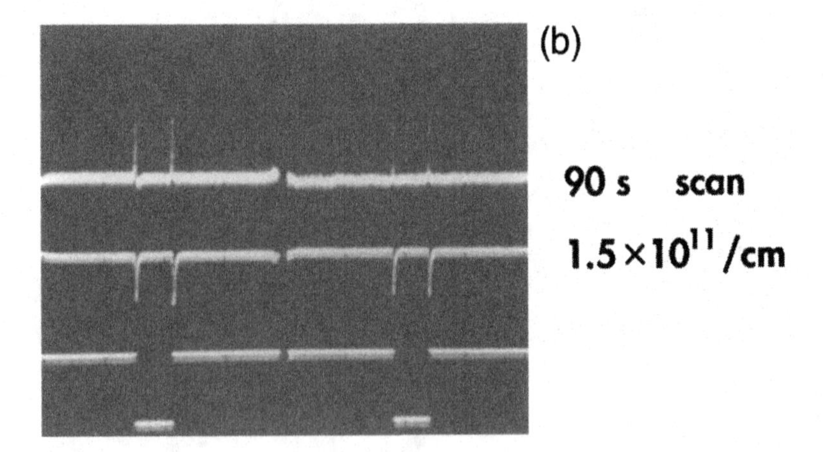

Figure 5.9 Mark-detection signals observed when 260-keV Si^{++} FIB was scanned across a mark covered with 0.5-μm-thickness PMMA resist.

keV Si^{++} ions (ion current $I_p = 6.3$ pA). After 90 s FIB scanning, the resist on the mark was removed, and the SE signal increased to the level sufficient for mark detection.

Double exposure using two ion species with different ion-scattering ranges has been applied to make mushroom-shaped cross-section gates in a GaAs FET (see Subsection 5.2.4). The overlay accuracy in between Si^{++} and Be^{++} FIB at 260 keV was measured. The first exposure was carried out using a 260-keV Be^{++} FIB, and the second exposure was carried out using a 260-keV Si^{++} FIB. The registration mark used was cross-shaped, 10-μm wide, 100-μm long, and 0.8-μm high. The PMMA resist was 0.1-μm thick. The registration accuracy was measured using verniers. Most of the overlay errors obtained for both the x- and y-directions were less than 0.1 μm. Figure 5.10 demonstrates the good overlay accuracy obtained. The etched depths were 0.6 μm for the 260-keV Si^{++} and 1.2 μm for the Be^{++} FIB. Overlay accuracies for various combinations of ion species are listed in Table 5.1. For each combination, the accuracy obtained was better than 0.26 μm at 3σ.

5.2.3 Fabrication of 0.1-μm-linewidth nMOS gate patterns

nMOSFETs were fabricated using a hybrid process combining a FIB and an optical stepper. FIB lithography was applied for gate formation.[3] Figure 5.11 shows the process flow. In order to increase adhesion between the resist and the substrate, a 1.0-μm-thick bottom layer of MP2400 (Shipley Co. Ltd) was first spin-coated on a substrate and then baked for 1 h at 250 °C. Then, a 0.2-μm-thick top layer of silicone-based negative resist SNR™ (Toyo Soda Manufacturing Co.) was spin-coated on

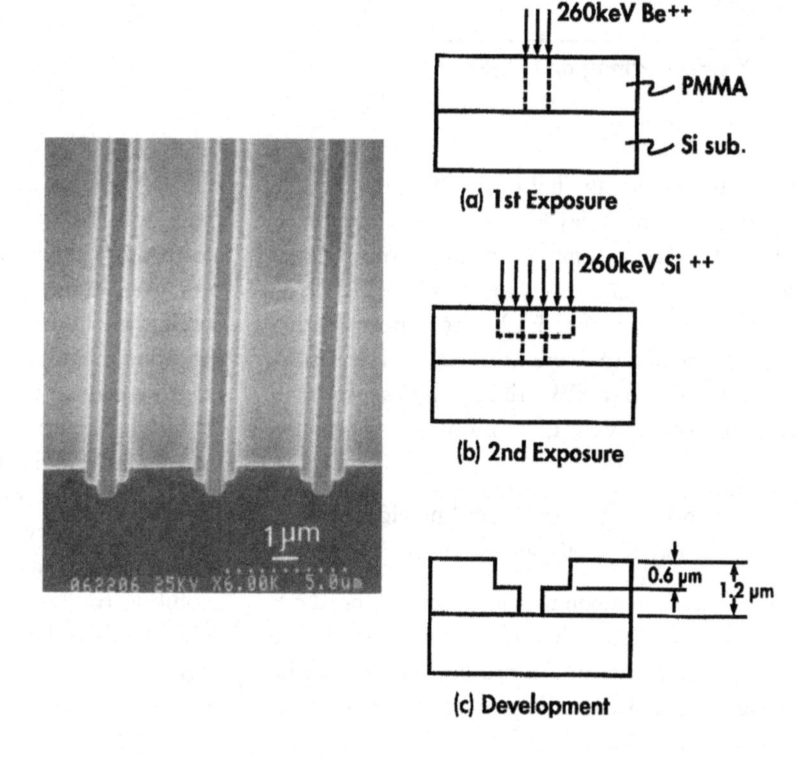

Figure 5.10 T-shape cross-section profile fabricated by double exposure of 0.1-μm-thick PMMA resist using 260-keV Be^{++} FIB at a dose of 3.0 $\times 10^{13}$ ions/cm^2 followed by 260-keV Si^{++} FIB at a dose of 1.0×10^{13} ions/cm^2.
Left: SEM photograph;
Right: Process route.

Table 5.1 *Overlay accuracy for several combinations of 1st and 2nd FIB exposure.*

Exposure 1st/2nd (keV)	3σ (x, y over-all) (μm)
260 Be^{++}/260 Si^{++}	0.26
300 Be^{++}/160 Be^{++}	0.21
260 Si^{++}/260 Si^{++}	0.26

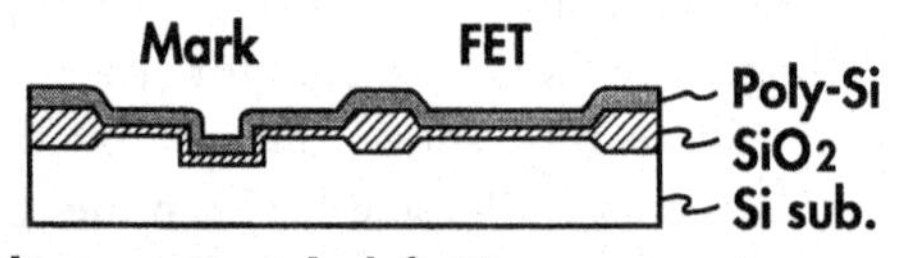

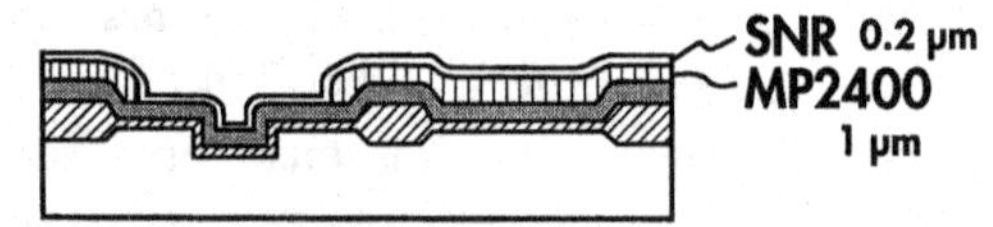

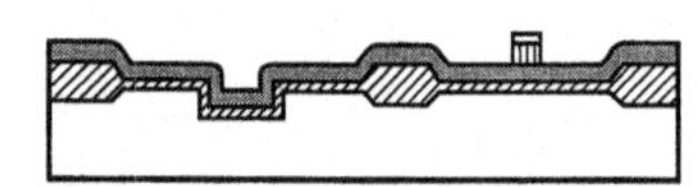

Figure 5.11 Process of FIB lithography with bi-layer resist for making 0.1-μm gate patterns.

the bottom layer and baked for 30 min at 80 °C. After the top resist layer was exposed and developed, it served as a mask when the bottom layer was etched by O_2 reactive ion etching (RIE).

As shown in Figure 5.12, gate patterns with a 0.1-μm linewidth were formed. These results demonstrate that FIB lithography is useful in making devices with dimensions of 0.1 μm.

5.2.4 Fabrication of 0.25-μm-linewidth mushroom-shaped gates

To fabricate semiconductor devices we need not only to make sub-micron patterns but also to control the resist pattern profiles. When making a GaAs microwave FET with agate length less than 0.5 μm, for example, to keep the gate resistance low enough we need to fashion a gate electrode with a mushroom-shaped cross-section. A multilayer resist structure and a side-etching method might be used to make such a gate, but it is not easy to obtain an accurate gate length when using these methods. An advantage of FIB lithography is that it enables both the lateral and longitudinal dimension of patterns to be controlled precisely. Figure 5.13 shows the detailed process flow for FIB fabrication of a mushroom-shaped gate of a high-electron-mobility transistor (HEMT).[9,10] A 1.05-μm-thick PMMA layer was exposed successively with 192-keV Be^{++} and 260-keV Si^{++} ion beams (each to a dose of 2.0×10^{13} ions/cm^2).

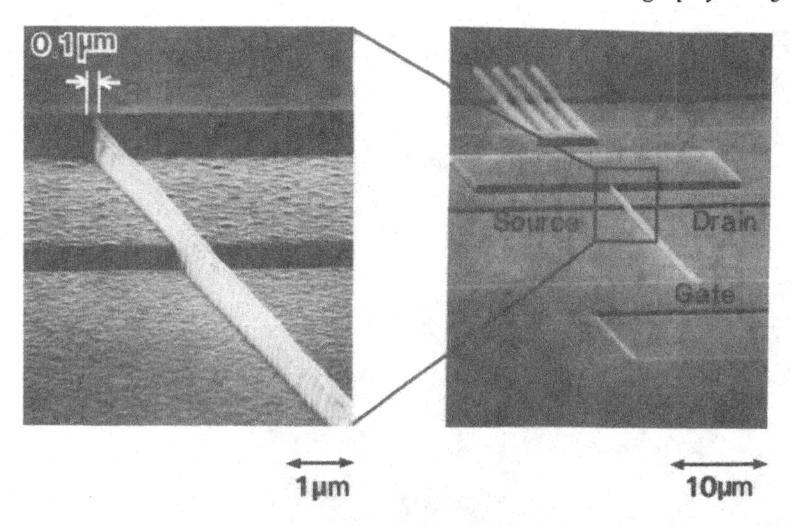

Figure 5.12 SEM photographs of 0.1-μm gate pattern replicated in 1.0-μm-thick MP 2400 resist.

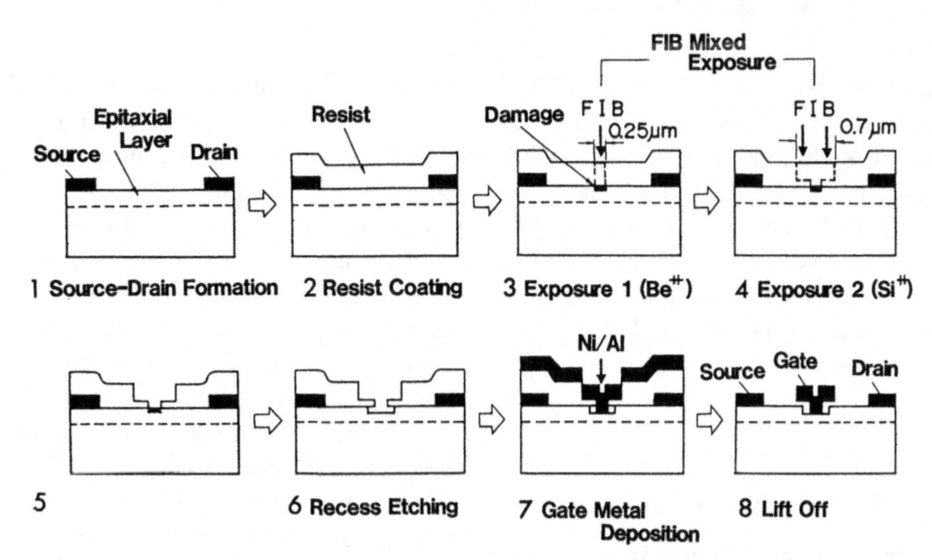

Figure 5.13 Process flow of mushroom-shaped gate fabrication using FIB. (1) Fabrication of source and drain; (2) Resist coating; (3) Be^{++} ion-beam exposure; (4) Si^{++} ion-beam exposure; (5) Developing; (6) Recess etching; (7) Metal deposition; and (8) Lift off.

The resist was developed by immersing the wafer in a 2:3 mixture of methylisobutylketone, MIBK, and isopropylalcohol, IPA, (resist profiles were almost independent of developing conditions). The gate region was then recessed to the optimum depth by wet chemical etching, and Ni–Al alloy metal gate about 0.6-μm thick was deposited. A lift-off procedure finished the gate fabrication process.

A SEM photograph of the mushroom-shaped cross-section gate fabricated this way is shown in Figure 5.14. A 0.25-μm gate having a mush-

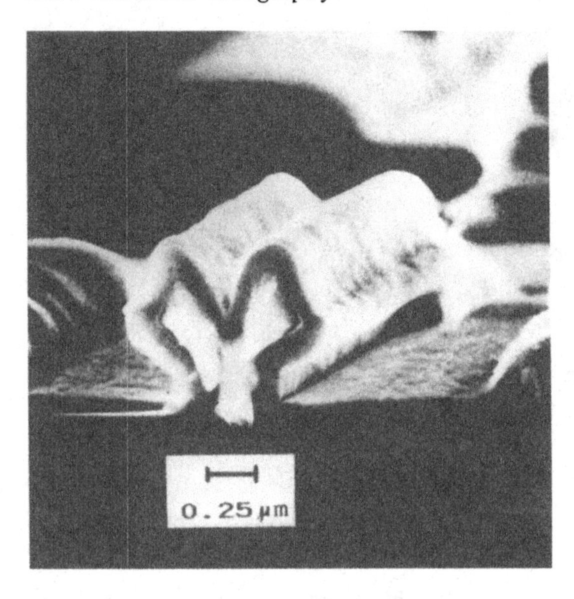

Figure 5.14 SEM photograph of mushroom-shaped cross-section gate of HEMTs.

room-shaped cross-section with a cap width of 0.7 μm was obtained. The size of the fabricated chip was 350 × 380 μm^2 and the gate width was 200 μm. For passivation, a plasma CVD SiN layer about 0.1-μm thick covered the gate metal and the epitaxial wafer surface. Excellent microwave performance, with minimum noise figures of 0.68 dB at 12 GHz and 0.83 dB at 18 GHz, was obtained.

5.3 Masked-ion-beam lithography

5.3.1 Proximity ion-beam lithography

Figure 5.15 shows the concept of a proximity ion-beam lithography system developed by Hughes Research Laboratory.[29] In this system, the mask is located 10–30 mm above the wafer with an exposed area of 1–2 cm^2, and alignment is achieved by an interferometer technique. The patterns are thus replicated by step-and-repeat. The most important factor limiting the resolution of the replicated pattern is the beam divergence after the ions pass through the membrane. A 0.1-μm blur, for example, can be produced by a 0.3-degree divergence when the proximity gap is 25 μm.

Various mask structures proposed so far are summarized in Figure 5.16(a)–(d). The thin amorphous mask[29] shown in Figure 5.16(a) uses an alumina (Al_2O_3) membrane only 0.1–0.2-μm thick, and this membrane produces a divergence of 1.4–2 degrees when incident protons have an initial energy of 250 keV. This angular divergence corresponds to 0.2–0.25-μm blurring even when the proximity gap between the mask and the wafer is only 10 μm. To decrease the angular divergence without sacrificing the mechanical stability of the mask membrane, researchers proposed a channeling mask (Figure 5.16(b))[29] in which a (100) crystalline Si film about 0.7-μm thick is used and the surface normal of the membrane is aligned parallel to the incident ion direction so that the penetrating ions are subjected to less scattering. Measurements of the angular distribution for 250-keV protons yield a divergence of about 0.9 degree for a 0.7-μm-thick membrane.

In the dechanneling mask shown in Figure 5.16(c),[30] a thin gold (Au) pattern is deposited on the upper surface of a rather thick (110) Si crystalline film. This gold pattern acts as a scattering film. The angular divergence of the dechanneling mask is similar to that for the channeling mask, and the highest pattern resolution is achieved by using a stencil or open-hole mask. This is because there are essentially no scattering events in this kind of mask, so the minimum linewidth is limited only by the ion beam itself. Films of crystalline Si or Si_3N_4 are used as membranes in the stencil masks (Figure 5.16(d)).[29,31] A severe problem with this mask

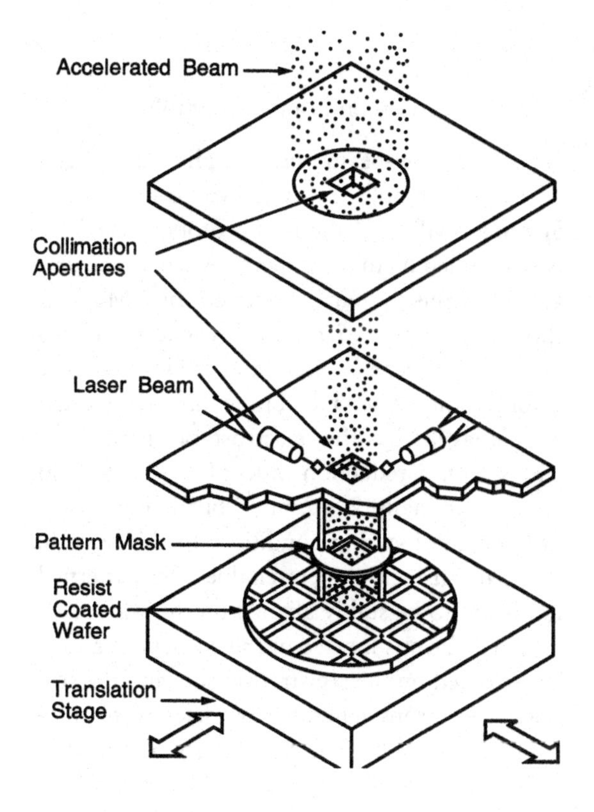

Figure 5.15 Schematic diagram showing step-and-repeat proximity ion-beam lithography system.

Figure 5.16 Diagrams of several mask structures proposed for masked-ion-beam lithography. (a) Thin amorphous mask (hyperthin membrane mask); (b) channeling mask; (c) dechanneling mask; and (d) stencil mask.

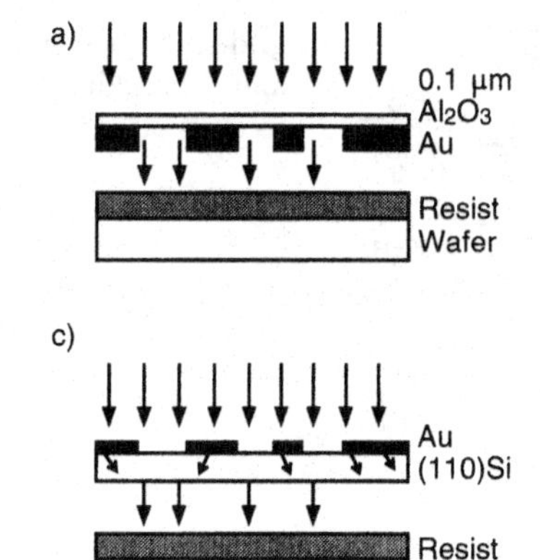

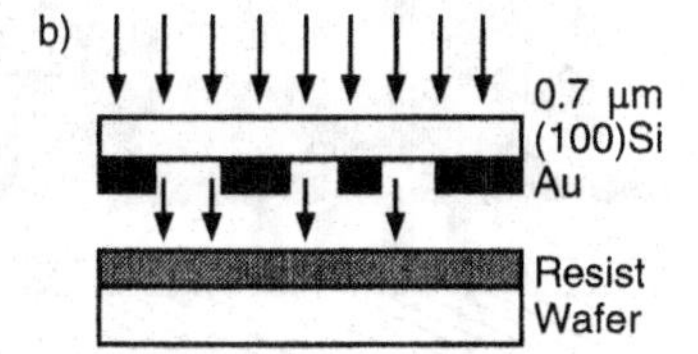

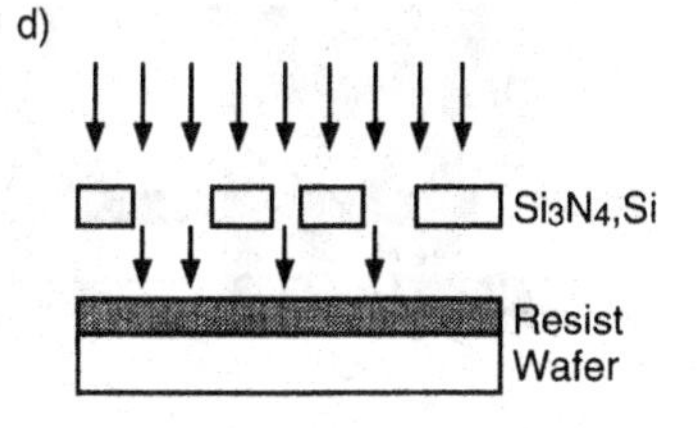

structure is that it cannot produce a donut-like pattern, so two complementary masks must be prepared when such a pattern is needed. A stencil mask supported by a grid pattern was therefore proposed.[32,33] As seen in Figure 5.17, the incident ion beam is sometimes deflected ('rocked') during the exposure in order to eliminate the shadows of the supporting grid patterns. There are other systems where a conformable mask made of a polyimide membrane and/or a carbon thin-film has been used for contact printing.[16,34]

A problem with all the mask configurations proposed is that ion-beam-induced heating results in in-plane distortion of absorber patterns. The amount of distortion depends on ion energy and current density, membrane thickness, membrane area, absorber coverage, and the type of mask used. The results of computer simulations of the in-plane distortion for various types of masks are shown in Figure 5.18 as a function of mask-membrane thickness.[35] Since a higher power density results in a lower exposure time, there is some trade-off between mask size and incident power.

A group at Hughes Research Laboratories has demonstrated the fabrication of a nMOS test chip using proximity ion-beam lithography.[36] The channeling mask and the wafer were aligned using an optical microscope and were then clamped 50 μm apart. An nMOS transistor with a sub-micrometer gate length was produced using only ion-beam lithography. A group at MIT has also reported a 0.3-μm-gate-length GaAs FET[37] made using proximity ion-beam lithography to replicate the gate pattern.

5.3.2 Projection ion-beam lithography

Projection ion-beam lithography has been developed mainly by Ion Micro-fabrication Systems (IMS) and the first prototype machine was supplied by that company. A schematic diagram of the system developed by IMS[38] is shown in Figure 5.19. A low-energy (5–10 keV) proton beam is used, and the ions passing through the mask membrane are accelerated and imaged by the electrostatic immersion lenses with a reduction ratio of 1 : 5. The field size is nominally $2 \times 2\,\mathrm{mm}^2$, but it depends a little on the focusing condition. In this system, the 'pattern lock' system enables the patterned ion beam to be stably aligned on the wafer surface. The beams located at the corners of a chip are scanned on the registration slits and the secondary electrons generated are detected to compensate the positional fluctuation of the chip

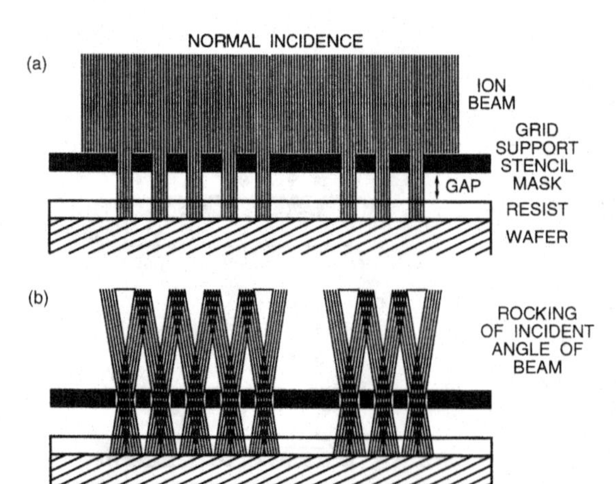

Figure 5.17 Schematic of (a) grid-supported stencil mask and (b) rocking-beam method to eliminate the grid image.

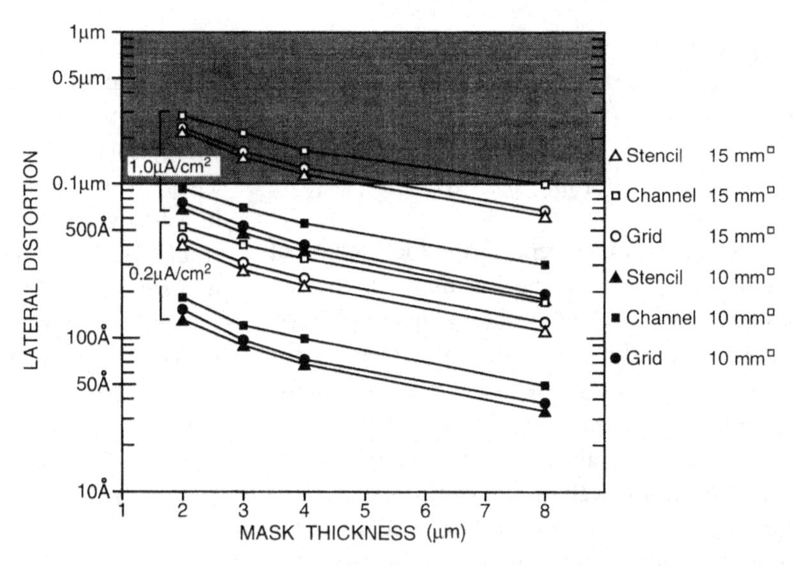

Figure 5.18 Dependencies of calculated maximum in-plate distortion of mask due to ion-beam heating on mask-membrane thickness for two ion current-densities and mask sizes of 10-mm and 15-mm square.

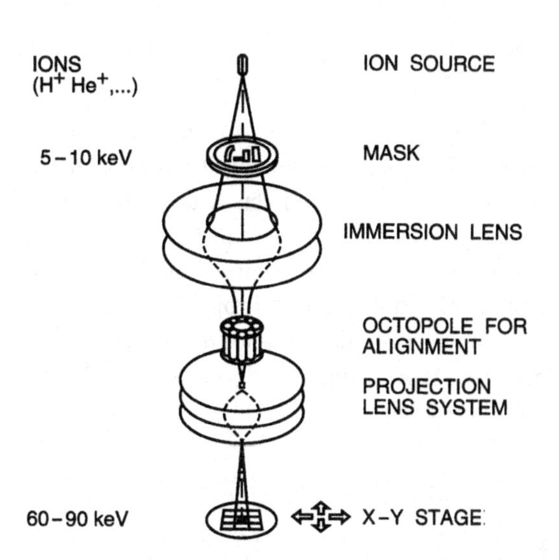

Figure 5.19 Schematic of demagnifying projection ion-beam lithography system.

beam. Thus, the position of the chip beam can be maintained to within less than 10 nm by feeding the detected signal to the multipole deflector, focusing lens, and solenoid coil.

A stencil-type mask made of a Si crystalline membrane or a Ni foil is used in this system, but this type of lithography is also limited by the donut-problem and in-plane distortion due to ion-beam heating. It has been proposed that forcibly heating the mask will reduce the distortion due to ion-beam heating, and a new system with a field size of 20×20 mm^2 has been developed.[39]

The stochastic space-charge effect has been closely examined because it is the most important factor limiting the resolution when the ion current is increased in order to shorten the exposure time. The stochastic blurring diameter

d_{stoc} due to the space charge is generated mainly at the crossover in the ion optics and is given by:[40]

$$d_{stoc} = kI^{1/2}F_2m^{1/2}/E^{5.4}\alpha D^{1/2} \qquad (5.1)$$

where I is the ion current, F_2 the focal length of the imaging lens, m the ion mass, E the ion kinetic energy, α the ion-beam-envelope crossover angle, and D the ion-beam diameter at the crossover. The most effective way to reduce the blurring is to increase the ion energy. A space-charge effect obtained experimentally is illustrated in Figure 5.20, which compares the resolution obtained with a projector (ALPHA 5x) in which the ion kinetic energy at the crossover was about 7.5 keV with the resolution obtained with a projector (IPLM 02) in which the energy at the crossover was 50 keV.[41] When the ion-beam current is reduced to a level at which the space charge is not effective, the chromatic aberration restricts the pattern resolution. The system used a duo-plasmatron ion source which gave rise to an ion kinetic energy spread of about 6 eV for He^+. A multi-cusp ion source which has an energy spread of about 2 eV was recently developed by the Lawrence Livermore Laboratory,[42] and a computer simulation on chromatic aberration and on intradistortion as a function of axial wafer plane position was carried out assuming the use of such a low-energy-spread ion source (Figure 5.21).[29] Thus, the modernized system which will appear in a few years is expected to provide a pattern resolution better than 20 nm within a field 20-mm square.

Several test devices have demonstrated the value of projection ion-beam lithography. Good step coverage can be seen in Figure 5.22, which shows a 0.18-μm-wide resist pattern (Ray PN(AZ114PN) resist) on 0.5-μm-high SiO_2 steps.[41]

5.4 Summary

FIB lithography has several advantages over EB lithography, such as high sensitivity and negligibly small proximity effects, but damage is a serious problem and throughput for device fabrication is lower than attained with variably shaped EB lithography. Projection ion-beam lithography is under development for use in the mass-production of ULSIs with 0.1-μm-level dimensions.

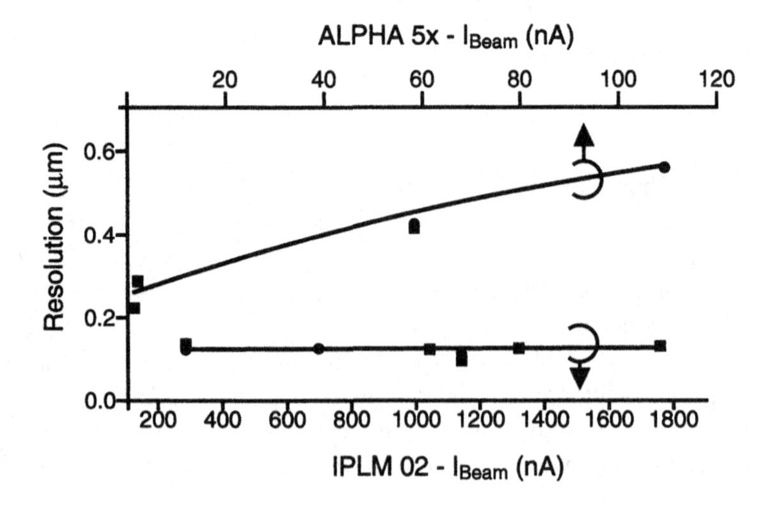

Figure 5.20 Experimental results on influence of space-charge blur on resolution using He^+ for two different projections, ALPHA5x, in which the crossover was formed at 7.5-keV ion energy; and IPLM 02 with the crossover at 50 keV.

Figure 5.21 Calculated maximum distortion and maximum chromatic aberrations within a 20 × 20 mm^2 exposure field as a function of wafer position along the ion-optical axis of a recently designed projector.[42]

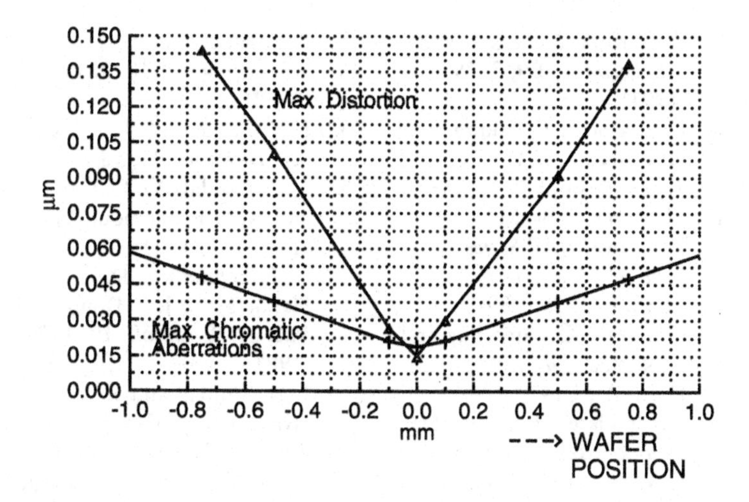

Figure 5.22 SEM photograph of 0.18-µm-wide resist lines crossing 0.5-µm-high SiO_2 steps replicated by ion projector.[41]

5.5 References

1. M. Komuro, N. Atoda and H. Kawakatsu, *J. Electrochem. Soc.*, **126**, 483 (1979)
2. G. Brault and L. J. Miller, *Polymer Eng. & Sci.*, **20**(16), 1064 (1980)
3. T. M. Hall, A. Wagner and L. F. Thompson, *J. Vac. Sci. Technol.*, **16**, 1889 (1979)
4. L. Karapiperis, I. Adesida, C. A. Lee and E. D. Wolf, *J. Vac. Sci. Technol.*, **19**, 1259 (1981)
5. M. Komuro, H. Hiroshima, H. Tanoue and T. Kanayama, *J. Vac. Sci. Technol.*, **B1**, 985 (1983)
6. T. Ohta, T. Kanayama, H. Tanoue and M. Komuro, *J. Vac. Sci. Technol.*, **B7**, 89 (1989)
7. T. Kanayama *et al.*, *J. Vac. Sci. Technol.*, **B9**, 295 (1991)
8. T. Venkatesan *et al.*, *J. Vac. Sci. Technol.*, **19**, 1379 (1981)
9. I. L. Berr and A. L. Caviglia, *J. Vac. Sci. Technol.*, **B1**, 1059 (1983)
10. R. L. Seliger *et al.*, *J. Vac. Sci. Technol.*, **16**, 1610 (1979)
11. G. Stengl, H. Losschner, W. Maurer and P. Wolf, *J. Vac. Sci. Technol.*, **B4**, 194 (1986)
12. R. L. Seliger, J. W. Ward, V. Wang and R. L. Kubena, *Appl. Phys. Lett.*, **34**, 36 (1979)

13. R. L. Kubean *et al.*, *J. Vac. Sci. Technol.*, **B9**, 3079 (1991)
14. K. Horiuchi, T. Itakura and H. Ishikawa, *J. Vac. Sci. Technol.*, **B6**, 937 (1988)
15. M. Konishi and M. Takizawa, *JJAP Series 3, Proc. 1989 MicroProcess Conf.*, p. 139 (1989)
16. K. Gamo, K. Yamashita and S. Namba, *Jpn. J. Appl. Phys.*, **23**, L141 (1984)
17. I. Adesida *et al.*, *J. Vac. Sci. Technol.*, **B3**, 45 (1985)
18. S. Matsui *et al.*, *J. Vac. Sci. Technol.*, **B4**, 845 (1979)
19. S. Matsui *et al.*, *J. Vac. Sci. Technol.*, **B5**, 853 (1987)
20. Y. Kojima, Y. Ochiai and S. Matsui, *Jpn. J. Appl. Phys.*, **27**, 1780 (1988)
21. S. Matsui, Y. Kojima and Y. Ochiai, *Appl. Phys. Lett.*, **53**, 868 (1988)
22. T. Shiokawa *et al.*, *Jpn. J. Appl. Phys.*, **23**, L232 (1984)
23. Y. Ochiai *et al.*, *Proc. SPIE,* **923**, 106 (1988)
24. H. Morimoto *et al.*, *J. Vac. Sci. Technol.*, **B4**, 205 (1986)
25. K. Nakagawa *et al.*, *Electronics Letter,* **24**, 242 (1988)
26. H. Ryssel, K. Haberger and H. Krang, *J. Vac. Sci. Technol.*, **19**, 1358 (1981)
27. L. Karapiperis, I. Adesida, C. A. Lee and E. D. Wolf, *J. Vac. Sci. Technol.*, **19**, 1259 (1981)
28. Y. Ochiai, Y. Kojima and S. Matsui, *J. Vac. Sci. Technol.*, **B6**, 1055 (1988)
29. D. B. Rensch *et al.*, *J. Vac. Sci. Technol.*, **16**, 1897 (1979)
30. L. Csepregi, F. Iberl and P. Eichinger, *Appl. Phys. Lett.*, **37**, 630 (1980)
31. P. E. Mauger *et al.*, *J. Vac. Sci. Technol.*, **B10**, 2819 (1992)
32. J. N. Randall *et al.*, *J. Vac. Sci. Technol.*, **B3**, 58 (1985)
33. J. N. Randall, D. C. Flanders and N. P. Economou, *Proc. SPIE,* **471**, 47 (1984)
34. N. P. Economou, D. C. Flanders and J. P. Donnely, *J. Vac. Sci. Technol.*, **19**, 1172 (1981)
35. J. N. Randall and R. Sivasankar, *J. Vac. Sci. Technol.*, **B5**, 223 (1987)
36. C. W. Slayman, J. L. Bartelt, C. M. McKenna and J. Y. Chen, *Optical Eng.*, **22**, 208 (1983)
37. S. W. Pang *et al.*, *J. Vac. Sci. Technol.*, **B5**, 215 (1987)
38. G. Stengle, H. Loschner and P. Wolf, *Nucl. Instrum. Meth.*, **B19/20**, 987 (1987)
39. A. Chalupka *et al.*, *J. Vac. Sci. Technol.*, **B12**, 3513 (1994)
40. E. Hammel *et al.*, *J. Vac. Sci. Technol.*, **B12**, 3533 (1994)
41. W. H. Brunger *et al.*, *Microcircuit Eng.*, **27**, 323 (1995)
42. K. N. Leung *et al.*, *J. Vac. Sci. Technol.*, **B13**, 2600 (1995)

Resists 6

Hiroshi Ban, Tadayoshi Kokubo,
Makoto Nakase and Takeshi Ohfuji

6.1 Principles

6.1.1 Introduction

One of the major factors determining lithographic performance is the resist: the photosensitive material that forms a relief image after an imaging exposure to light (or other radiation source) and subsequent development. The resist pattern protects portions of the underlying substrate during the etching process, and the term 'resist' came from this function (i.e., 'resistant' to etching). Many photosensitive chemistries for resist materials have been proposed and studied[1] but only a few are actually used in device manufacturing. Most of the practical resist materials are organic polymers.

Figure 6.1 outlines the basic process of resist pattern formation. A resist layer coated on a substrate is exposed to light or to some other radiation source, and a photochemical reaction takes place in the resist layer. Figure 6.1(b) shows the resist layer after the chemical changes due to the exposure. The amount of the chemical change is generally dependent on the flux of light entering the layer, and the chemically changed portion of the layer constitutes what is called a 'latent image'. The latent image is then converted to a 'relief image' by using a developer to selectively dissolve either the chemically changed part or the unchanged part. The resulting relief image, now termed the resist image, as shown in Figure 6.1(c), is used as an etching mask. A resist image is therefore formed by two steps: formation of the latent image, and development.

A positive resist is one in which the image is formed by dissolving out the exposed area, as in Figure 6.1(c), and a negative resist is one in which the image is formed by dissolving out the unexposed area. The diazonaphthoquinone (DNQ) resists widely used in industry are positive photoresists. The diazonaphthoquinone resist itself is insoluble in aqueous alkaline solutions, but the photochemical reaction that occurs during exposure makes the exposed areas soluble in the aqueous alkaline developer. The positive photoresist relief image is thus obtained by treating the resist layer with an aqueous alkaline developer which selectively dissolves the exposed photoresist.

Other photochemical reactions can also be used to change a resist material's solubility (dissolution rate) in a developer. For example, reactions causing molecular-weight degradation or a significant polarity change resulting in a solubility increase have been used in positive resists. And the photochemical changes that occur in negative-tone resists reduce solubility in the developer. A dual-tone resist (i.e., one which can be positive or negative tone) can in some cases be obtained by choosing the proper developer polarity.

The latent image can also be formed without using liquid solutions. In a 'dry development' technique, the developer is in the gas phase. Generally, either the exposed parts or the unexposed parts of the resist layer are selectively etched away in an oxygen plasma.

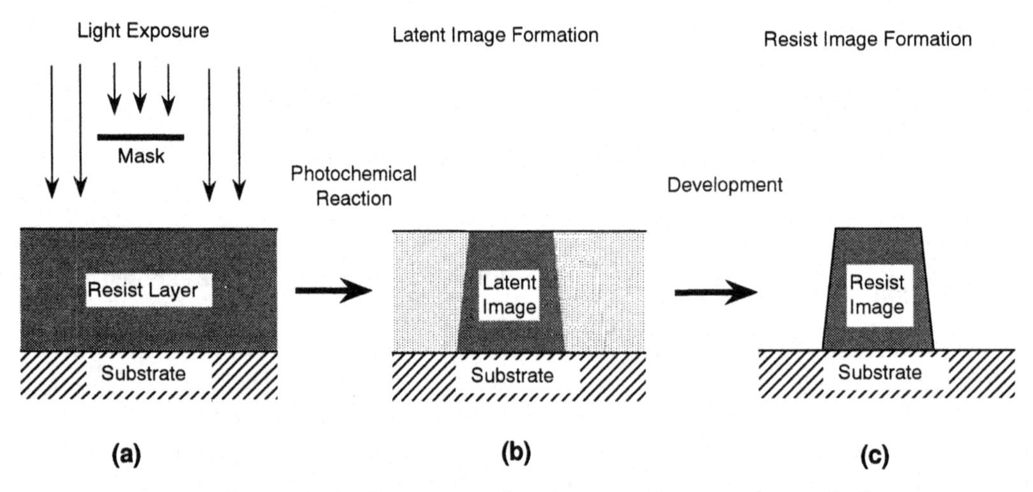

Figure 6.1 Basic process of forming a positive-tone photoresist image.

6.1.2 Kinds of resists

Resists for sub-half-micron lithography are classified into three categories depending on the wavelength of the light used in their imaging.

(a) Near-UV photoresists: g-line, i-line, and broadband (including g, h, and i lines);
(b) Deep-UV photoresists: KrF and ArF excimer lasers, and Hg-lamp emissions shorter than 300 nm;
(c) High-energy-beam resists: electron beams, X-rays, and ion-beams.

Shorter wavelengths are generally preferred for high-resolution imaging because of optical principles (see Section 2.3), but the suitability of any particular system is determined by resist parameters and the specific application. Resist technologies for the near-UV region are well established, whereas resist technologies for the shorter-wavelength regions are not matured and are therefore used less in practical applications. Details of each type of resist will be discussed in Sections 6.2, 6.3, and 6.4.

Resist materials usually consist of a photoactive component and a polymeric component that serves as a binder resin to improve the film-forming properties. Although many chemistries have been proposed for sub-half-micron lithography resists, few are actually used in practical wafer processes. The only chemistry used for near-UV photoresists is a positive resist system composed of a diazonaphthoquinone photoactive compound and a novolac resin (DQ/N resist, hereafter). Chemical-amplification (CA) resist chemistries, which utilize a catalytic reaction for latent-image formation, are used in the deep-UV region for the reasons of yielding high transparency to deep-UV light and high sensitivity to compensate the weak light intensity in this region. Both positive and negative CA resists are available, and they typically consist of a photo-acid generator and a polymeric component that is decomposed, cross-linked, or polarity-changed by a catalytic amount of the photo-acid. As with a DQ/N resist, it is possible to obtain a large difference between the dissolution rates of exposed and unexposed areas (dissolution-rate contrast) during development in aqueous alkali. For a high-energy-beam resist, high sensitivity is more important for achieving practical throughput because the exposure is performed with beam scanning. Fortunately, this same CA chemistry, which can yield high photospeed, can also be used for a high-energy-beam resist with some slight modification.[2]

Another category of resist uses only the aerial image produced in the upper portion of the resist layer. Two types of chemistry can be employed, both of which utilize silicon atoms and an O_2 plasma-etching step to complete the formation of the relief image. In top-surface imaging (TSI), a photochemical reaction changes the reactivity (to silylating reagents) of the resist material at or near the surface.[3] The subsequent silylation step thus results in a gradient of silicon incorporation between the exposed and unexposed areas, a gradient that forms the latent image and provides the etching-rate discrimination during the subsequent O_2 plasma development. The other type of resist chemistry uses a thin layer of a resist over a thicker nonphotosensitive layer. The thin resist layer already contains silicon atoms, and a relief image of the resist is formed by wet development. The subsequent O_2 plasma treatment then etches away the newly exposed underlying nonphotosensitive layer. From an optical point of view, confining the aerial image to the thin layer in the resist surface results in better latent images and thus, if an equivalent transformation of the latent image into a relief image after the dry development is guaranteed, higher resolution.

6.1.3 Physics and chemistry of resist processes

Figure 6.2 shows the flow of the photolithographic processing typically used in IC device fabrication. This section will describe each step of the resist process and each of the materials required and will describe the chemical and physical changes that occur during the processing steps. The DQ/N resist chemistry is cited for showing the chemical and physical changes.

(1) *Substrate pretreatment*: The adhesion between the resist and the substrate is usually improved by dehydrating the substrate and priming its surface before it is coated with a resist. If any hydrophilic sites remain at the substrate surface, a hydrophilic molecular orientation will be formed at the resist–substrate interface and will allow water to penetrate there during the wet processes (e.g., development and wet-etching), causing loss of adhesion. The substrate is typically primed by exposing it to hexamethyldisilazane (HMDS) gas. This is a silylating reagent which reacts with OH or equivalent polar groups on the surface to covert them to hydrophobic ones and thereby makes the van der Waals bonding effective at the resist–substrate interface. This makes the interface more resistant to water penetration.

(2) *Resist coating*: Resist is spin-coated onto the wafer using a casting solvent. Spin-coating easily gives a very uniform and well-controlled thickness across the wafer and is used very commonly in IC manufacturing. The casting solvent is chosen to provide a viscosity and an evaporation rate within process boundaries. If the evaporation rate is too high, spontaneous convective flow of solvent in the wet film will form a Bénard Cell and will thus result in striation defects in the coating.[4] Most practical resists contain a surfactant to minimize the chance of forming striations.

(3) *Pre-bake*: Most of the casting solvent evaporates during the spin-coating process, and the resist film is baked to remove most of the remaining solvent. A little solvent, however, remains in the resist film even after the baking process and can influence the diffusion and dissolution properties of the film.[5] If the resist surface is dried too fast during the pre-baking

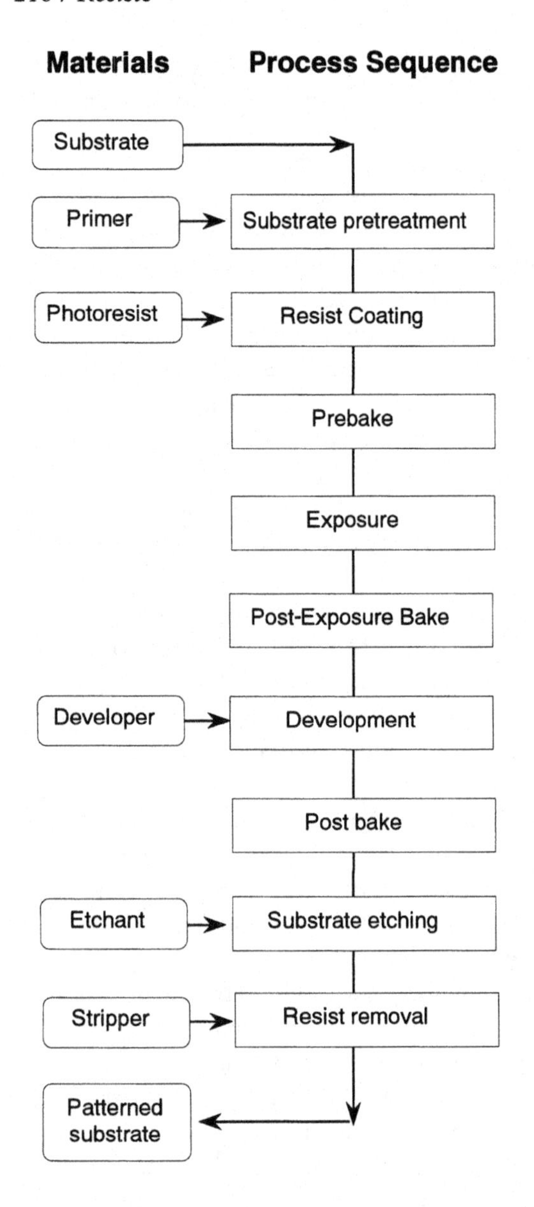

Figure 6.2 Typical photolithography process flow.

process, it will become dense, and this will make it difficult for solvent in the bulk of the film to evaporate through the surface. This densification would also make the resist surface dissolve more slowly than the bulk resist during development and thereby distort the resist cross-sectional profile.[6]

(4) *Exposure*: In the case of DQ/N resist, a photochemical reaction that takes place in the film completes the formation of the latent image. Specifically, the diazonaphthoquinone (DNQ) photoactive compound (PAC), decomposed into indenecarboxylic acid (ICA), distributes to form the latent image. In CA resists, on the other hand, exposure produces only an acid

distribution. The latent image is formed through a subsequent reaction catalyzed by the acid, a reaction that largely occurs after the exposure.

The optical image determined by the mask is, in practice, not necessarily transformed into the latent image accurately. As shown in Figure 6.3(b), the depth of the latent image generally has a wavy outline. This is the result of standing-wave interference between the incident light and the light reflected from the substrate. The period of the cycle, from node to node, is $\lambda/2n$, where λ is wavelength of the exposure light and n is refractive index of the resist film. No standing wave is formed in non-optical exposures (electron beam or X-ray), but beam scattering or the generation of secondary electrons in the layer can prevent the formation of an accurate latent image.

(5) *Post-exposure bake*: A resist pattern with wavy side walls is formed if the latent image is developed without additional processing, and such a pattern is not suitable for high-resolution lithography. A latent image with smooth side walls like those shown in Figure 6.3(c) can be obtained when local diffusion of the DNQ or ICA compound is driven by heating the resist layer.[7] A post-exposure baking (PEB) step is optionally applied for this purpose. The diffusion length depends on the size of the molecule, polymer properties (particularly the glass-transition temperature T_g), and residual solvent.[8] Too much diffusion degrades the contrast of the latent image and results in resolution loss. The temperature and time of the PEB therefore needs to be optimized for each resist. PEB is indispensable for CA resists because heat is required after exposure to complete the acid-catalyzed reaction required for latent-image formation.

(6) *Development*: An aqueous alkaline solution of a quarternary ammonium salt, particularly of tetramethylammonium hydroxide (TMAH), is normally used as the developer for DQ/N and CA resists. Some additives (e.g., surfactants) are optionally used. Organic solvents may also be used for other types of resists that utilize a solubility change caused simply by molecular-weight change after the exposure. The latent image in the resist layer is changed to a real resist image by dissolving out the exposed or unexposed parts selectively (Figure 6.4). The

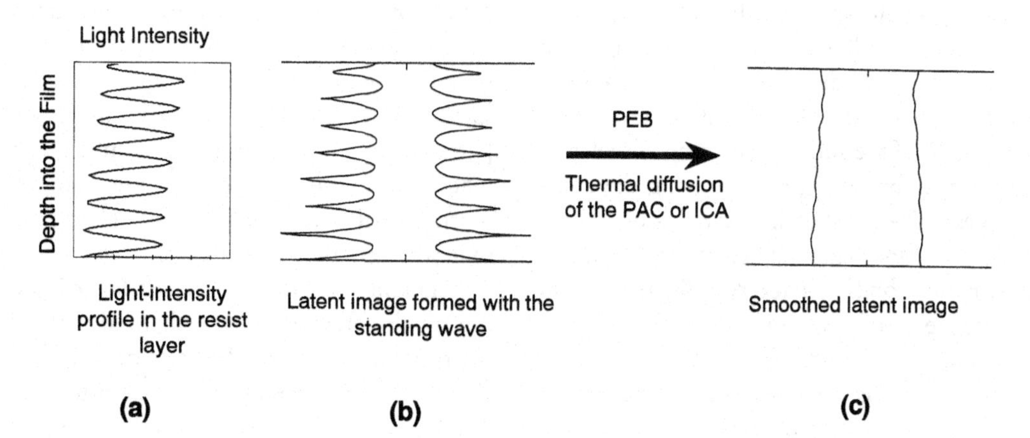

Figure 6.3 Standing wave in latent image and its smoothing by post-exposure baking (PEB).

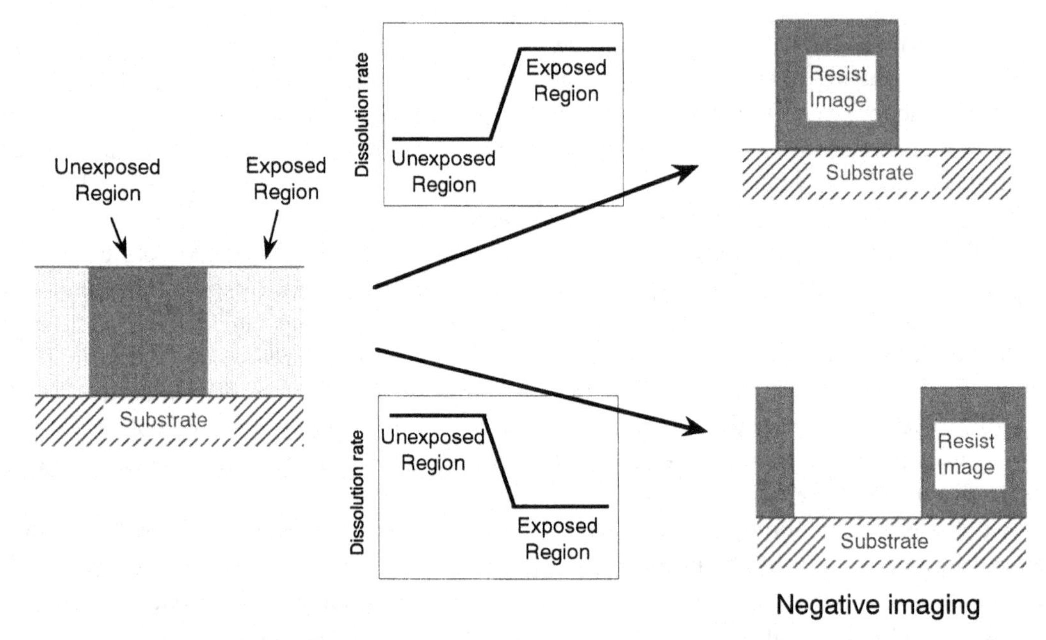

Figure 6.4 Resist-image development: positive-tone and negative-tone images.

dissolution-rate difference between the two parts determines the contrast of the resist image and thus the resolution. This is explained in more detail in Section 6.1.4. The developed resist image is then baked (post-bake) to remove residual solvent and moisture. Then it is utilized as the mask for etching or ion-implantation.

(7) *Etching*: In dry-etching, the resist is eroded by chemical and physical processes induced by the high-energy plasma and ion bombardment. The rate of the resist etching must be sufficiently less than that of the substrate etching in order to ensure proper pattern transfer during the etching process. The resist etch-rate is mainly determined by the binding energy of the components, that is, by its bonding stability.[9] Table 6.1 lists the energies of several organic bonds. Compounds with multiple C—C bonds have the largest binding energies, and a compound with a high density of multiple C—C bonds, such as aromatic phenolic compounds, therefore has good etch stability. Novolac resin and polyhydroxystyrene (PHS) resin have very high aromatic : aliphatic carbon ratios and are very resistant to etching. Those polymers are respectively used for DQ/N and CA resists binders. For the same reason, many phenolic compounds are preferably used for the other components, like PAC and additives. Alicyclic compounds also give low rates of etching, similar to those given by aromatic compounds,[10] probably because of the multiple bonds in the same ring. Compounds of this kind are studied particularly for use in 193-nm lithography,[11] where aromatic compounds are difficult to work with because their optical absorption at that wavelength is too strong.

The temperature of the resist layer increases during dry-etching, so the thermal stability of the resist is important. The rate of resist etching shows a sharp increase at temperatures above T_g, decreasing the resist–substrate etching selectivity. In addition, the pattern may be deformed due to a softening of the resist film occurring at

Table 6.1 *Energies of various organic bond scissions.*

Type of bond	Energy (kJ/mol)	Type of bond	Energy (kJ/mol)
CH_3-H	434	$HO-H$	499
CH_3-CH_3	368	CH_3O-H	436
CH_3-COOH	403	$C_6H_5-CH_3$	417
CH_3-COCH_3	355	C_6H_5-OH	458
CF_3-F	543	$C_6H_5-OCH_3$	409
CF_3-CF_3	410	$C_6H_5-NH_2$	405
CH_3-OH	383	$CH_2=CH_2$	718
CH_3-NH_2	333	$CH\equiv CH$	960

the high temperature. Such a situation can degrade the transfer from the original resist pattern to the etched substrate.

In wet-etching, on the other hand, the resist is eroded mainly through chemical reactions with the etchant, and a high selectivity can thus be achieved by choosing a suitable etching chemistry. If the etchant penetrates into the resist–substrate interface, however, the resist may peel off and thus never work as a mask. Adhesion at the interface is therefore much more important in wet-etching than it is in dry-etching.

(8) *Resist removal*: The resist pattern is removed after the etching or ion-implantation process. One of two kinds of methods is generally used for resist removal: a dry process or a wet process. In the former, oxygen plasma is used to burn off the resist. A problem with this process, however, is that inorganic contaminants or metal-containing etching byproducts can form metal oxides that often remain on the substrate.[12]

In a wet process, solutions containing an amine and an organic solvent are generally used for resist stripping. A resist that was subjected to high temperature or high energies during etching generally has a chemically hardened surface and is often difficult to dissolve. In practice, resist removal uses both processes in a complementary manner.

6.1.4 Parameters governing the formation of high-resolution resist images

The main components of the resists are polymeric materials, but the size of the molecules is still sufficiently small compared to that of the pattern to be printed. The size of the molecules is therefore not the dominant factor determining the resolution limit. The major factor is how the resist can recover the contrast of the image that was degraded by optical diffraction, focus failure, beam scattering, and other physical factors. Figure 6.5 shows schematically how image contrast is lost as a result of optical principles but can be recovered by exploiting resist chemistry.

The resolution capability R of optical projection lithography is basically given by the equation: $R = k_1 \times \lambda/NA$. The term λ/NA correlates to the contrast of the aerial image projected to the resist layer. This equation, however, says nothing about how the resist resolution can be determined from the predetermined aerial image because the coefficient k_1 includes various parameters other than the optics and cannot be treated theoretically. The

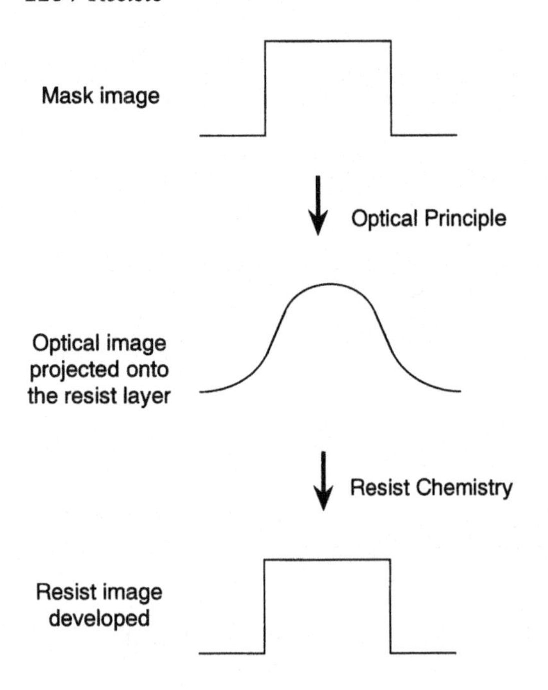

Figure 6.5 Conceptual image of the contrast changes occurring in projection lithography.

resolution of resist imaging is determined through the parameters involved in the following steps:

(1) Projection of the mask aerial image onto the resist layer.
(2) Latent-image formation by the photochemical reaction induced by the projected light.
(3) Modification of the latent image by diffusion or a subsequent imaging reaction.
(4) Conversion of the latent image into a real image by development.

This section explains the parameters in each step and also explains how they interact to determine the resolution.

(1) *Aerial image projection*: This is determined by the optics, and the contrast of the projected aerial image is the parameter that correlates to the resolution as shown with the above equation. The details are described in Section 2.3.

(2) *Latent-image formation*: For a given projected image, the latent image is determined by the properties of the projected light and by the optical properties of the resist. Figure 6.6 shows schematically the image projections into a resist layer of a certain thickness. Images projected to the resist surface and bottom can differ from each other, because the depth of focus of the optics can be smaller than the dimension of the layer thickness. If the projection is focused onto the resist surface, the projected image at the bottom of the resist is degraded compared to that at the surface because of loss of the focus. The image at the bottom is also influenced by the optical properties of the resist because the resist layer itself is in the light projection path and absorbs light. The latent image is determined as a function of light intensity distribution through the whole thickness of the film.

If we take DQ/N resist as an example, the imaging reaction is the photochemical decom-

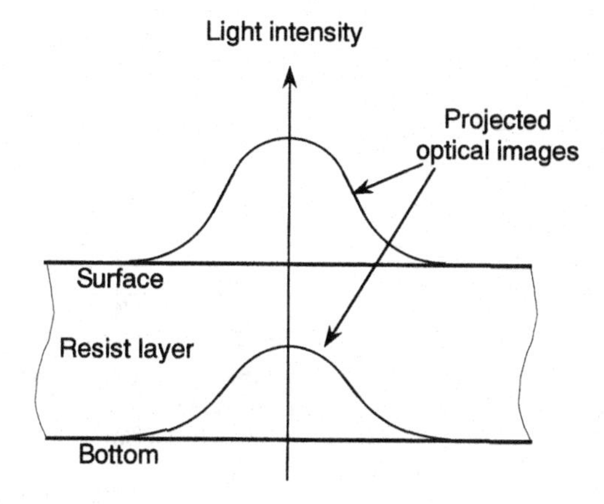

Figure 6.6 Aerial images projected into resist layer. The top and bottom images are not identical because of the intensity loss due to optical absorption and because of the focus difference.

position of the DNQ-PAC. This is a first-order reaction with regard to the DNQ group concentration, and the progress of the reaction is therefore represented by

$$-\mathrm{d}M/\mathrm{d}t = CIM, \tag{6.1}$$

where M represents the normalized concentration of the remaining DNQ groups, C represents the quantum yield of the DNQ decomposition, and I and t respectively represent light intensity and exposure time. With this equation, the light intensity distribution in the resist layer is transformed to the M-value distribution after a certain exposure time. The M-value distribution determines the latent image.[13] Most conveniently, the latent image is expressed with the contour of the M-value across the resist layer. Figure 6.7 shows examples of M-value latent-image contours for line-and-space patterns, calculated using a lithographic simulation program PROLITH/2.

If the resist has a high optical absorption, the latent-image profile becomes trapezoidal because of the light intensity attenuation in the bulk of the film. On the other hand, if the resist layer is transparent, a strong standing-wave appears in the latent image because of thin-film interference. Both problems interfere with the precise reproduction of the aerial image. The ideal condition for precise reproduction of the aerial image is that the resist is transparent and the substrate reflection is minimal.

(3) *Modification of the latent image*: Resolution is significantly degraded if the standing-wave is retained in the latent image. This is because the development is blocked by the 'dark' node, where the dissolution rate becomes slow. The initially wavy profile is therefore converted to a smooth one by thermal diffusion of the molecules as described in Section 6.1.3.(5). Figure 6.8 shows an example of the latent image obtained after the smoothing bake. Figure 6.9 shows actual resist images processed with and without PEB.

The M-value difference, shown with the number of the contour lines, in a certain horizontal length represents the contrast of the latent image. The contrast, particularly at the mask edge, is correlated with the resolution of the developed image. The contrast generally decreases because of the diffusion that occurs during PEB, because it averages the M-value differences. A non-diffusion process coupled with a non-reflective substrate is consequently the best condition in terms of resolution.

When CA resists are used, the latent image is formed through an acid-catalyzed reaction wherein the acid species need to wander from

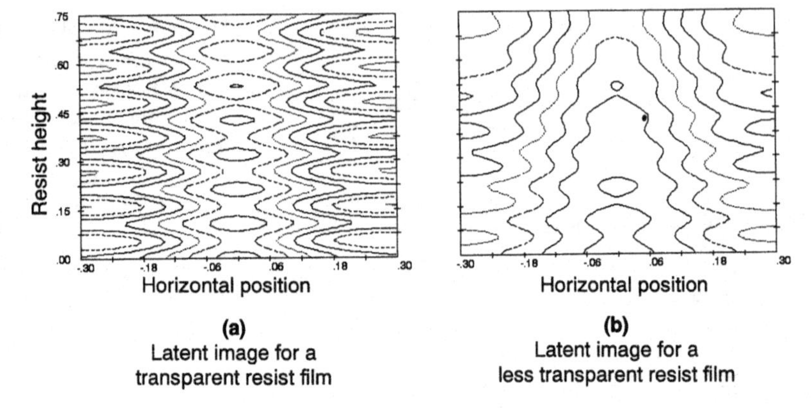

Figure 6.7 Examples of latent images drawn with *M*-value contours (calculated with PROLITH/2 resist simulator for i-line resists). Units are μm.

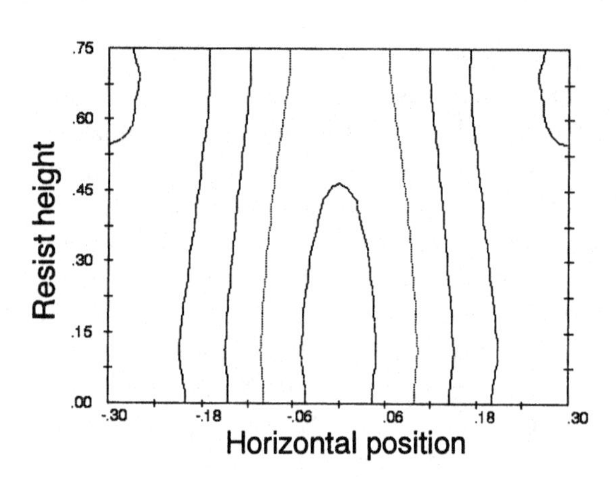

Figure 6.8 Latent image after PEB for a transparent resist film (the same resist as in Figure 6.7(a); diffusion length = 5 nm). The standing-wave is smoothed. Units are μm.

(a)
Resist image processed without PEB

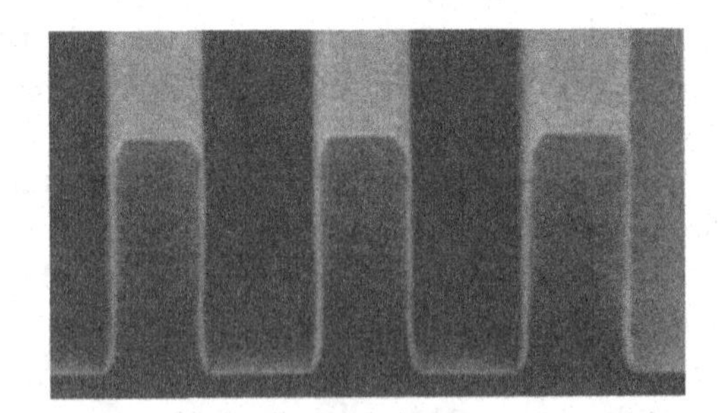

(b)
Resist image processed with PEB

Figure 6.9 SEM photographs of resist images of 0.35-μm L/S pattern replicated in 0.98-μm thickness i-line resist developed with and without PEB.

molecule to molecule. Molecular diffusion is thus critically involved, and the light intensity profile is thus not transformed directly into the latent-image profile. Diffusion is thus one of the key factors to influence resolution, particularly in CA resists.

(4) *Conversion to a real resist image*: The resolution of each resist material is determined mainly by this step. DQ/N resist is again used here for a detailed explanation. Figure 6.10 shows a typical plot of dissolution rate against *M*-value. The slope of this curve represents the dissolution contrast, which is the key factor in this step. During the development step, each small portion of the resist layer dissolves into the developer at a rate determined by its *M*-value. The dissolution starts from the surface of the layer and then propagates into the film to dissolve out the bulk, and a resist image is eventually developed as shown in Figure 6.11. In a simplified development model, the dissolution front propagates from top to bottom and then laterally to reach the mask-edge position. If the dissolution contrast is sufficiently high, the dissolution front propagates rapidly to the bottom and then to the mask edge and finally stops

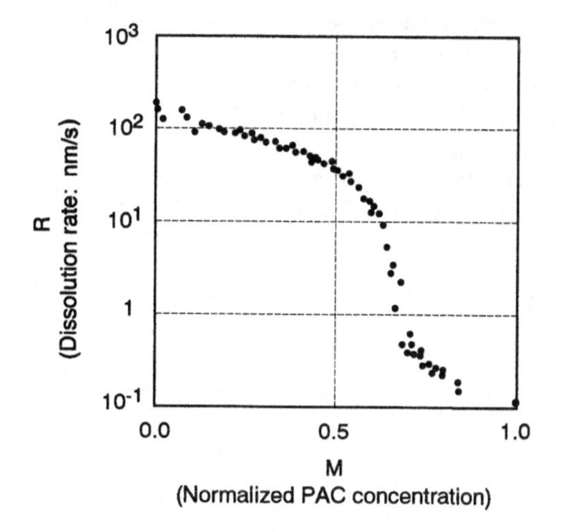

Figure 6.10 Dissolution-rate curve for DQ/N i-line resist.

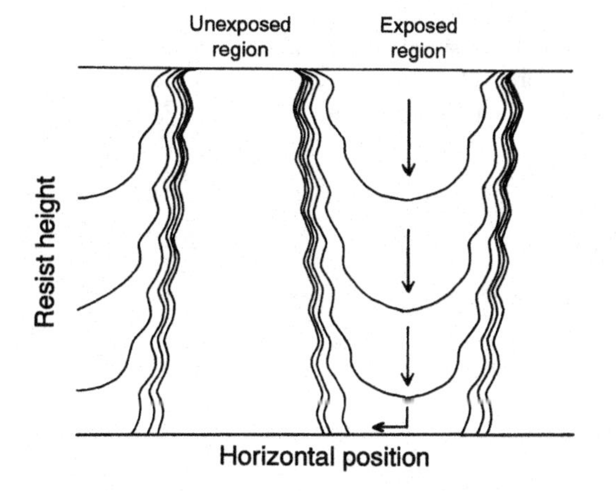

Figure 6.11 Progress of positive resist image development. Results from PROLITH/2 simulation: development times are 0, 10, 20, 30, 40, 50, and 60 s.

there because of the rapid decrease in the dissolution rate at the mask edge. A resist image with exact dimension as defined by the mask will thus be obtained. On going to a smaller geometry, the latent image is generally given a lower contrast (i.e., a smaller difference in M-value), and therefore yields a smaller differentiation in the dissolution rates. The dissolution front does not reach the mask edge in this case and therefore the resolution is lost. If a resist has a higher dissolution contrast, the small difference in M-value can yield a larger differentiation in the dissolution rates, and thereby the resist can develop a smaller image.

This can be clearly shown with a mathematically derived resist simulation. The two dissolution-rate curves in Figure 6.12 are for resists having (a) high dissolution contrast and (b) low dissolution contrast. The rate drops rapidly at a certain M-value for resist (a), whereas it changes gradually for resist (b). Figure 6.13 shows the results of the simulation for the two imaginary resists as defined with the dissolution-rate curves in Figure 6.12. The plots for the resist

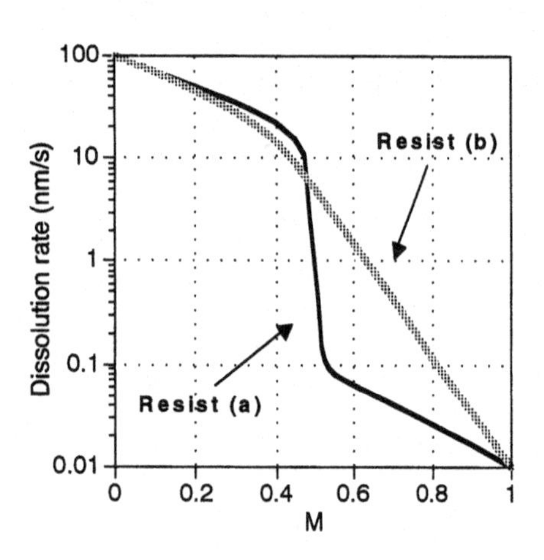

Figure 6.12 Dissolution-rate curves for two imaginary resists having different curve shapes. (a) A high dissolution contrast with rate switching. (b) A low dissolution contrast with monotonic rate change.

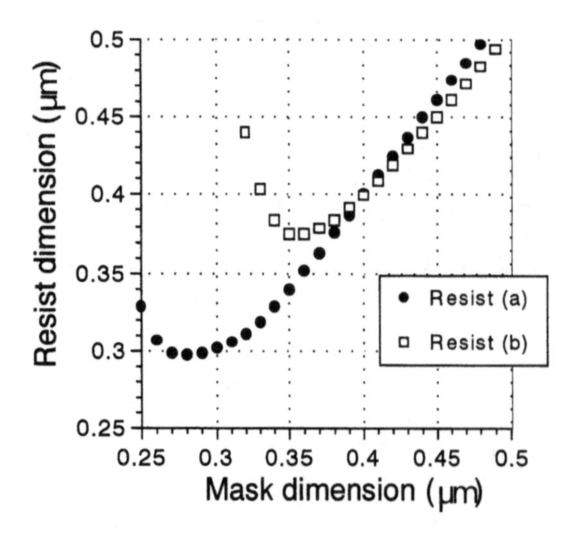

Figure 6.13 Mask linearity of the two imaginary resists in Figure 6.12. Results of PROLITH/2 simulation: i-line exposure, 0.57 NA ($\sigma = 0.6$) with normal illumination.

dimension vs mask dimension in Figure 6.13, which are termed 'mask linearity curves', are a good measure of resist resolution. The plots for both resists clearly show that resist (a) can print images of smaller dimension than resist (b); in other words, resist (a) gives a higher resolution than resist (b).

The dissolution-rate curve varies from resist to resist depending on molecular design, composition, and other chemical and physical parameters. The curve for a real resist is almost impossible to predict theoretically, but it can be measured. Many mathematical equations have been used in attempts to precisely fit the actual dissolution-rate data.[14] Recent high-resolution resists have very high dissolution contrast values. Figure 6.14 plots the resolution against the dissolution contrast of actual i-line resists to show their relation. The figure shows that a set of imaginary resists having different dissolution contrast also fits the same trend. It is seen that the resolution is improved dramatically by increasing the dissolution contrast.

To summarize this section, the optics and the resists are the factors that determine the resolution of optical lithography. The optics determines the contrast of the optical image, and the resist determines the contrast of both the latent image and the developed resist image. Assuming a certain contrast of the latent image, the key parameter determining the resolution is the contrast in the dissolution-rate curve. In non-optical lithography (like electron-beam or X-ray lithography), diffraction does not limit the contrast of the latent image, so high resolution can be achieved easily. Other mechanisms, however, can degrade the contrast of the latent image and prevent accurate transfer of the image into the resist. A resist with high dissolution contrast is therefore desirable even for the non-optical lithographies to compensate for such a degradation in the contrast of the latent image.

6.2 **Resists for i-line lithography**

6.2.1 Introduction

Almost all of the positive photoresists used for g-line and i-line lithography in the IC manufacturing industry are DQ/N resists, and the chemical structures of the major components in those resists are all very similar. Specifically, the photoactive compounds (PACs, hereafter) are

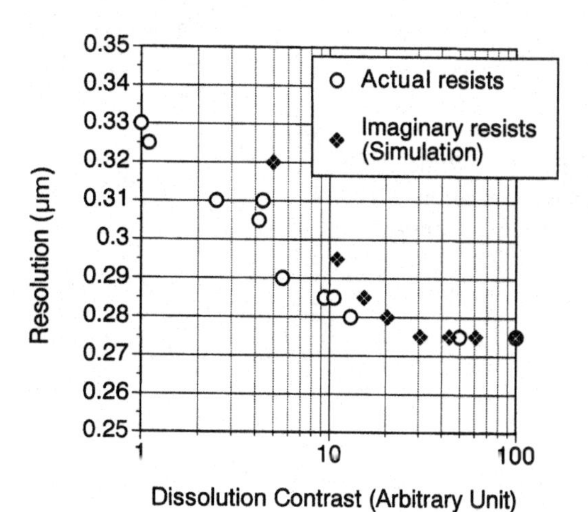

Figure 6.14 Resolution as a function of dissolution contrast for actual and imaginary resists (i-line exposure, 0.57 NA ($\sigma = 0.6$) with normal illumination).

1,2-naphthoquinonediazide-5-sulfonyl esters of aromatic polyhydroxy compounds, and the resins are novolac polymers made by polycondensation of phenol derivatives (e.g., cresols or xylenols condensed with formaldehyde). Although this basic composition has not been changed since the DQ/N resist was introduced in the early 1970s, their resolution capability has been improved significantly. This section describes the technology behind the improved performance of DQ/N resists and discusses their potential for use at the limits of i-line lithography.

6.2.2 Characteristics and features of DQ/N resist materials

6.2.2.1 *Baseline chemistry*

The photochemical reaction of the 1,2-naphthoquinonediazide (DNQ) group has been known since the 1940s. As shown in Figure 6.15, when the DNQ is irradiated it produces an unstable intermediate, a ketene, through the Wolf rearrangement reaction.[15] The ketene (indene in our example) immediately reacts with a water molecule in the resist matrix to form indenecarboxylic acid (ICA).[16] The lifetime of the ketene is normally in the order of microseconds and the quantum yield of the ICA formation from DNQ is about 0.2 to 0.3, which is fairly efficient.[17] Figure 6.16 explains how this reaction results in the formation of a positive image. The novolac resin itself is basically soluble in the alkaline developer, but the DNQ-PAC inhibits its dissolution from the resist matrix. When the resist composition is exposed to light, DNQ is converted into ICA and this inhibition is lost because of the high alkali solubility of ICA. In fact, ICA is so soluble in alkali, that the fully exposed resist dissolves faster than the unexposed resin.

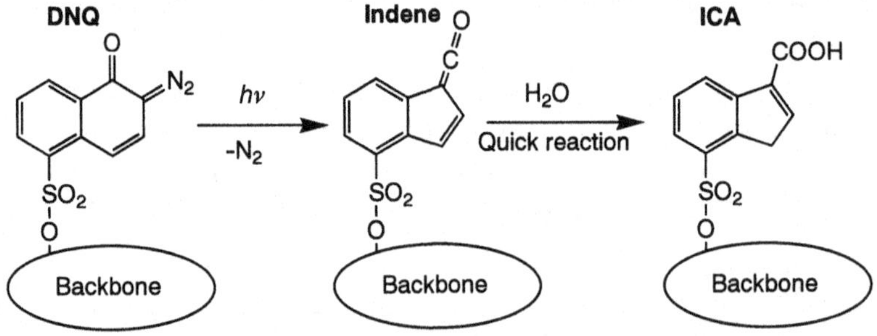

Figure 6.15 Photochemical reaction of the DNQ group.

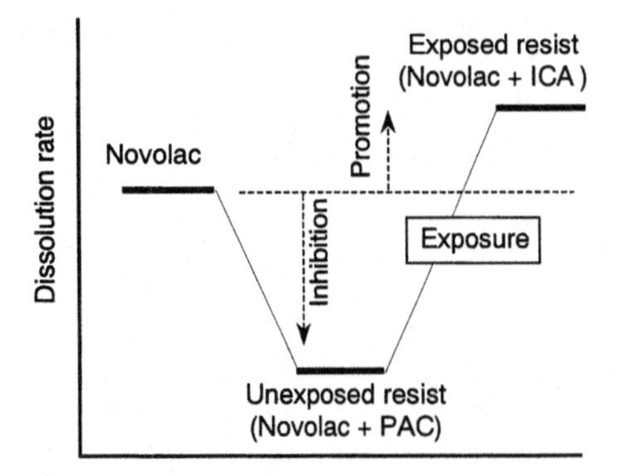

Figure 6.16 Principles of image formation of DQ/N positive photoresist.

The magnitude of the change in dissolution rate is a function of two factors. The major factor is the degree of dissolution-rate inhibition by the DNQ-PAC, and the other is the degree of dissolution-rate promotion by the ICA. The mechanism of dissolution inhibition has been studied extensively, and several molecular models have been proposed.[18,19,20,21,22] Figure 6.17 shows an example in which hydrogen bonding between the DNQ group and novolac OH groups creates a domain structure[19] wherein the hydrophilic sites of both molecules are oriented to the domain interior and the hydrophobic outer shell of the domain protects the domain from the aqueous developer. These structures sit in hydrophilic channels of the novolac, blocking water penetration and thus making the whole matrix insoluble in the developer.[23] A resist matrix structure like 'stonewall' may be formed between the PAC and resin,[22] which makes the dissolution inhibition efficient. Once the DNQ is converted to ICA, the hydrogen bonding is lost and the domain is easily dissolved. This is illustrated schematically in Figure 6.18. However, the dissolution of DQ/N resist cannot be so simply explained and many other models are proposed besides these for discussing the mechanisms.[24]

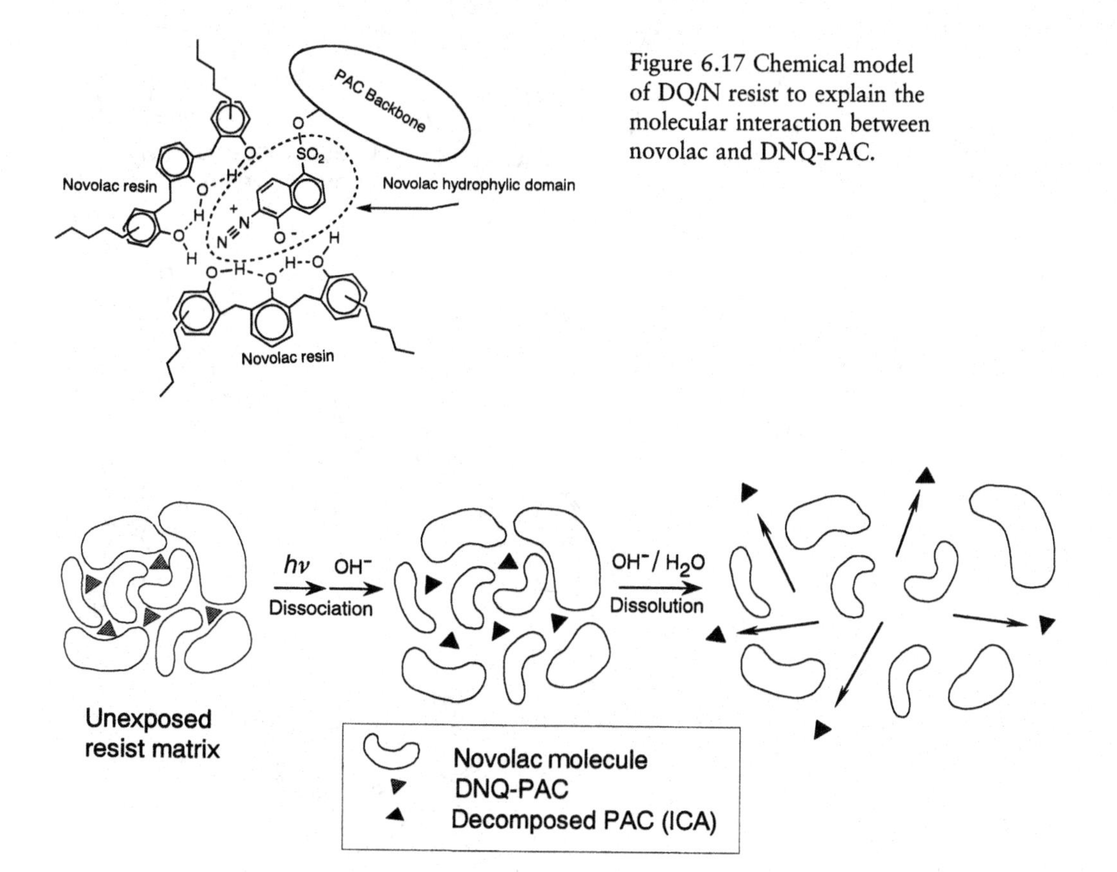

Figure 6.17 Chemical model of DQ/N resist to explain the molecular interaction between novolac and DNQ-PAC.

Figure 6.18 Physical model of DQ/N resist to explain the dissolution inhibition and promotion.

6.2.2.2 *Features of DQ/N resist*

(1) *Good resistance to dry-etching*: As described in Section 6.1.3(7), the resistance to dry-etch is primarily determined by the content of aromatic C—C bonds. Novolac resin has many such bonds and thus resists dry-etching. Note that the other components in DQ/N resist are also aromatic. Figure 6.19 shows typical examples of the type of compounds used in a DQ/N i-line resist.[25,26] Each compound consists of phenolic or similar aromatic groups modified with a very small number of aliphatic groups.
(2) *Aqueous development*: The resist can be developed using an aqueous alkaline developer, which is environmentally safer than one based on an organic solvent.
(3) *High dissolution contrast*: A high dissolution contrast is essential for high resolution, and the primary factor determining the contrast is the magnitude of the difference in dissolution rate. A significantly large difference can be

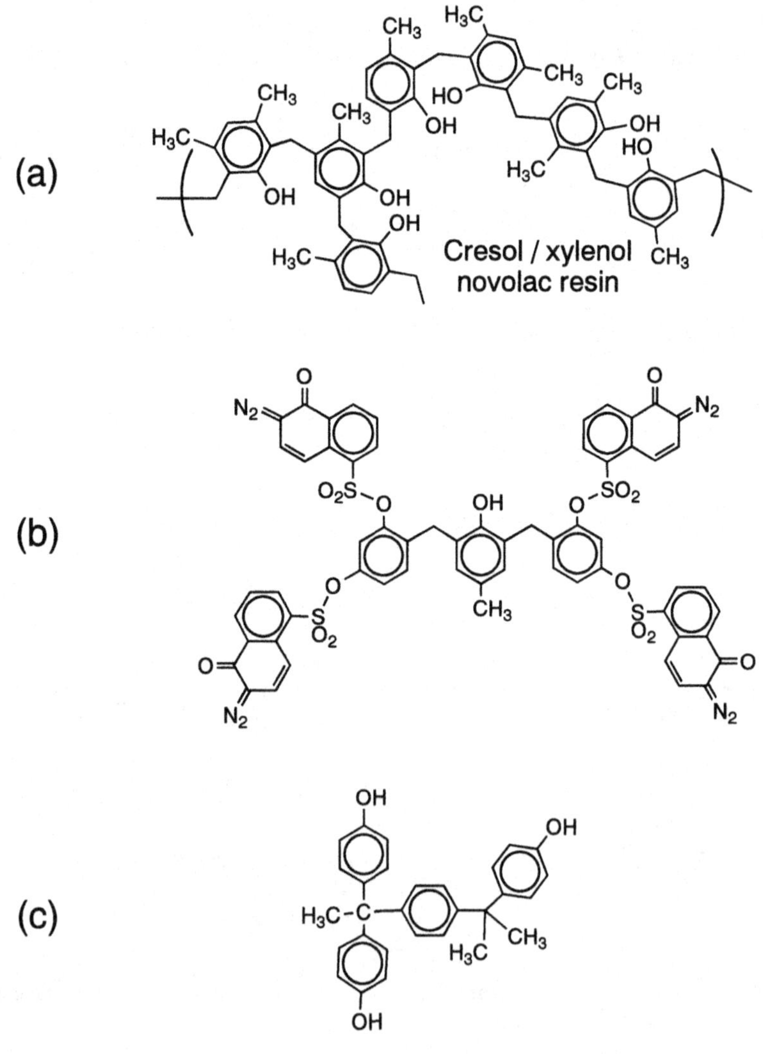

Figure 6.19 Chemical structures of some i-line resist components: (a) novolac resin, (b) DNQ-PAC, (c) additive dissolution promoter.

(a) Cresol / xylenol novolac resin

(b)

(c)

obtained in the DQ/N resist system by designing the molecular structures of the PAC and novolac to interact with each other efficiently. While dissolution rates differing by a factor of ten would be just sufficient for a relief image to develop, a factor of 10^4–10^5 can easily be obtained with DQ/N systems,[27] which therefore can be used in high-resolution lithography.

(4) *Simple imaging reaction*: The imaging reaction is completed during the exposure step because at room temperature the ketone intermediate reacts immediately with the water that is available in a sufficient amount from the resist matrix.[16] Therefore, given a certain quantum yield, there is a quantitative correlation between the number of photons absorbed and the extent of the DNQ-to-ICA conversion reaction that causes the change in dissolution rate. Process control for DQ/N resists is relatively easy because of this simplicity. This contrasts favorably with CA chemistries, which are more complex and for which the simple quantitative correlation does not apply. In CA resists, several parallel reactions, such a quenching of the photo-acid, take place during the imaging.[28] Furthermore, the imaging reaction is catalytic and therefore the degree of the reaction is determined not only by the amount of photo-acid but also by temperature, reaction time, and other conditions. The CA and DQ/N resist systems are compared schematically in Figure 6.20.

(5) *Additional reaction modes*: The DNQ and the novolac can react with each other in several different modes when a resist is treated with heat or alkali without exposing to light. These various reaction modes are used in the resist process to make it more sophisticated. Figure 6.21 shows the reaction modes feasible between DNQ and novolac.

Mode (a) is the normal imaging reaction induced by exposure to light. When the resist is exposed not to light but to alkali, however, cross-links may form through mode (b).[29] The reaction spontaneously takes place during the development and makes the dissolution contrast greater by making the unexposed portion less soluble.[29] The same reaction can also be used, prior to exposure to light, for hardening the resist surface to make the resist profile more nearly vertical.[30,31] The resist also can form cross-links without being exposed to light through mode (c)[32] when it is heated to a high

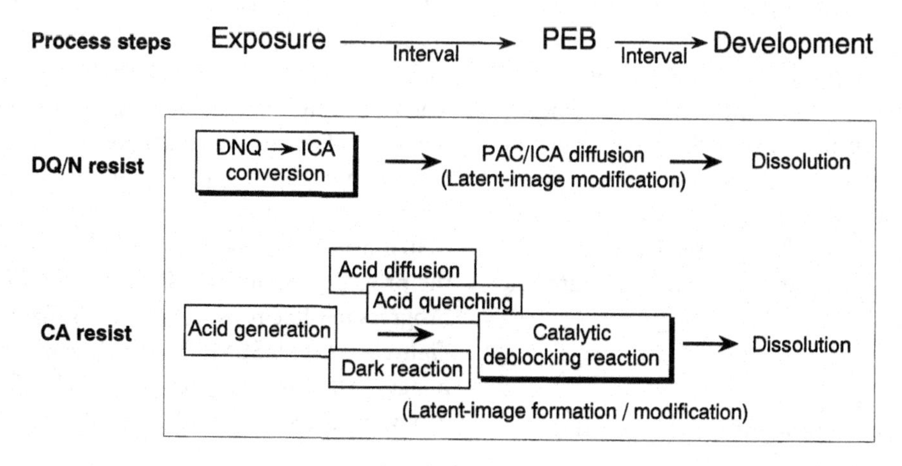

Figure 6.20 Chemical and physical events influencing the imaging of DQ/N and CA resists.

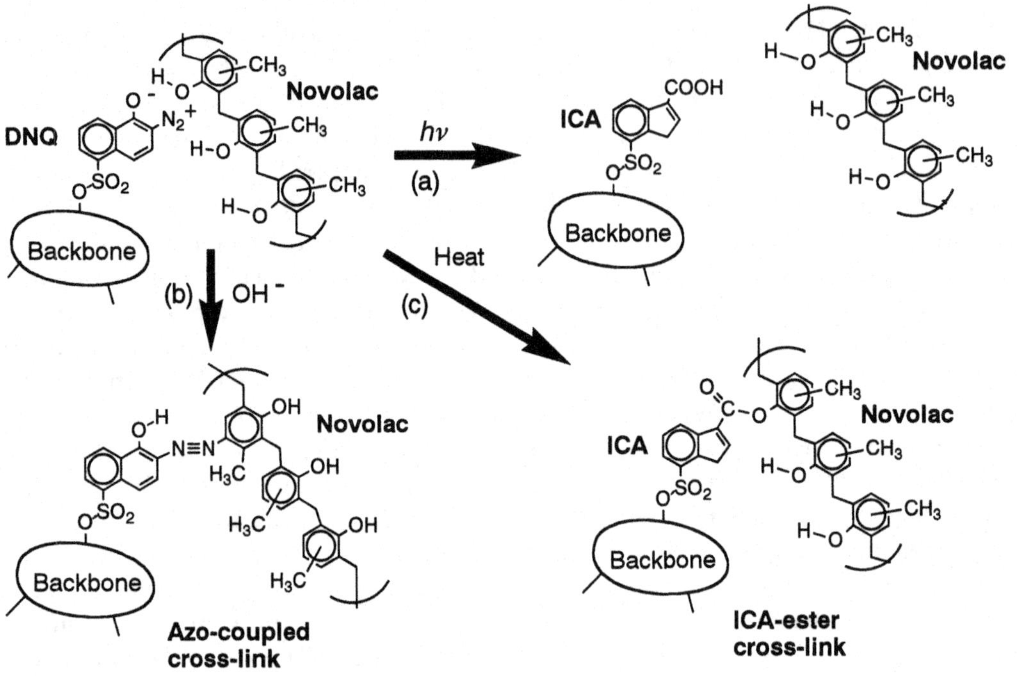

Figure 6.21 Various reaction modes feasible for DNQ group and novolac molecule exposed to (a) light, (b) alkali, and (c) heat.

temperature. This reaction mode is used for thermal curing.

(6) *Raw material availability*: A variety of phenol derivatives are commercially available at low cost because they are intermediates for many chemical products. As shown in Figure 6.19, the bases for each component are typically phenolic compounds. A small difference in the molecular structure (e.g., substituents, linkage, configuration) can dramatically change the dissolution rate and other physical properties of the resist and, hence, the lithographic performance.[33]

6.2.3 Progress in resist materials technology

The improvement in performance of g-line and i-line resists, both based on essentially the same DQ/N resist chemistry, has been substantial and has resulted in high-resolution resists. The improvement is continuing and the end is not clearly in sight. The resolution of some i-line resists is approaching a quarter-micron (0.25 μm) which is even below the image wavelength of 365 nm (0.365 μm). Although the minimum feature size that can be used in LSI devices has to be larger than this resolution limit, i-line lithography targeting below 0.30 μm has now become realistic based upon the progress in the DQ/N resist resolution. The progress in i-line resist resolution is illustrated in Figure 6.22 with SEM photographs of real resist images. The following key resist technology factors were the source of this improvement.

(1) *Increase in dissolution contrast*: The most influential factor on resist resolution is dissolu-

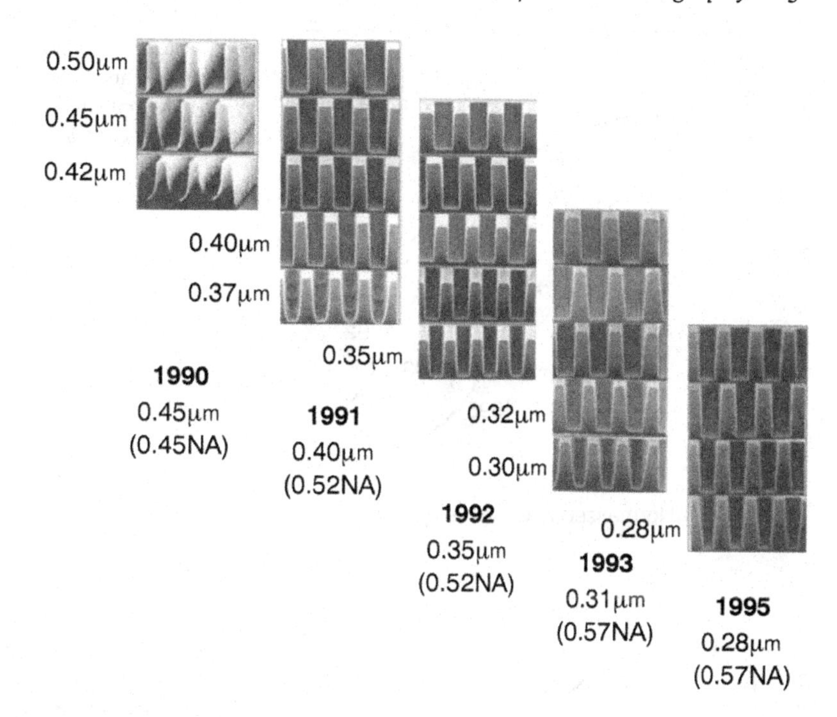

Figure 6.22 Cross-sectional SEM photographs of commercial i-line resists showing year-to-year improvement of resolution.

tion-rate contrast.[34] Figure 6.23 demonstrates the typical dissolution-rate curves of i-line resists, represented as *M–R* curves (see Subsection 6.1.4(4), Figure 6.10) to show the trend of this contrast increase over the first five years of the 1990s. Recent resists have a 'switching' dissolution-rate response, in which the rate increases very rapidly over a small range of *M*-value.[35] A very high dissolution-rate contrast can be obtained with this 'switching' response. The continuous increase in dissolution-rate contrast is the major factor that has contributed to the remarkable improvement in the resist resolution.

(2) *Use of unique surface dissolution properties*: The resist surface sometimes dissolves more slowly than the bulk resist during development. This phenomenon is called surface induction.[36] If the surface induction is optimized appropriately, we can obtain a more nearly vertical resist profile.[37] The optimized surface induction is also effective in increasing the resist's depth of focus (DOF) by suppressing the thickness loss at a defocused exposure. Figure 6. 24 shows an example of dissolution curves of a resist film that has surface induction.

(3) *Optimization of the optical absorption*: The optical absorption of a DQ/N resist is determined largely by the concentration of DNQ groups in the film. The magnitude of the absorption gives both positive and negative impacts to resist resolution. The absorption generally decreases the resolution by weakening the light intensity at the bottom of the film. On the other hand, it increases the resolution by suppressing the formation of a standing-wave. This impact has become more important because the standing-wave is now more pronounced than before due to the smaller geometries and stronger interference caused by the use of a thinner film on moving to sub-half-micron lithography. Another positive impact is the so

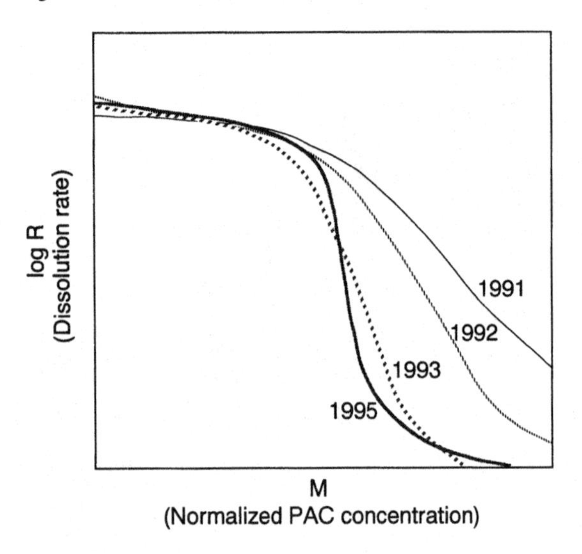

Figure 6.23 Dissolution-rate curves of i-line resists showing year-to-year increase in the dissolution contrast.

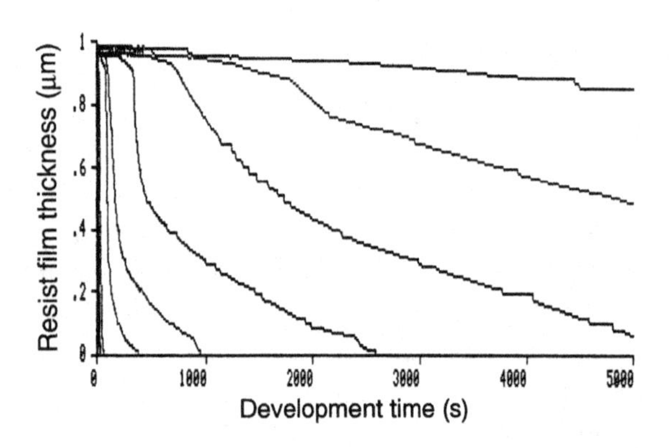

Figure 6.24 A resist film dissolution in development showing surface induction. Different curves represent the film exposed with different dose levels.

called 'Internal Contrast Enhancement Layer' effect.[38] The absorption by DNQ bleaches upon exposure. This bleaching yields an *in-situ* optical mask in the resist layer (Figure 6.25), thereby enhancing the optical image contrast at the bottom of the resist and resulting in higher resolution. A high optical absorption is therefore applied to a high-resolution resist in general. In application-specific resist developments, resist absorption has been optimized to meet the individual process requirements by taking all these different influences into account.

6.2.4 Towards the ultimate i-line resolution

Resolution of i-line resists has been improved year after year and no end is yet clearly in sight. Where is the theoretical limit? This question is not easy to answer, because resolution depends on many lithographic parameters other than resist material, i.e., optics and resist processes (e.g., resist thickness, reflection form substrate, and kinds of developer). Figure 6.26 illustrates the resolution limit predicted using a lithographic simulation that assumed an ideal resist and an ideal process but no special optical

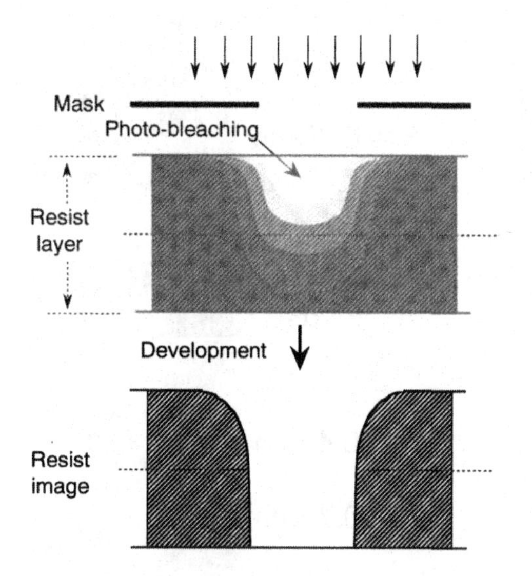

Figure 6.25 Schematic diagram explaining *in-situ* optical mask formation through photo-bleaching of DNQ in the resist layer. The mask increases the aerial image contrast at the bottom of the resist.

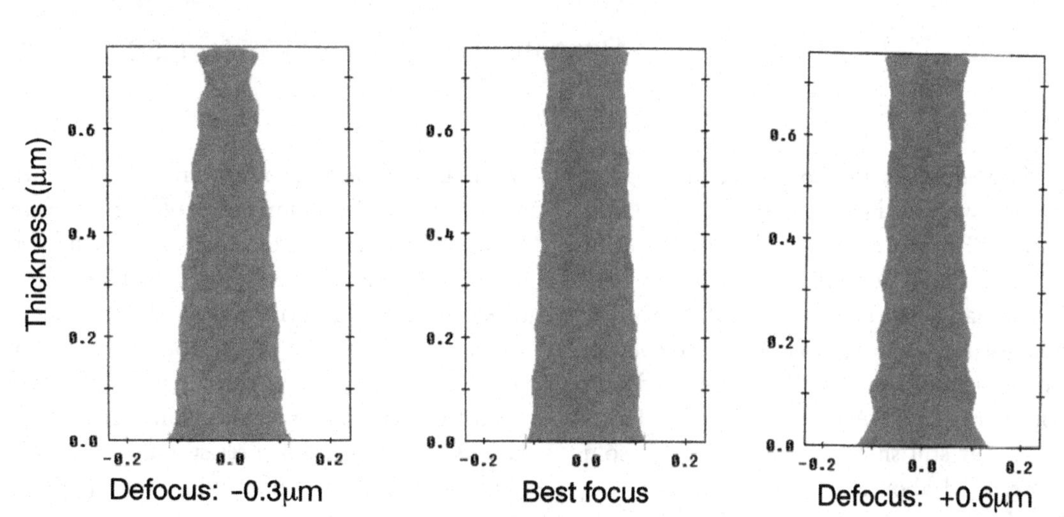

Figure 6.26 Simulated resist images showing the best resolution and DOF with i-line exposure: 0.24-μm-L&S patterns with 0.76-μm film thickness. A transparent resist imaged on a non-reflective substrate was assumed. Patterned with 0.57 NA ($\sigma = 0.6$) optics without using resolution-enhancement techniques.

tricks (e.g., oblique illumination or a phase-shift mask). An ideal resist is defined as one whose *M–R* curve shows a rectangular dissolution-rate response (infinite dissolution contrast). Other assumptions are: no substrate reflection, no PAC/ICA diffusion, and a sufficiently transparent resist. The results indicate that a 0.24-μm-L&S pattern can be resolved with ordinary 0.57 NA ($\sigma = 0.6$) optics. If the NA is increased to 0.7, resolution is further improved to nearly 0.2 μm. Of course those values give only the ultimate resolution and do not imply practical use.

Figure 6.27 shows actual examples of the resolution obtained using a commercially avail-

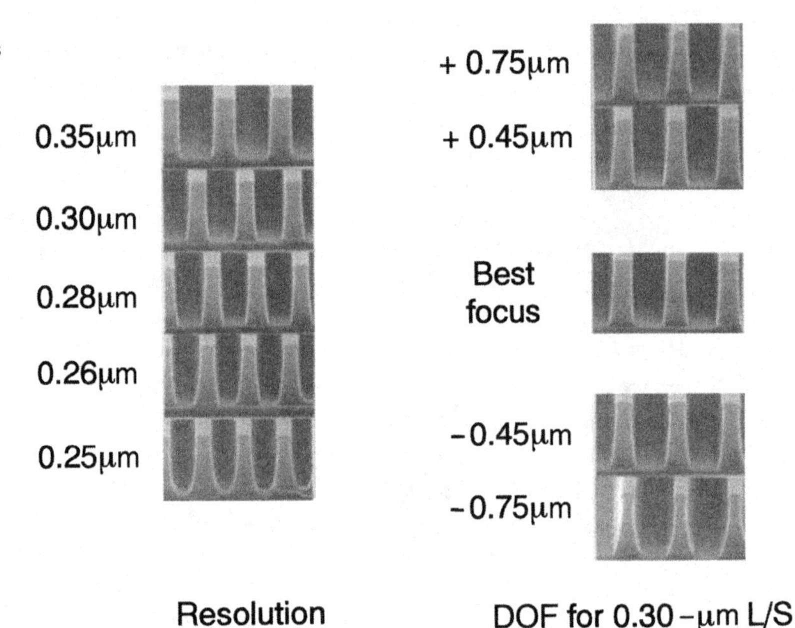

Figure 6.27 SEM photographs showing examples of resolution and DOF of a commercially available i-line resist. Imaged on a non-reflective substrate with 0.57 NA ($\sigma = 0.6$) optics without using resolution-enhancement techniques. Thickness was 0.86 μm.

able i-line resist. In this example, the imaging was made with an ideal situation, i.e., without using the PEB process (to avoid causing diffusion), and on a non-reflective substrate that can eliminate the negative influence of the standing-wave on resolution. Resolution down to 0.26 μm was actually obtained with the ordinary 0.57 NA ($\sigma = 0.6$) optics. This is close to, but still short of, the ultimate resolution predicted above.

Resolution is a good measure of other imaging performances (e.g., DOF and exposure latitude). However, increasing performance requirements go beyond simply resolution, particularly when optical tricks are applied to push the limit of i-line lithography. For example, the DOF for isolated lines becomes critical when annular illumination is used. Resistance to resist surface erosion becomes important to prevent the unfavorable sub-pattern formation that can be induced by side-lobe exposure when an attenuated PSM is used. Fast photospeed is another issue for newer resists because the optical tricks used to improve the resolution generally decrease the illuminance of the exposure tool. Resist technologies required for achieving these performance criteria are not identical to these required for resolution improvement, which has been the major objective in the past. Therefore, a new resist has to be developed to meet each criterion. Considering these factors, progress in i-line resist materials will be made not only by improving the materials themselves in resolution, but also by designing resists optimized for use with specific optical tricks as illustrated in Figure 6.28.

6.3 **Resists for deep-UV lithography**

6.3.1 Introduction

Deep-UV lithography using a Xe–Hg lamp as a light source was proposed in 1975.[39] Because of the difficulty of manufacturing projection lenses, however, an optical system equipped with a chromatic quartz lens and with a spectral narrowing excimer laser has generally been used. At present, deep-UV lithography essentially means excimer laser lithography. This section describes recent progress in resist material for KrF and ArF excimer laser exposures, that is, for exposure wavelengths of 248 nm and 193 nm, respectively.

6.3.2 Principle

6.3.2.1 *Transparency of polymer*

To obtain a high-resolution resist pattern with a steep profile, we need to use a resist film that is transparent. Figure 6.29 is a schematic diagram which explains the relation between transparency and resist profile. When the transparency of the resist film is low, the exposure dose at the

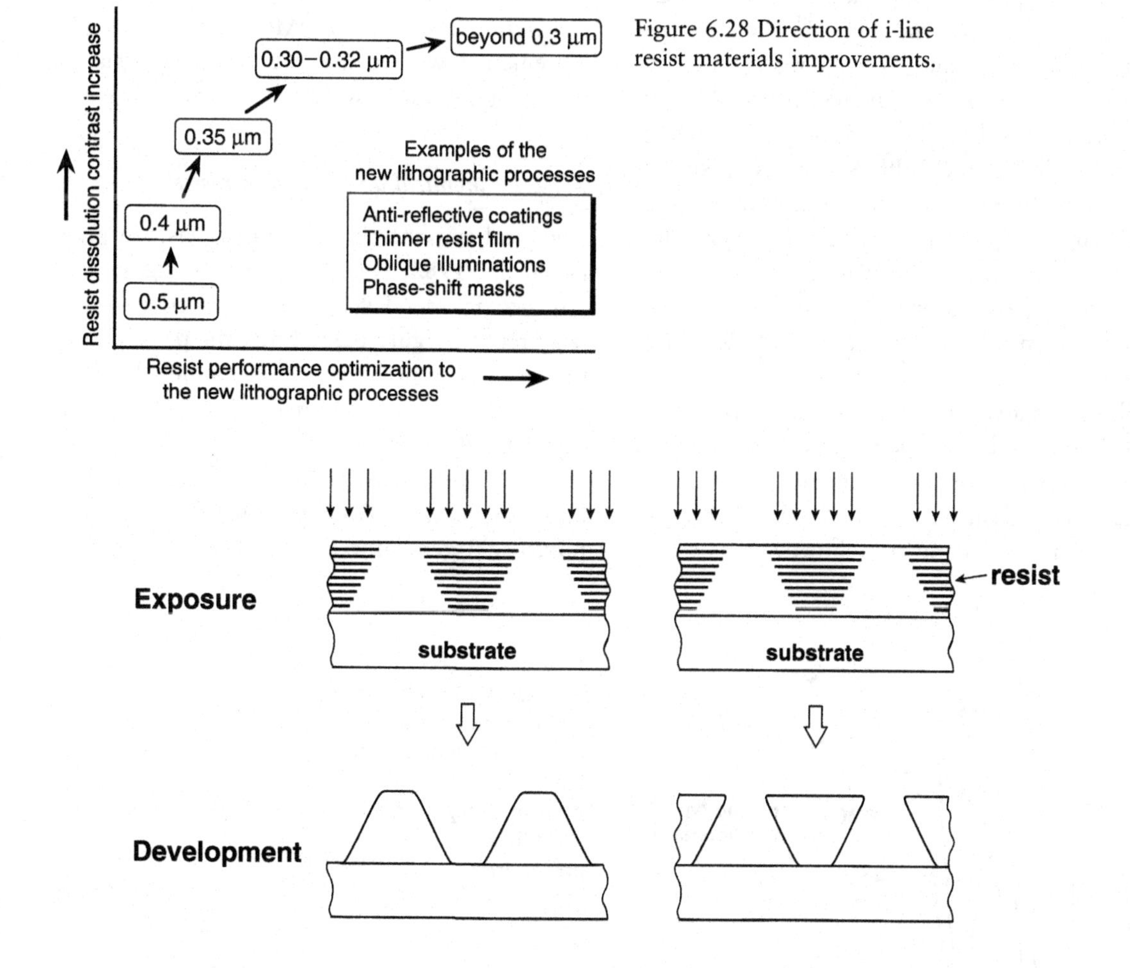

Figure 6.28 Direction of i-line resist materials improvements.

Figure 6.29 Schematic diagram of the relation between the transparency of a resist film and resist profile after development.

top of the resist is greater than it is at the bottom. Consequently, the surface region is overexposed and the substrate region is underexposed. The resist profiles after development are thus excessively narrow at the top (footing profile) for positive-tone resists and excessively narrow at the bottom (undercut profile) for negative-tone resists.

Figure 6.30 shows the transmission spectra of DQ/N resist, polymethylmethacrylate (PMMA), polydimethylglutarimide (PMGI) and polyhydroxystyrene (PHS).[40] The transmissivity, sensitivity, and dry-etching resistance[41] of three of these resist materials for 248-nm exposure, as well as the resist profiles obtained, are given in Table 6.2. PMMA has a sensitivity in the exposure wavelength of less than 300 nm, and thus is applicable to deep-UV exposure. PMGI has been developed to serve as an aqueous-developable deep-UV resist for a bilayer resist system.[42] These resists are of the main-chain-scission type, having no aromatic ring, and exhibit low absorption around 248 nm.

As described in the previous section, a DQ/N photoresist is a high-performance resist for g- and i-line exposures. Because of the absorption band of the aromatic ring, however, its transmissivity at 248 nm is inadequate: less than 10% at 1-μm thickness. A good resist profile thus cannot be obtained.

Since the profile of the DQ/N (novolac type) photoresist exhibits a triangular shape due to its high absorption, linewidth control on a substrate with varying topography is very difficult. PMGI and PMMA, by contrast, exhibit a good resist profile. This suggests that a transmissivity of more than 30% is necessary for obtaining a steep resist profile. However, the sensitivity and dry-etching resistance of PMGI and PMMA are extremely low, because they are main-chain-scission-type resists and have no aromatic ring.

6.3.2.2 *Chemical amplification resist*

Chemical amplification of a resist was first proposed in 1982,[43] and enables a significant advance in the development of KrF excimer laser resists. Figure 6.31 shows the principle of the positive-tone chemically amplified (CA) resist in comparison with that of the conventional DQ/N photoresist.

A CA resist consists of an alkaline-soluble base polymer, a dissolution inhibitor, and a photo-acid generator (PAG). The PAG gener-

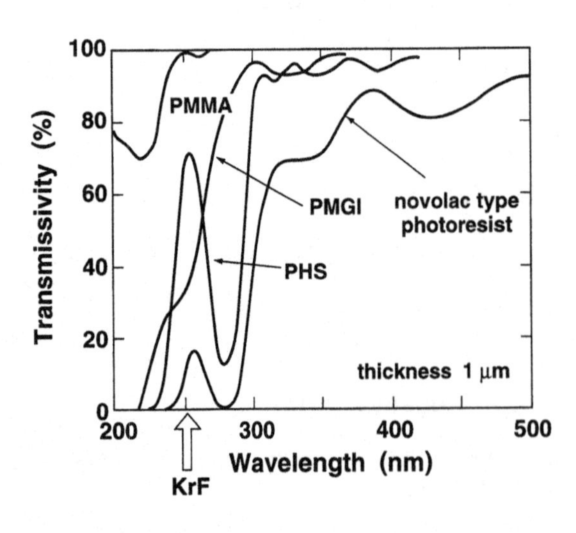

Figure 6.30 Transmission spectra of a DQ/N (novolac type) photoresist, polymethylmethacrylate (PMMA), polydimethylglutarimide (PMGI) and polyhydroxystyrene (PHS).[40]

Table 6.2 *Transmissivity at 1-μm thickness, sensitivity, dry-etching resistance and resist profile obtained by 248-nm exposure of DQ/N photoresist, polydimethylglutarimide (PMGI) and polymethylmethacrylate (PMMA), respectively.*

Chemical structure	Transmissivity (%)	Sensitivity (J/cm^2)	Dry-etching resistance (nm/min)	Resist profile (0.4-μm L/S)
novolac type photoresist	6	0.1	170	
PMGI (R:H, alkyl)	32	10	270	
PMMA	98	32	330	

The etching resistance is expressed as etching rate under the condition of CF_4, 0.6 Torr, 100 sccm, 500 W.

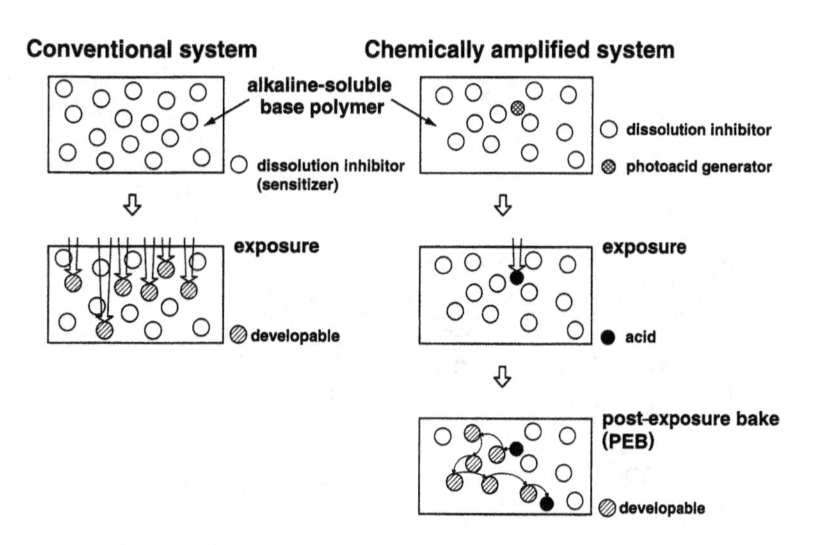

Figure 6.31 Pattern formation with conventional type (DQ/N photoresist) and CA resists.

ates an acid during exposure, and then a post-exposure bake (PEB) is carried out. During the PEB, the acid diffuses and catalytically decomposes the dissolution-inhibitor molecules one after another. This catalytic amplification increases the resist sensitivity beyond the limitation of the quantum yield.

6.3.3 KrF excimer resist

6.3.3.1 *Technical issues of resist design*

The technical issues of the KrF excimer laser resists are summarized as follows: (1) high transparency; (2) high sensitivity, (3) high dry-etching resistance, and (4) aqueous developable. In the early stage of KrF excimer laser stepper development, a high-power illumination was expected using the KrF excimer laser as a light source, but such a source has not been obtained, since the spectral narrowing of the excimer laser decreases the light power: light intensity at the image field of an excimer laser stepper is less than 50% of that of an i-line stepper. Therefore, a resist sensitivity better than 50 mJ/cm^2 is required to achieve the same level of throughput. Furthermore, an aqueous-developable resist is desirable for ease of handling and for safety.

To solve these technical issues, PHS was used as an alkaline-soluble base resin. It has a window of absorption (transmissivity peak) which just aligns with 248 nm as shown in Figure 6.30, and is resistant to dry-etching because it includes an aromatic ring. Additionally, a CA system was used in order to obtain high sensitivity.

6.3.3.2 *Positive-tone KrF excimer laser resist*

The most popular positive-tone KrF excimer laser resist is a CA resist based on the acid-catalyzed deprotection of tertiary-butoxycarbonyloxystyrene (tBOC-PHS). The resist chemistry is shown in Figure 6.32.[43] Before exposure, the phenol OH group of PHS is protected by the t-butoxycarbonyl (tBOC) group. Thus tBOC-PHS resist has no polarity and the base resin is insoluble in the polar solvent used as the alkaline developer such as tetramethylammonium hydroxide (TMAH) solution. An onium salt acts as a photo-acid generator and generates the acid during the exposure. The acid-cata-

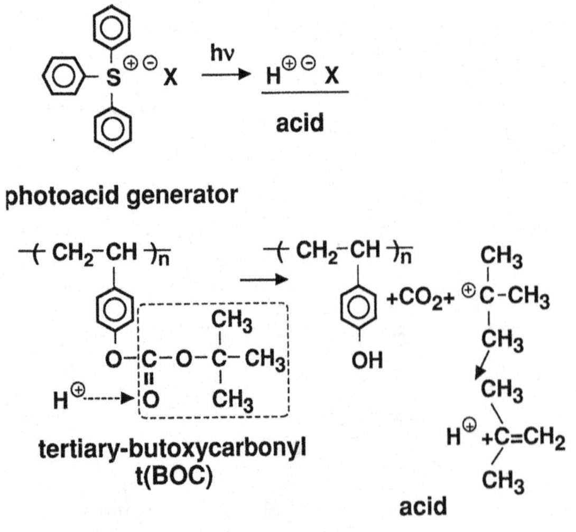

Figure 6.32 CA resist chemistry based on the acid-catalyzed deprotection of tertiary-butoxycarbonyloxystyrene (tBOC-PHS).

Table 6.3 *Photo-acid generators used for chemically amplified KrF excimer laser resists.*

Photoacid generator	Chemical structure	
onium salt	S^+ OTf^-	I^+ OTf^-
halogen compound	XH_2CXHCH_2C N O N CH_2CHXCH_2X O N O CH_2CHXCH_2X	X_3C N CX_3 N N CX_3
sulfonic acid ester	NO_2 CH_2OSO_2Ar ; O N_2 SO_2OR	OSO_2CH_3 OSO_2CH_3 OSO_2CH_3 ; $Ar - SO_3 - N = R$
sulfonyl compound	$Ar - SO_2 - SO_2 - Ar$	$Ar - SO_2 - \overset{N_2}{\overset{\parallel}{C}} - SO_2 - Ar$

lyzed reaction induces chain reactions and deprotects tBOC during the PEB process. As a result, the phenol OH groups are generated in the exposed area and can be dissolved in the alkaline developer. Since the system takes advantage of the polarity difference between the exposed and unexposed areas, a negative pattern can be obtained when a non-polar organic solvent is used as the developer.

The photo-acid generator is an important factor in the designing of CA resists, where a higher quantum yield of acid generation is desired so that a higher sensitivity can be obtained. The principal acid generators that have been reported are listed and shown in Table 6.3. The ones that are generally used are onium salts, such as triphenylsulfonium triflate (TPS-OTf) and diphenyliodonium triflate.[44]

The protection groups that have been reported are shown in Figure 6.33. A resist consisting of PHS partially protected by trimethylsilyl group can be used in a bilayer resist because it contains Si. Most of today's commercially available resists use PHS partially protected with 1-ethoxyethyl group. Since 100% protected PHS shows poor developability and an insoluble layer at the surface is likely to

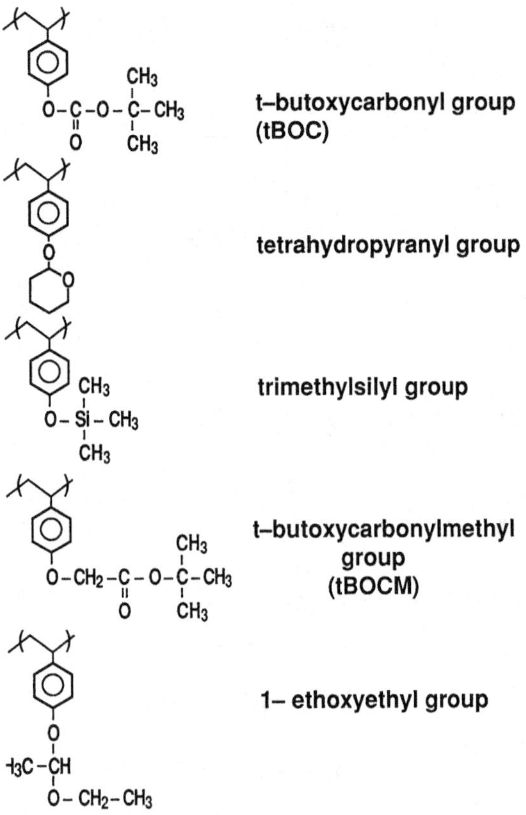

t–butoxycarbonyl group (tBOC)

tetrahydropyranyl group

trimethylsilyl group

t–butoxycarbonylmethyl group (tBOCM)

1– ethoxyethyl group

Figure 6.33 Protection groups used for positive-tone CA KrF excimer laser resists.

arise, the degree of protection has been optimized to be about 20%.[45]

Several concepts have been used in designing high-resolution resists. One is to increase the dissolution-rate contrast (difference) between exposed and unexposed areas described in Subsection 6.2.2.2. As illustrated schematically in Figure 6.34, the dissolution rate of partially protected PHS is lower than that of the original PHS. After exposure and PEB, the protection group is decomposed and the dissolution rate increases.

When tBOC is used as the protection group, tBOC-PHS only returns to the original PHS by the deprotection reaction (dashed line in Figure 6.34). In this case, the dissolution rate of the exposed area is equal to that of PHS. When the PHS is, instead, partially protected with t-butoxycarbonylmethyl group (tBOCM-PHS), the deprotection reaction generates a carboxyl group. Since the polarity of the carboxyl group is higher than that of the hydroxyl group, the dissolution rate of polyhydroxycarbonylmethyl-oxystyrene is greater than that of the original PHS. As a result, higher dissolution-rate contrast can be obtained (solid line). Figure 6.35 shows the dissolution rate as a function of PEB temperature for tBOC-PHS and tBOCM-PHS.[46] It is evident that the dissolution-rate contrast of tBOCM-PHS is higher than that of tBOC-PHS.

Figure 6.36 shows examples of resist patterns obtained by a 0.5-NA excimer laser stepper

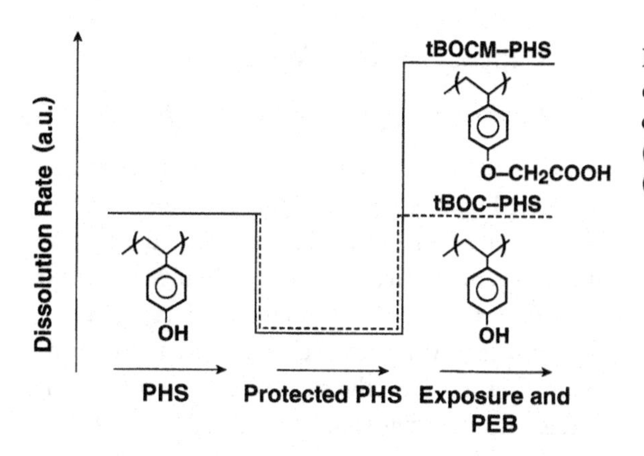

Figure 6.34 Schematic diagram of dissolution-rate characteristics for tBOC-PHS (dotted line) and tBOCM-PHS (solid line).

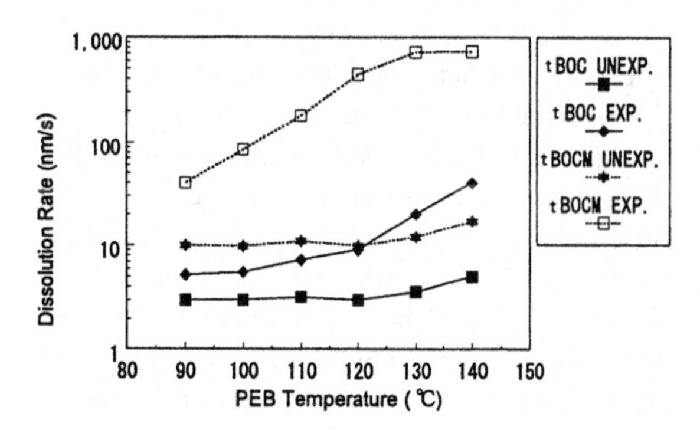

Figure 6.35 Dissolution rate vs PEB temperature for exposed and unexposed areas of tBOC-PHS and tBOCM-PHS. Protection degree is 10%, PAG is TPS-OTf of 0.5 wt%, exposure dose is 50 mJ/cm^2 and developer is 0.27N TMAH solution.[46]

Figure 6.36 SEM photographs of resist patterns obtained by positive-tone KrF excimer laser resist consisting of tBOCM-PHS and TPS-OTf.[46] Exposure dose is 38 mJ/cm^2, resist thickness is 0.85 μm and PEB is 90 s at 98 °C.

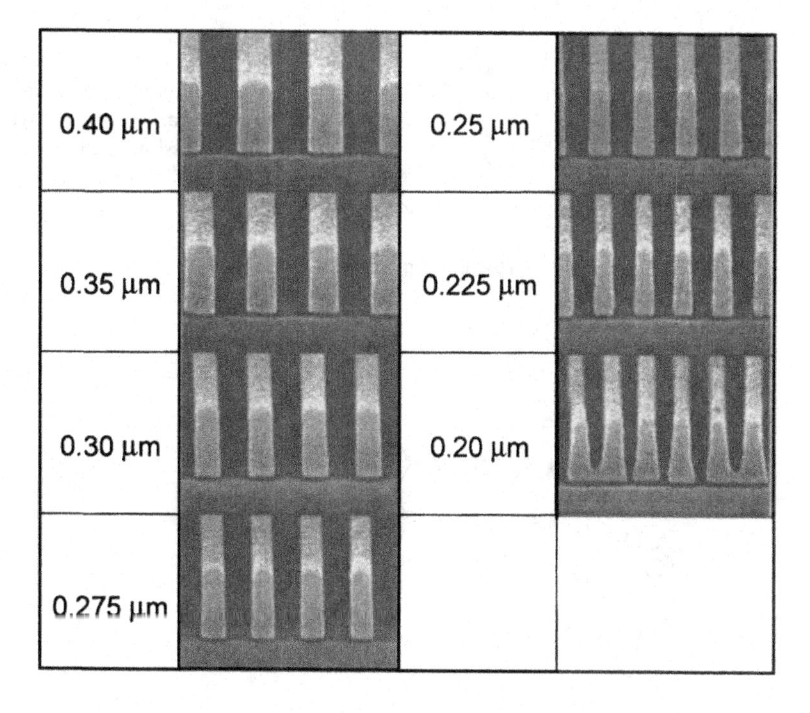

with a positive-tone CA resist consisting of tBOCM-PHS and TPS-OTf.[46] A 0.225-μm line-and-space (L/S) pattern is resolved with a steep profile.

6.3.3.3 *Negative-tone KrF excimer laser resist*

Negative-tone KrF excimer laser resists consist of a photo-acid generator, a crosslinking agent and PHS. Although many crosslinking agents for the phenol polymer have been reported, the one usually used is hexamethoxymethylmelamine. The crosslinking mechanism of a well-known commercially available negative-tone resist is shown in Figure 6.37.[47] As a result of the catalytic crosslinking reaction occuring during the PEB process, the exposed area becomes insoluble to the alkaline developer. Reduction of the phenol OH group of PHS by etherification also contributes to the insolubility.

6.3.4 Physics and chemistry of CA resists

6.3.4.1 *Behavior of acid during PEB*

Acid behavior during PEB is shown schematically in Figure 6.38. (a) Photogenerated acid at the resist surface may be deactivated (neutralized) by basic airborne contaminants. (b) Acid at the resist–substrate boundary may also be deactivated by hydroxyl group or basic contaminants absorbed on the substrate. (c) Acid evaporates into the atmosphere. (d) Acid at the resist–substrate boundary may also diffuse to the substrate and be trapped. These phenomena lead to the loss of the acid and formation of a thin soluble or insoluble layer within the resist film.

As a result, the resist profile abnormalities like those shown in Figure 6.39 occur. A thin insoluble layer is formed at the surface of a positive-tone CA resist and results in a T-top resist profile (a). When a negative-tone CA resist is spin-coated onto a SOG (spin-on glass) or an aluminium substrate, an undercut at the resist–

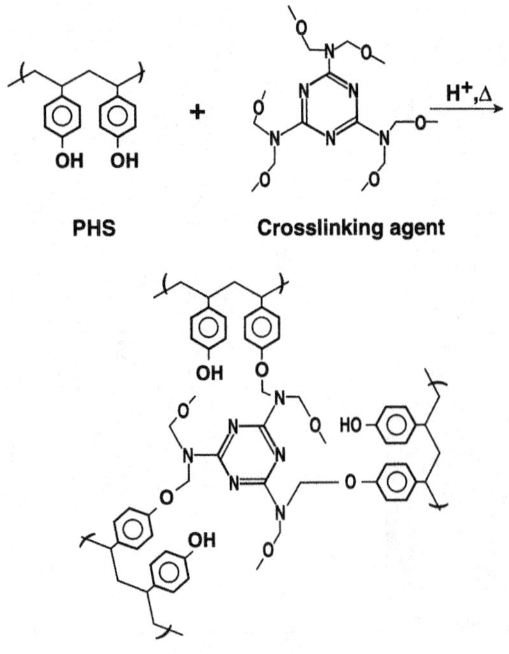

Figure 6.37 Crosslinking mechanism of PHS by hexamethoxymethylmelamine crosslinker in negative-tone CA KrF excimer laser resist.

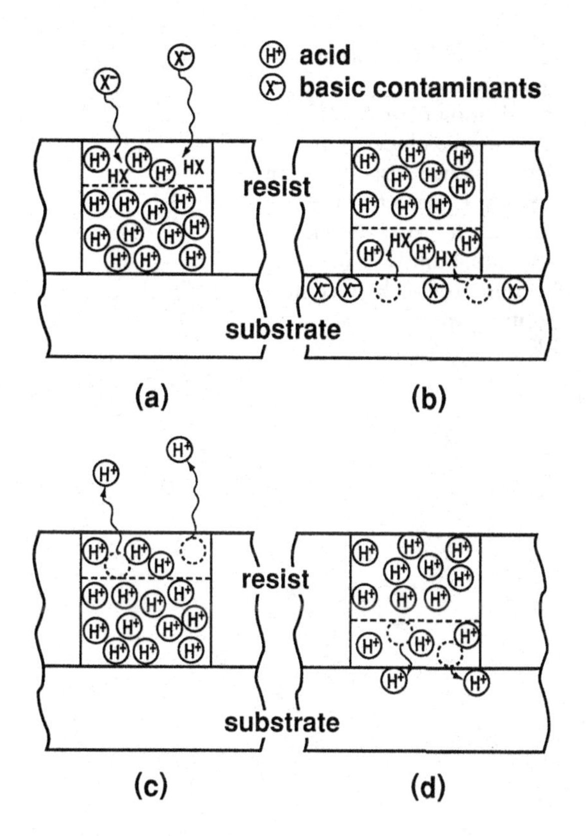

Figure 6.38 Schematic representation of acid behavior during PEB. (a) Deactivation of acid by basic airborne contaminants. (b) Deactivation by basic contaminants absorbed on the substate. (c) Evaporation. (d) Diffusion to the substrate.

substrate boundary occurs because of the thin soluble-layer formation at the boundary (b). In contrast, a positive-tone resist on BPSG (boron–phosphorus silicate glass), Al and SiN exhibits a footing profile because of the thin insoluble-layer formation (c). Several approaches to preventing these abnormalities have been proposed: (1) use an overcoating to protect the resist surface;[48] (2) filter the air to remove the basic contaminants;[49] (3) use an additive that counteracts the basic contaminants.[50,51,52] and (4) use an appropriate protection group which is easily decomposed at room temperature.[53]

6.3.4.2 *Diffusion of acid*

The diffusion of acid during PEB determines the resolution, sensitivity and resist profile. Many investigators have examined the diffusion range of an acid and have proposed several methods for measuring it.[54,55,56] For example, the diffusion range can be estimated by measuring the extent of the region within which the dissolution rate of the resist is changed after PEB, as shown in Figure 6.40.[57] A water-soluble polymer containing PAG is spin-coated onto a resist without PAG. Photogenerated acid in the water-soluble polymer diffuses into the resist during PEB and changes the dissolution rate of the exposed resist material, the extent of which can be measured by a development-rate monitor (DRM). The diffusion range of the acid can be defined by the depth to which the dissolution rate has been changed.

Figure 6.41 shows the diffusion range (filled circles) and the amount of solvent (ethyl cello-

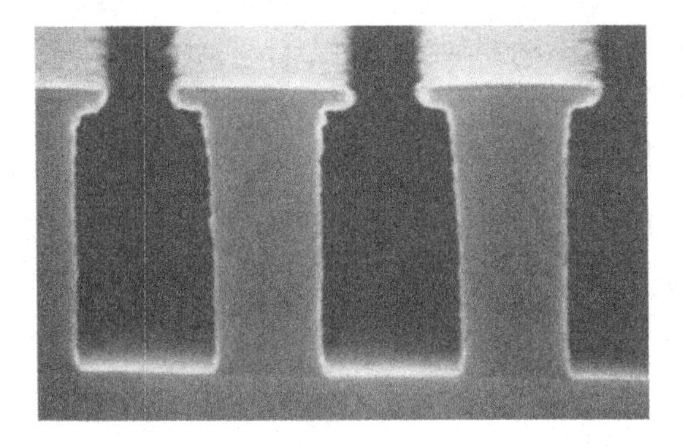

(a)

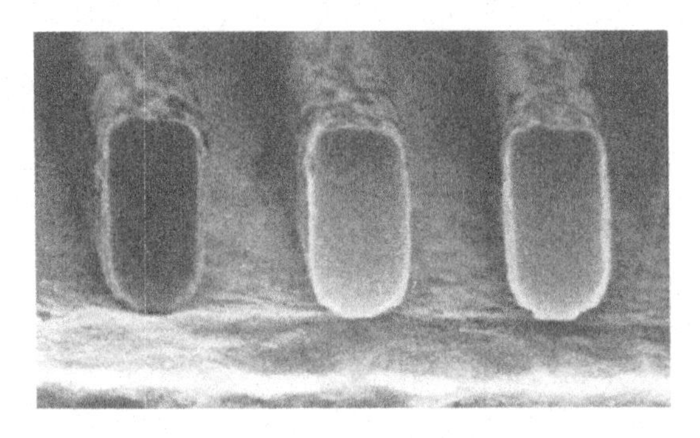

(b)

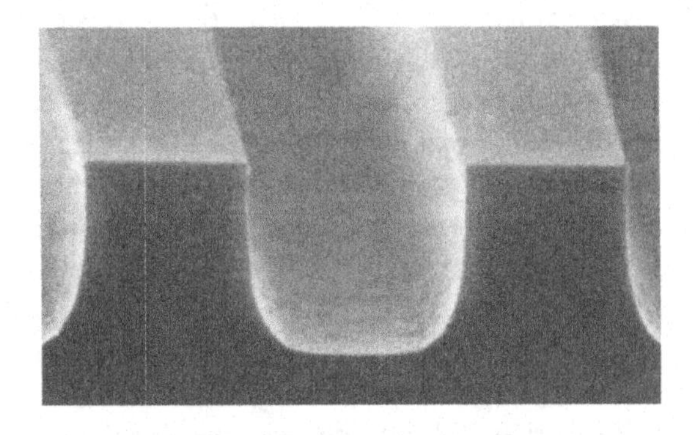

(c)

Figure 6.39 Resist profile abnormality due to deactivation of acid. (a) T-top resist profile observed in positive-tone resists (0.35-μm L/S). (b) Undercut profile observed in negative-tone resists (0.4-μm L/S). (c) Footing profile observed in positive-tone resists (0.8-μm L/S).

solve acetate) remaining after pre-bake (PB) (open circles) vs PB temperature. The data plotted here were obtained when tBOCM-PHS and TPS-OTf were used as components of a model resist.[57] The diffusion range changes markedly with changes in PB temperature below 110 °C but has a constant value at temperatures above 110 °C. This corresponds clo-

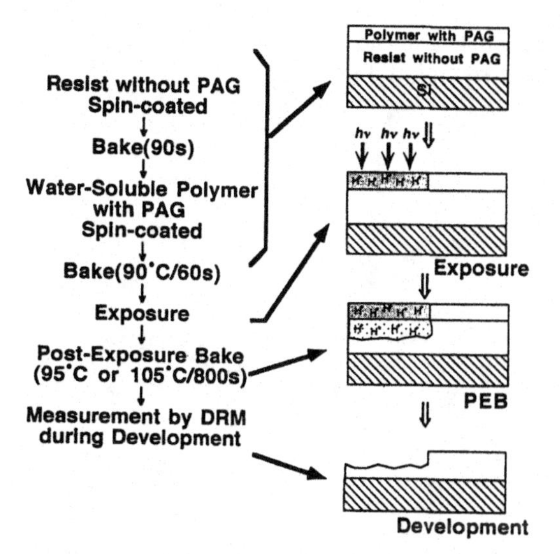

Figure 6.40 Measurement of diffusion range of acid.[57]

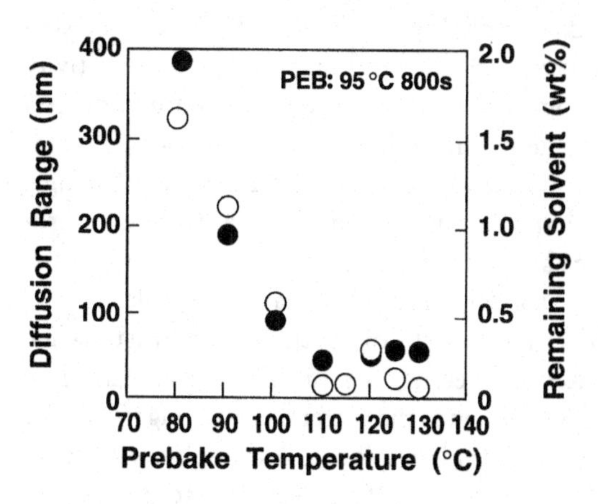

Figure 6.41 Diffusion range of acid, and remaining solvent, vs PB temperature.[57] Filled circles show the diffusion range of acid during PEB at 95 °C for 800 s, where PB time is 90 s. Open circles show the solvent remaining after PB for 90 s.

sely to the change in the amount of remaining solvent, and these results suggest that the acid diffuses through a solvent path and an open space derived from the free volume of the matrix polymer. The solvent makes open spaces in the polymer matrix and the polymer has a proper free volume, which determines the constant value of the diffusion range above 110 °C. These results can be well explained if we assume that the open space formed by the solvent is proportional to the concentration of the solvent and that the free volume does not depend on the solvent concentration.

Figure 6.42 shows the diffusion range plotted again the square root of PEB time, when the PB has been carried out at 95 °C for 90 s and the PEB temperature was 85 °C.[56] Since the diffusion range is proportional to the square root of PEB time, acid diffuses according to the diffusion equation. The diffusion range d_R is thus given by

$$d_R = 2\sqrt{Dt}, \qquad (6.2)$$

where D is the diffusion coefficient and t is PEB time. The diffusion range of acid during PEB

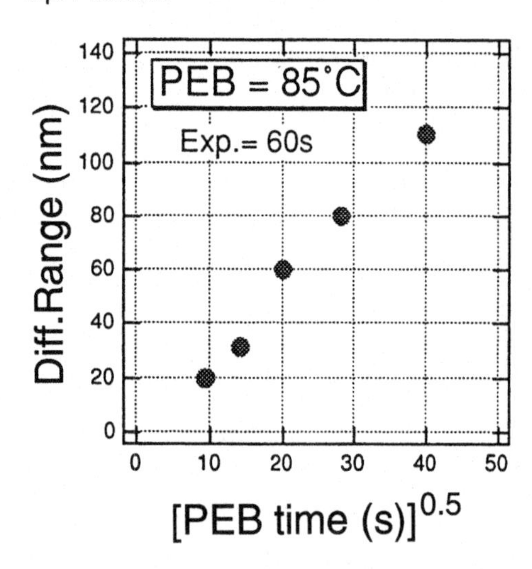

Figure 6.42 Diffusion range vs square root of PEB time.[56]

condition at 95 °C for 100 s is estimated to be about 25 nm.[56]

6.3.4.3 *Deactivation of acid by basic airborne contaminants*

If the resist-coated wafers are left to stand for a while before exposure or PEB, the resist performance is changed. This was a serious problem in the early stages of developing a CA resist. For a given exposure dose, the longer the coated wafer is left standing before PEB, the more serious are the phenomena occurring, such as image-size changes, T-top profile or footing profile (see Figure 6.39). This is called the post-exposure delay (PED) effect. A dynamic test has been performed to examine quantitatively the PED mechanism.[49] It revealed that the PED effect is caused by the photogenerated acid being neutralized by airborne basic organic contaminants present in amounts on the order of 15 ppb. Furthermore, paints, adhesive and sealants commonly used in a clean-room are sources of volatile amine contaminants.

Figure 6.43 shows the PED effect measured when a tBOC-PHS/iodonium-salt CA resist was exposed to an electron beam in the presence of various kinds of airborne contaminants.[58] It is evident that basic contaminants such as aniline and *N*-methylpyrrolidine (NMP) cause the sensitivity to change and that there is no sensitivity change when the exposed resists are kept in a nitrogen atmosphere (or clean air). Water (humidity) also causes the sensitivity to change, perhaps because, when no water is present, protons are transferred to the base polymer or casting solvent (in this case diethyleneglycol dimethyl ether), or both. In contrast, when water is present, the proton is transferred to the water, since the water has a stronger basicity than the casting solvent. Although the transfer of a proton to tBOC is necessary to decompose the tBOC group, protonated water is less acidic than protonated casting solvents. Thus, the efficiency of tBOC decomposition decreases.[59]

The thickness of the insoluble layer and the content of airborne contaminants absorbed in the resist film are approximately proportional to the square root of the delay time.[59] Thus, the movement of the basic contaminants is also governed by Fick's law, whereby the concentration:

$$C(x,t') = C_0\left\{1 - \mathrm{erf}\left[\frac{x}{2\sqrt{Dt'}}\right]\right\}, \qquad (6.3)$$

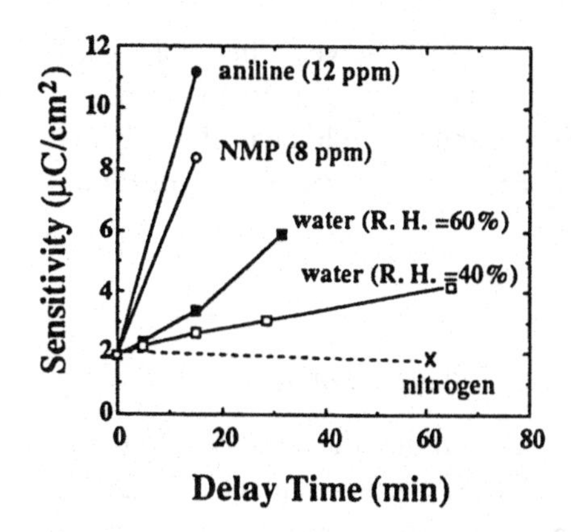

Figure 6.43 Relation between post-exposure delay time and sensitivity of exposed resist. Effect of several kinds of airborne contaminant.[58]

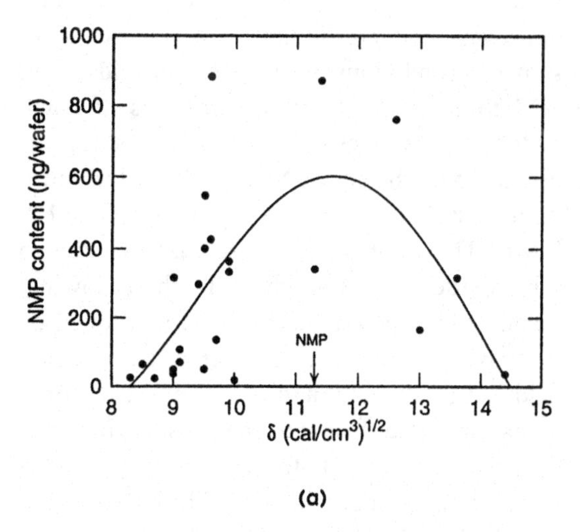

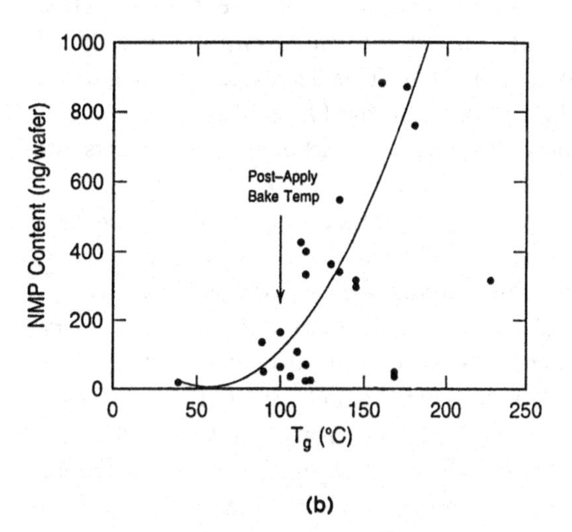

Figure 6.44 Contaminant (NMP) absorption on various polymers.[60] (a) Amount of NMP as a function of solubility parameter δ of the polymers. Solid line represents a fourth-order polynominal fit to the data. (b) Amount of NMP as a function of the glass transition temperature T_g of the polymers. Solid line represents a second-order polynomial fit to the data.

where C_0 is the concentration of basic contaminants, x is the depth of the resist, t' is PED time.

6.3.4.4 *Contaminant absorption and annealing theory*

Contaminant absorption in the resist surface has been examined by a radiochemical technique. NMP labeled with the radioactive isotope ^{14}C at the methyl carbon was used to ascertain the amount of absorption.[60]

Figure 6.44 shows the amount of NMP absorbed under equivalent conditions by various polymers as a function of both the solubility parameter δ of the polymer and the glass transition temperature T_g of the polymer, when all polymers had been pre-baked at 100 °C for 5 min.[61] Although there is much scatter, the equilibrium solubility of NMP in each polymer is characterized by its solubility parameter. Maximum NMP absorption occurs when the solubility parameter of the polymer is near to that of NMP. This result means that the solubility parameter is one factor influencing NMP absorption. The free volume in a polymer matrix strongly influences the diffusibility of small molecules, and the free volume can be reduced by annealing spin-coated film at a temperature above T_g of the polymer. The amount of NMP absorption increases with increasing T_g due to the excess of the free volume. These results suggest that the absorption of airborne contaminants can be decreased by using a low T_g polymer and that PB temperatures above T_g prevent diffusion of airborne contaminants into the bulk polymer.

An environmentally stable CA positive-tone resist (ESCAP) based on this annealing theory has been developed. It consists of a base copolymer of 4-hydroxystyrene with t-butyl acrylate P(HS-tBuA), instead of the tBOC-PHS, as shown in Figure 6.45.[62] tBOC-PHS undergoes thermal deprotection at 130 °C and its T_g is 180 °C. That is, its T_g is higher than the deprotection temperature. Therefore, the PB temperature cannot be raised above the T_g used for annealing. In contrast, P(HS-tBuA) undergoes thermal deprotection at 180 °C and its T_g is about 150 °C. The ESCAP polymer thus can be safely heated above its T_g of 150 °C without thermal deprotection. A dramatic reduction of NMP uptake – down to 1.8 ng/min (cf. tBOC-PHS, 18 ng/min) – is observed when the ESCAP is baked above its T_g. This reduction is due to the reduction of the free volume in the polymer. As a result, ESCAP is extremely insensitive to the PED effect.[62]

6.3.4.5 *Effect of basic additive*

It has been reported that the addition of organic base such as NMP, *o*-aminobenzoic acid, 2-benzylpyridine and diphenylamine to the CA resist reduces the severity of the PED effect and improves the resolution.[51,52] Figure 6.46 shows the relationship between linewidth and PED time for various concentrations of NMP added to the resist.[51] The PED effect is decreased by the added NMP, and the linewidth is very stable for the resist containing 0.1% NMP. The mechanism of stabilization due to a basic additive is thought to be the following. Some of the photogenerated acid is trapped by the basic additive and is deactivated by being held in a weak combination. When the airborne contaminants diffuse into the resist surface and deactivate the acid there, trapped acid is released to maintain the equilibrium between activated and deactivated acid. As a result, acid concentration in the resist film becomes uniform, and the resist profile is kept normal.

The behaviour of a basic additive has been simulated using the Monte-Carlo method.[63] It was assumed that acid and basic additive are random-walked by a certain probability corre-

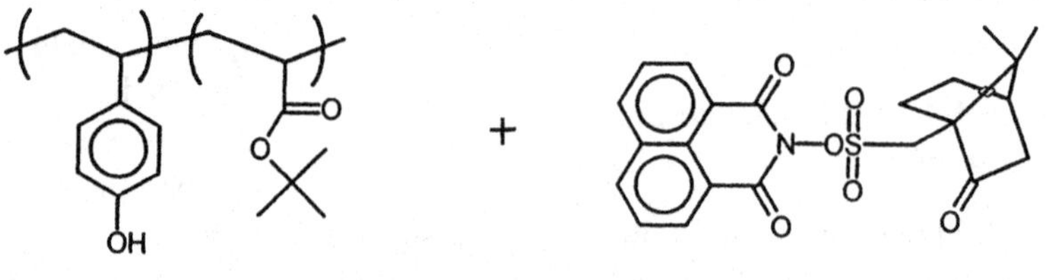

Figure 6.45 Chemical structures of base polymer (a) and PAG (b) of environmentally stable chemically amplified positive-tone resist (ESCAP).

(a) (b)

sponding to their diffusion coefficient D, which is given by:

$$D = \frac{p\Delta x^2}{2\Delta t}, \tag{6.4}$$

where p is the probability, Δx is the random-walk distance in a unit time Δt. When the acid collides with an inhibitor and the basic additive, the inhibitor is decomposed and acid is deactivated by a certain probability. Repetition of this calculation procedure enables the concentrations of acid and the basic additive after PEB to be simulated as shown in Figure 6.47, where the exposure distribution is given by the step function.[63] Note that the basic additive in the exposed area is rapidly deactivated and the concentration profile of the remaining basic additive is changed to a profile mirroring of that of the acid. This result means that the remaining basic additive acts as a barrier to acid and prevents the diffusion of acid from the exposed area to the unexposed area. The basic additive thus enhances the resist contrast and stabilizes the linewidth during the PEB process.

6.3.4.6 *Acid evaporation*

Since the absorption maximum at 629 nm of tetrabromophenol blue sodium salt (TBPB) changes in contact with the acid, the evaporation of acid during PEB can be evaluated by measuring its absorption spectrum.[64,65] An indicator film containing TBPB is spin-coated

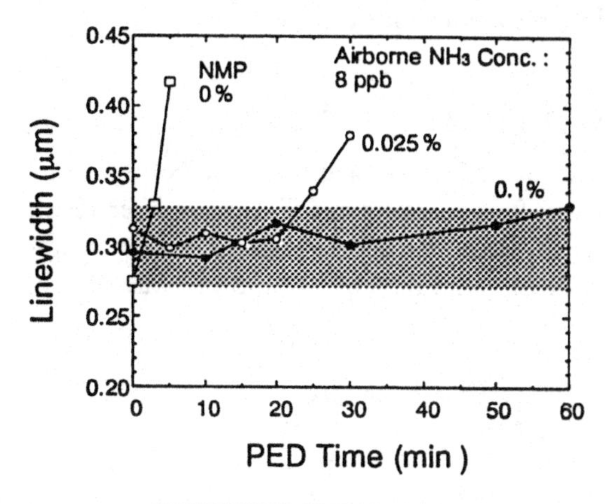

Figure 6.46 Relationship between linewidth and PED time for 0.3-µm L/S for NMP concentrations in the atmosphere of 0, 0.025 and 0.1%.[51]

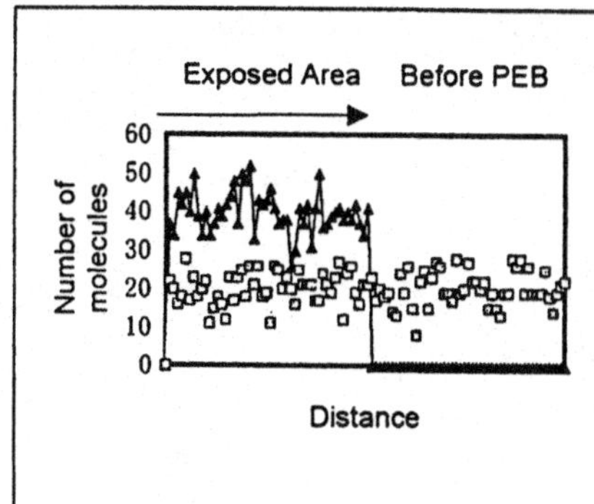

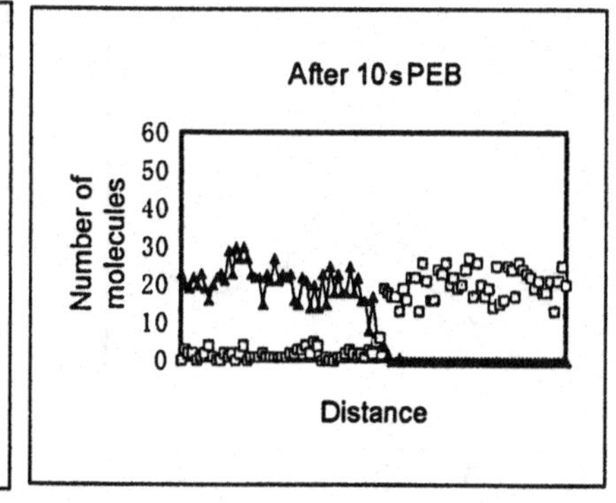

Figure 6.47 Concentration profiles of acid (▲) and basic additive (□) before and after PEB.[63]

onto a quartz wafer and is set, with a small gap, on the exposed resist sample during PEB as shown (inset) in Figure 6.48. The amount of acid evaporation can be expressed by the bleaching ratio of the indicator film, which is defined as the difference between the absorption coefficients measured before and after PEB.[65]

The relationship between the bleaching ratio and the PB temperature is shown in Figure 6.48. These data were gathered in experiments in which PHS without acid-labile group containing 5 wt% TPS-OTf was used as a resist sample. The casting solvent was ethyl cellosolve acetate, PB time was 4 min, the exposure dose was 100 mJ/cm^2, and the PEB temperature and time were 80 °C and 5 min. In the case of no exposure, the value of the bleaching ratio is zero, but the exposed sample shows some value. This result clarifies the acid evaporation during PEB. The bleaching ratio decreases with increasing PB temperature, and saturates at temperatures above 110 °C. This tendency is consistent with the relation between PB temperature and the amount of solvent remaining in the resist film (see Figure 6.41). The solvent remaining, which decreases with increasing PB temperature, provides the pathway for evaporating acid. Therefore, the PB temperature must be optimized to prevent excessive acid evaporation.

Figure 6.48 also shows the relationship between the bleaching ratio and the PEB temperature when the PB temperature and time were 120 °C for 4 min. It is clear that the acid evaporation increases dramatically with increasing PEB temperature.

Another experiment confirmed that the thickness of the surface insoluble layer (namely, the degree of the T-top profile) is in proportion to the value of the bleaching ratio.[65] These results suggest that T-top profiles can be prevented by using a high-temperature PB and a low PEB temperature.

6.3.5 ArF excimer laser resist

The need for transparency has been met by using PHS to make a KrF excimer laser resist, but it has an extremely high absorption coefficient of about 35 μm^{-1} at 193 nm wavelength.

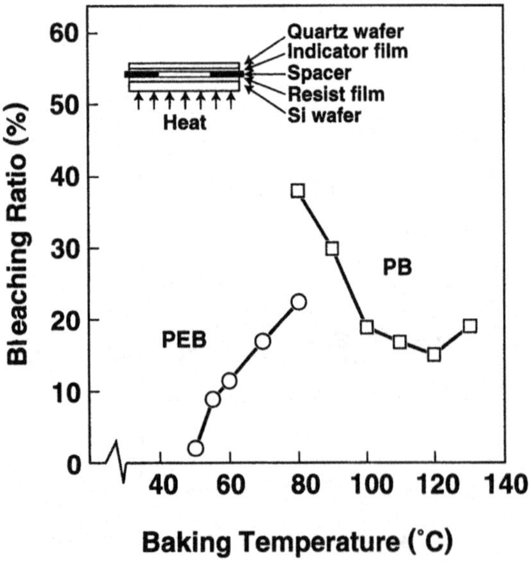

Figure 6.48 Bleaching ratio (relative amount of evaporated acid) versus PB (□) and PEB (○) temperatures.[65] See text for details.

This is due to the π–π^* electronic transition of the aromatic ring. As a consequence, the photon penetration depth into a resist is limited to less than several tens of nanometers. As a result, increasing the transparency of resist components while maintaining the dry-etching resistance has again become the most critical requirement in the development of ArF excimer laser resists.

An aliphatic polymer containing no aryl functionality has a much lower absorption than one with aryl groups, so a methacrylate terpolymer consisting of methylmethacrylate, t-butylmethacrylate, and methacrylic acid was developed for use as an alkaline-developable positive-tone CA resist.[66] The methylmethacrylate promotes adhesion and mechanical properties, the t-butylmethacrylate provides an acid-cleavable side group, and the methacrylic acid controls the solubility for the alkaline developer and adhesion.

To provide the required dry-etching resistance, alicyclic pendant groups such as adamantyl,[67] isobornyl,[68] tricyclodecanyl[69] and menthyl[70] were introduced into the methacrylate polymer. These chemical structures are shown in Figure 6.49. These alicyclic compounds have no strong absorption at 193 nm due to no aromaticity, and show high dry-etching resistance. Figure 6.50 shows the dry-etching rate normalized to that of DQ/N resist with respect to the polymerization ratio of adamantylmethacrylate and t-butylmethacrylate (poly(AdMa-tBuMA)).[67] The dry-etching resistance increases with increasing AdMA content and when the AdMA is 80% or more the dry-etching resistance of the methacrylate copolymer is comparable to that of DQ/N type resist.

The concept of alicyclic-compound introduction was also applied to the chemical structures of the PAGs and the dissolution inhibitor. PAG consisting of alkysulfonium salt and a dissolution inhibitor consisting of a steroid compound have been reported.[69,71] The use of an alicyclic compound is the main-stream approach to developing single-layer ArF excimer laser resists, but the hydrophobicity of the alicyclic group results in non-uniform dissolution in an alkaline developer. Therefore, the use of an alicyclic compound modified with a hydrophilic group and an alicyclic acid-cleavable group have been studied.[69,72].

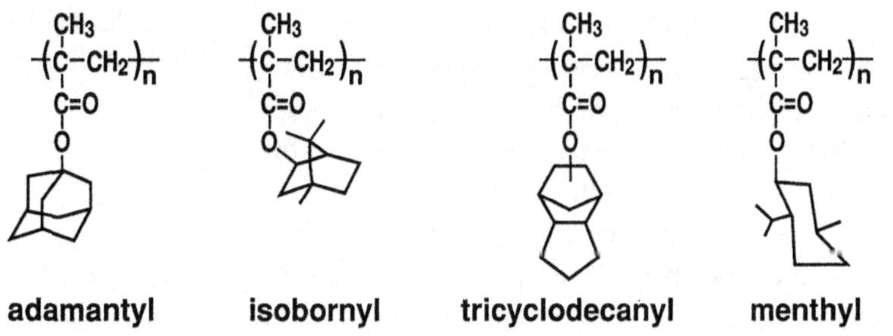

Figure 6.49 Chemical structures of ArF excimer laser resists.

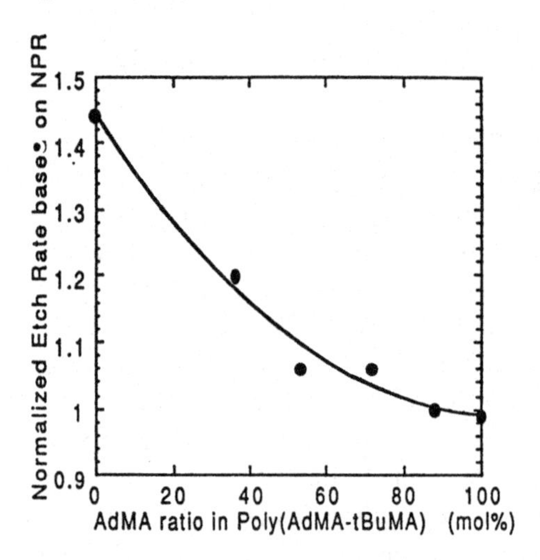

Figure 6.50 Dry-etching rate normalized to that of DQ/N resit (NPR) with respect to the polymerization ratio of adamantylmethacrylate and t-butylmethacrylate (poly(AdMA-tBuMA)).[67] Etching condition is CF_4, 100 sccm, 0.02 Torr, 200 W.

An alternative approach which maintains both high transparency and dry-etching resistance has been reported. The VUV (vacuum ultraviolet)-absorption spectrum of aromatic compounds can be red-shifted toward longer wavelengths and the window of absorption made to align with 193 nm by extending the conjugation length of the double bonds. Based on this observation, a naphthalene ring instead of a benzene ring is introduced into the resist components.[73]

6.4 Resists for electron-beam and X-ray lithography

6.4.1 Historical background

The first electron-beam (EB) resist, polymethylmethacrylate (PMMA), was studied by an IBM group in the 1960s,[74] and was used in the first trial of direct wafer writing fabrication of semiconductor devices in the mid-1970s.[75] Although PMMA provides high resolution, it is not sufficiently sensitive to EB. Various analogues of polyhalomethacrylates and polyolefinesulfones have therefore been investigated, and some of them have been commercialized for use in photomask fabrication. These resists, as well as PMMA, however, have little resistance to the reactive ion etching (RIE) required for LSI fabrication. Thus, the need for RIE-resistant high-sensitivity resists is increasing with progress in direct wafer writing.

X-ray lithography also began with PMMA in the 1970s.[76] Although EB resists have been used with X-rays for a long time, highly sensitive and highly resolving X-ray resists are exclusively in demand as X-ray lithography is maturing. The invention of chemical amplification in the early 1980s was an epoch-making development for EB and X-ray resists as well as for deep-UV applications[77] because it brought about unprecedented lithographic performance: high sensitivity, high resolution, good RIE durability, and compatibility with current LSI processes. Chemically amplified (CA) resists have invigorated the challenges involved in sub-quarter-micron fabrication using EB or X-ray lithographies.[78,79,80]

6.4.2 Excitation process

The energy of the electron beams commonly used in EB lithography ranges from 10 to 100 keV. As illustrated in Figure 6.51(a), an accel-

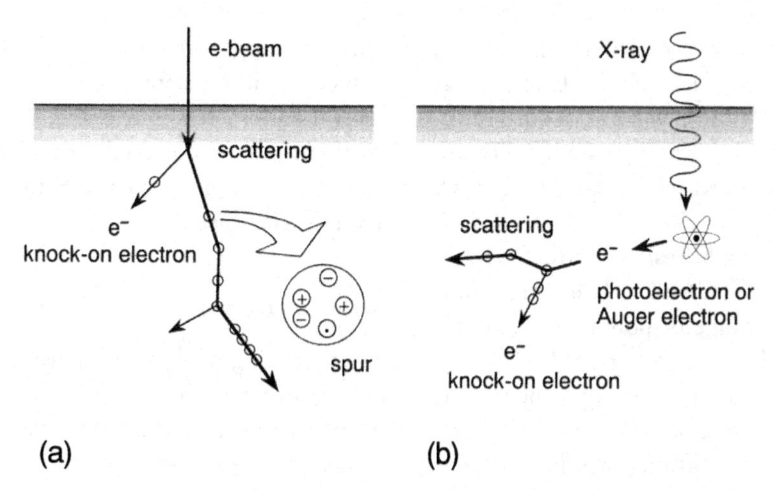

Figure 6.51 Schematic drawing of excitation processes for EB and X-ray exposures.

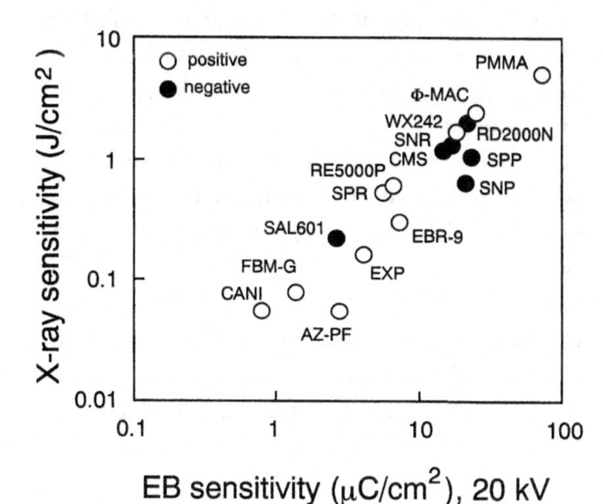

Figure 6.52 Correlation of the X-ray and EB sensitivities of various resists, including conventional and chemically amplified resists.

erated electron excites resist materials by inducing avalanches of highly excited species in its track, the result of scattering by the electron clouds of the materials. Such excited species (ions and radicals) are discretely condensed in small regions called spurs, which are several nanometers in size.

In a resist under the irradiation of soft X-rays with a wavelength of about 1 nm used for proximity (1:1) X-ray lithography, the first step proceeds through the photoelectric effect accompanied by the ejection of photoelectrons or Auger electrons. As illustrated in Figure 6.51(b), however, the subsequent process is the excitation induced by these electrons, since the ejected electrons still have keV-level energy. Thus, the dominant radiochemical reactions are caused by this electron excitation process. Experiments on the correlation between the EB and X-ray sensitivities of resists revealed an approximately linear relationship (Figure 6.52).[81] Most of the X-ray data were obtained at the BL-1B beam line of KEK Photon Factory, Tsukuba, in Japan, in 1990. Commercial resists: Φ-MAC (Daikin), WX242 (Olin Hunt), RD2000N (Hitachi Kasei), CMS (Toso), RE5000P (Hitachi Kasei), EBR-9 (Toray), FBM-G (Daikin), SAL601 (Shipley), AZ-PF (Hoechst). House-made resists: SNR,[82] SNP,[83] SPP,[84] SPR,[85] EXP,[86] CANI.[87] The correlations for a much wider range of polymeric materials have been studied by Lingnau *et al.*[88,89]

It should be noted, however, that the performance of X-ray resists depends on the lithographic system and various conditions. The exposure ambient sometimes affects resist sensitivity[90] because the radicals produced are easily quenched by atmospheric oxygen. Another issue is the absorption of X-rays by resists. Calculations based on atomic absorption data[91] have revealed that the absorption increases markedly with increasing wavelength (Figure 6.53). Absorption also depends on the molecular composition of resists. The optimum wavelength for proximity X-ray lithography is determined largely by resolution-limiting factors such as X-ray optical blur and the flight range of secondary electrons in resists. The choice of the wavelength substantially influences the absorption of X-rays by resists and thus the apparent resist sensitivity. If X-ray reduction lithography using a wavelength of several nanometers is adopted, the resist system itself should be changed since the X-rays are absorbed too much to reach the bottom of a 1-μm-thick resist.

6.4.3 Fundamental radiochemical reactions

The scattering of electrons in a resist is regarded as a function of the atomic numbers of the elements involved. This suggests that introduction of heavy atoms into resists will result in resists that are more sensitive. However, this physical contribution cannot be so useful practically, since the density of organic materials changes by a factor of two at most. Besides, the contribution is not always valid, as demonstrated experimentally with a series of chlorinated acrylates[92] and polysulfones.[93] More important is the efficiency of the subsequent radiochemical reactions. Table 6.4 lists the *G*-values of various polymers, some of which are typical EB resists.[94,95,96,97,98,99] The *G*-value is an index of radiochemical reaction efficiency approximated as the number of degradation or cross-linking sites per a deposition energy of 100 eV. Polyetherimide, a radiation-resistant material,

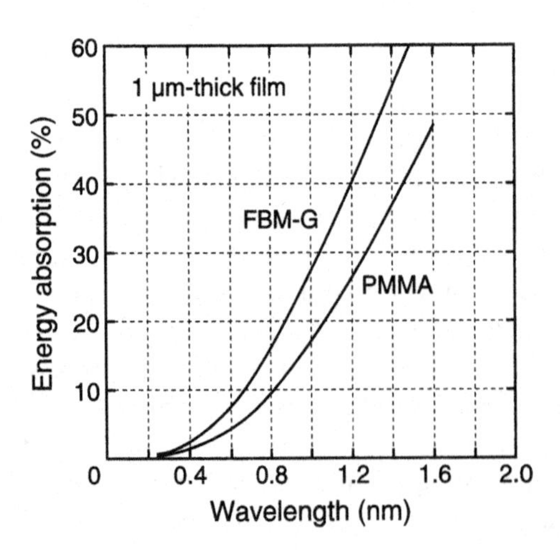

Figure 6.53 Calculated X-ray energy absorption by 1-μm-thick PMMA and FBM-G resists. FBM-G is a copolymer of hexafluorobutylmethacrylate and glycidylmethacrylate.

Table 6.4 *Polymer G-values.*

Polymers	G-value	Ref.
Scission (→ positive action)		
polymethylmethacrylate	2.23	94
polyphenylmethacrylate	0.86	94
polyethylmethacrylate	1.09	94
polytrichloroethylmethacrylate	2.14	94
polytrifluoroethylmethacrylate	3.28	94
polymethylisoprophenylketone	1.95	94
poly-αmethylstyrene	0.23	94
polybutenesulfone	12.06, 230[a]	95
polymethylpentenesulfone	3000 ± 300	96
Crosslinking (→ negative action)		
polystyrene	0.14	97
chloromethylated polystyrene	5.4	97
polybutadiene	14.3	98
polyetherimide[b]	0.014	99

[a] At the ceiling temperature of 64 °C
[b] ULTEM (General Electric Co.)

has a very small G-value of 0.014. Most EB-sensitive materials have a G-value of 0.1–10. The high G-values of polysulfones are due to chain reactions of depolymerization.[96] High G-value is also observed in polybutadiene, which undergoes crosslinking through the polymerization of vinyl groups.

Whatever the matrix elements initially excited by the EB, the deposited energy more or less transfers to the most reactive sites by means of physical effects or through several chemical steps. The dominant radiochemical reactions are, therefore, usually described as rather simple schemes and are very often similar to those caused by deep-UV excitations.[100] Typical radiochemical reactions useful for resist formulation are listed in Table 6.5.

Polymers have quarternary carbons in their mainchain repeating units, such as PMMA and poly-αmethylstyrene (PαMSt), degrade upon EB irradiation and provide positive resist action. Polyolefinesulfones undergo depolymerization upon irradiation and provide high sensitivity, of the order of 1 $\mu C/cm^2$. Polymethylpentenesulfone can be used, not only as a one-component resist, but also as a sensitizer in a novolac-based two-component resist system.[101]

Resist sensitivity can be enhanced by halogen substitutions because carbon–halogen bonds can be easily dissociated and the halogens have large cross-sections to electron beams. This technique is widely used to create sensitive EB resists: positive resists such as EBR-9, FBM-G, and ZEP (Nippon Zeon), and negative resists such as CMS.[102]

Epoxy and vinyl functional groups are easy to polymerize upon irradiation so they are useful substituents for making highly sensitive negative resists. An example of an epoxy-containing type is polyglycidylmethacrylate copolymer (COP)[103] which has been already used in photomask production lines. However, the use

Table 6.5 *Typical radiochemical reactions.*

Functional groups		Products
depolymerization (→ positive type)		
$\left[CH_2-\underset{-\underset{\mid}{C}-}{\overset{CH_3}{\overset{\mid}{C}}}-CH_2\right]_n$	$\leadsto$	$-CH_2-\underset{-\underset{\mid}{C}-}{\overset{CH_2}{\overset{\|}{C}}}+CH_3-$
$\left[CH_2-\underset{-\underset{\mid}{C}-}{\overset{\mid}{C}}-SO_2\right]_n$	$\leadsto$	$CH_2=\underset{-\underset{\mid}{C}-}{\overset{\mid}{C}}+SO_2\uparrow$
polymerization (→ negative type)		
$R-\underset{\diagdown O \diagup}{CH-CH}-R'$	$\leadsto$	$\left[CHR-CHR'O\right]_n$
$R-CH=CH_2$	$\leadsto$	$\left[CHR-CH_2\right]_n$
reactive radicals or ions		
$-N_3$	$\leadsto$	$-N\cdot + N_2\uparrow$
$-\underset{\mid}{\overset{\mid}{C}}-Cl$	$\leadsto$	$-\underset{\mid}{\overset{\mid}{C}}\cdot + Cl\cdot$
$M^{+\,-}X$	$\leadsto$	$M^{+} + X^{-}$

of such chain reactions brings a problem of post-polymerization in that the chain reaction continues for a relatively long time, typically more than one hour, so the sensitivity and resolution capability depends on the interval between exposure and the development processes.

Explosive materials such as nitro or azide compounds are sometimes good sensitizers to radiation. Azide groups produce nitrene radicals upon radiation, which are very reactive, to combine with hydroxyl groups. A bi-functional azide sensitizer is formulated as a cross-linker in RD2000N negative resist.

Onium salts decompose to produce strong acids upon radiation, similar to a photoreaction, although the details of the excitation process are rather complicated.[104] The acids produced can be used as a dissolution promoter of positive resists for base development.

Most of the resists described above are classified as one-step-reaction or 'conventional' types, meaning that the yield of intermediates produced by excitation essentially determines the resist characteristics. In CA resists, however, of more importance than the initial acid-generation reaction (Equation (6.5)) are the acid-catalyzed chain reactions in the second step

(Equation (6.6)) which can be independently controlled through a post-exposure bake.

$$(C_6H_5)_3S^+\ {}^-O_3SF_3 \xrightarrow{h\nu} (C_6H_5)_3S^+ + {}^-O_3SF_3 \quad (6.5)$$

$$\left[CH_2-CH(C_6H_4-O-C(=O)-O-t\text{-}C_4H_9)\right]_n \xrightarrow[\text{PEB} \sim 100\ ^\circ\text{C}]{H^+} \left[CH_2-CH(C_6H_4-OH)\right]_n + CO_2 + CH_2{=}C(CH_3)_2 \quad (6.6)$$

The most famous CA reaction is the decomposition of tert-butoxycarbonyloxy (tBOC) pendant groups in poly-(t-butoxycarbonyloxy) styrene (PBOCST) resist (Equation (6.6)) reported in earlier works;[77] it is still a major scheme used in current positive CA resists. The chain length of this catalytic reaction can reach 1000, depending on the post-exposure bake condition.[105] For negative CA resists, acid-catalyzed crosslinking of melamine derivatives and phenolic resins is such an excellent system that, as reported for the SAL series[106] and AZ-PN (Hoechst)[107], most commercially available resists utilize this type of reaction.

The CA resist scheme allows great flexibility in matching requirements to material design. A point of special distinction is that the CA resist formulations used with deep-UV radiation can also be used with electron beams and with X-rays. Various acid labile groups have been reported: tert-butyl ester,[108] polyphthalaldehyde,[109] tetrahydropyranylether,[110] acetal,[111] epoxy,[77] and melamine.[112] Details of CA formulation using these functional groups, and their chemical schemes, are described in a review.[113] Recently, resist manufacturers have been aggressively commercializing CA resists for EB because many LSI makers believe that EB direct writing will be a priority system of lithography for future-generation LSIs.

Although CA resists have various advantages, their sensitivity to environmental contamination during lithographic processes is a big problem.[114,115] This is because the catalytic acids generated in the resists are gradually neutralized by environmental base compounds such as ammonia and *N*-methylpyrrolidine even when their concentration is of the order of ppm. More details of the environmental contamination is described in Subsection 6.4.6(a).

6.4.4 Applications of EB resists

As shown on the resolution (pattern size) vs sensitivity map in Figure 6.54, there are three major application fields for EB lithography. The one that is already commercially viable is photomask production, in which the primary concerns are high sensitivity (about 1–5 $\mu C/cm^2$), good stability in the coated-film state, and reproducibility of lithographic performance. RIE durability is not needed in the conventional photomask process because wet-etching is used there. The requirement for resolution is about 0.5 μm at best, which corresponds to 0.1 μm in 5× reduction photolithography. To meet these specifications, one-component resists are suitable. Polybutenesulfone (PBS), EBR-9, and COP have been used to make photomasks. Recently, however, there is a growing need for high-resolution RIE-resistant materials that can be used to make precise phase-shift masks.

Much finer patterns are demanded in research and in X-ray masks. Resolution has

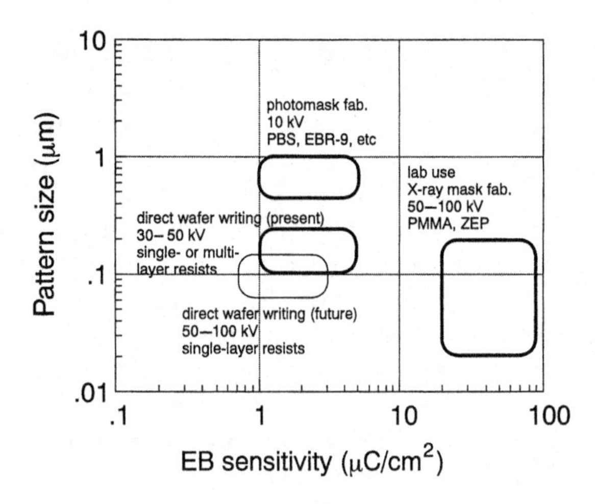

Figure 6.54 Applications of EB lithography plotted on a map of pattern size vs sensitivity.

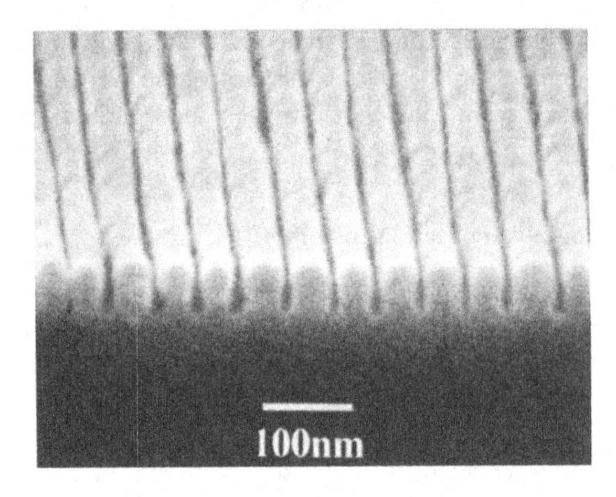

Figure 6.55 SEM photograph of ultrafine patterns delineated in ZEP resist.

priority, even at the cost of lower sensitivity. PMMA has long been used in research, but ZEP has recently been favored because of its excellent resolution, with 2–3 times higher sensitivity than PMMA, and good RIE durability. As demonstrated in Figure 6.55, 0.03-μm patterns have been fabricated in ZEP.[116]

A potential major field is LSI production, although the EB direct wafer writing technology is currently limited to the laboratory or to the fabrication of specific gate layers in GaAs FETs. Well-balanced resists achieving a resolution of 0.1 μm with 1 $\mu C/cm^2$, alkali development, and good RIE durability would push this technology to reliable use in LSI production lines. Conventional resists can no longer meet all these demands, so resist manufacturers have been investigating CA resists. An example, demonstrated in Figure 6.56, is a SEPR (Shinetsu Kagaku) positive resist. It has a D_0 sensitivity (minimum dose needed for a large exposed area of a resist film to be clear by development) of 1.2 $\mu C/cm^2$ at 30 kV and resolved a 0.2-μm-line-and-space pattern. As negative resists, the SAL series and AZ-PN appear promising, although further improvement in both sensitivity and resolution may be needed for this application field.

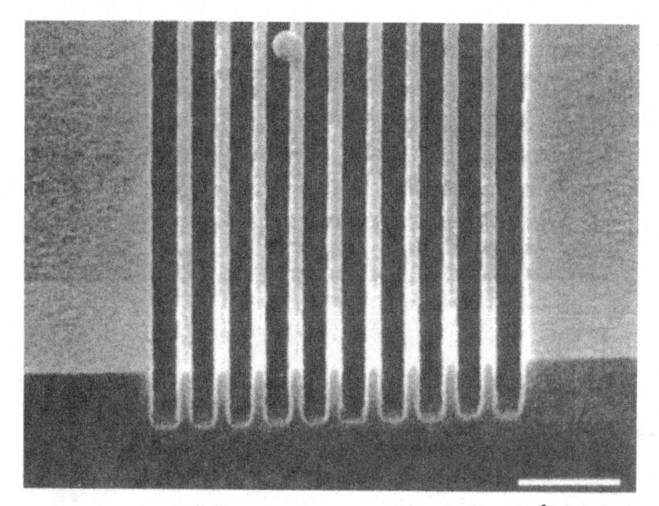

Figure 6.56 SEM photograph of 0.2-μm pattern delineated in 0.8-μm-thick SEPR positive resist when using 30-kV EB lithography. Its D_0 sensitivity is 1.2 μC/cm².

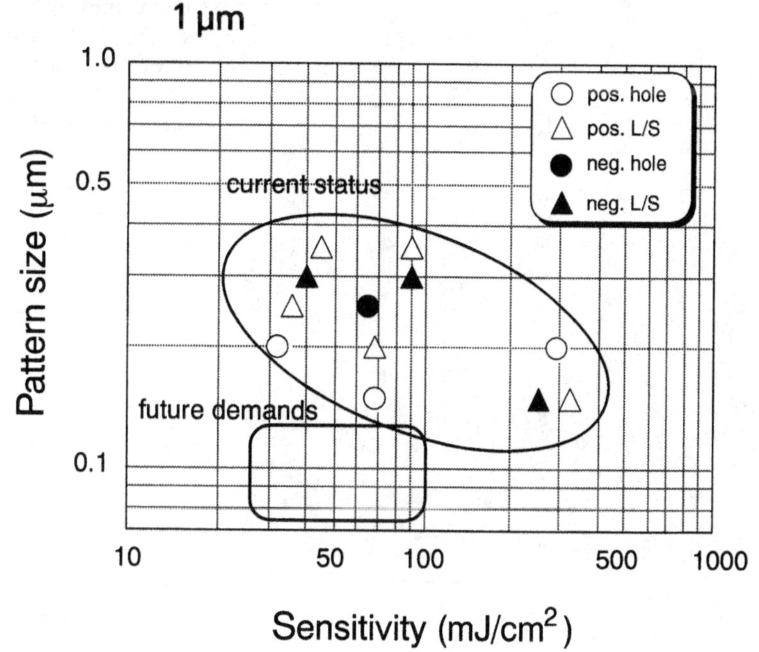

Figure 6.57 Specifications of recent X-ray resists. (Data from refs. 78, 86, 87, 118–122.)

6.4.5 Applications of X-ray resists

The major target for X-ray lithography is use in LSI production other than in micromechanics fabrication through the LIGA (Lithographie Galvanoformung Abformtechnik) process.[117] There are two technical approaches: one is 1 : 1 proximity printing using soft X-rays of about 1-nm wavelength, and the other is reduction printing using X-rays of 10–100-nm wavelength. Here, only the proximity method is focused on because the reduction system is still in quite basic research. The specifications required for proximity X-ray lithography used to be 0.2-μm resolution and 100-mJ/cm² sensitivity, and they were satisfied by specially tuned CA resists (Figure 6.57).[78,86,87,118,119,120,121,122]

As demonstrated in Figures 6.58 and 6.59, the positive resist CANI has 68-mJ/cm^2 sensitivity and 0.2-μm resolution.[87] CANI resist was used for contact and through-hole layers in 0.2-μm-LSI test fabrication.[123] However, 0.2-μm resolution is nowadays achievable with photolithography using ArF excimer laser deep-UV source ($\lambda = 193$ nm). The utmost limit of photolithography seems to be 0.13-μm fabrication using ArF. Therefore, the recent target of X-ray lithography goes down to 0.1-μm resolution. This new specification for X-rays suggests that even more improvement is needed, similar to that of the EB resists used in direct wafer writing.

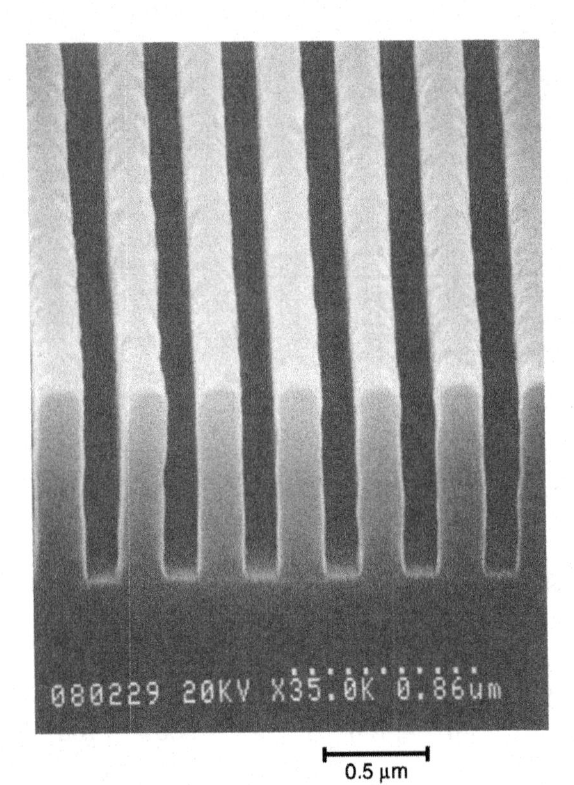

Figure 6.58 SEM photograph of 0.2-μm level patterns replicated in 1-μm-thick CANI resist when using X-ray lithography at NTT's synchrotron radiation facility (the peak of the wavelength is about 0.7 nm). Its D_0 sensitivity was 68 mJ/cm^2.

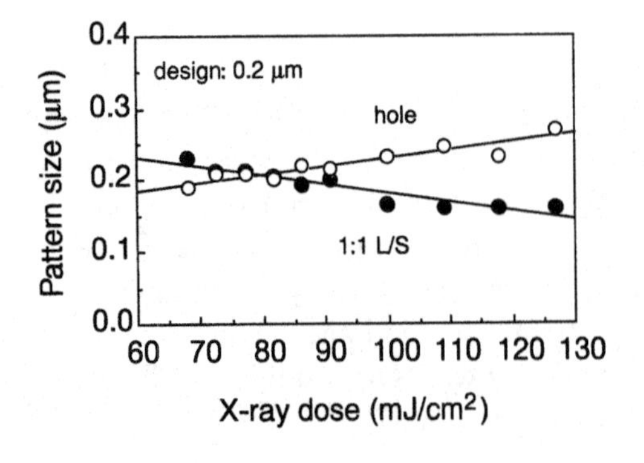

Figure 6.59 Dose margin of 0.2-μm-level patterns in a 1-μm-thick CANI resist. The tolerance ranges allowing ±10% deviation of the coded size were 40% for holes and 34% for L/S patterns.

6.4.6 Recent topics concerning CA resists

6.4.6.1 *Contamination from the environment and the substrate*

The catalytic acids in CA resists are sensitive to environmental base contaminants, and the extremely high sensitivity required of EB and X-ray resists makes this contamination problem more serious than it is for the deep-UV CA resists. To avoid airborne base contaminants, photolithography-process-manufacture lines usually adopt the solution of enclosing the whole stepper–developer system within a chemically clean booth equipped with charcoal or ion-exchangeable filters. Another simple solution is to use an insensitive overcoat on a resist. But these methods are not effective against contaminants from the substrate, such as ammonia from a TiN substrate. There are several approaches to reduce the contaminants from the substrate, such as a bottom barrier coat and addition of base compounds in the resist. Recently, a new resist scheme named ESCAP has been proposed by Ito *et al.*, to realize fundamental stability to environmental base contaminants.[124] The scheme is based on the concept that diffusion of contaminants in a resist should be controlled by its free volume. The free volume is minimized by heating the resist over its glass transition temperature (T_g) and annealing it.

6.4.6.2 *Diffusion of catalytic acids*

The catalytic acids inevitably diffuse from place to place and activate reactive sites during chemical amplification. This causes the initial latent image, which is formed by the concentration profile of exposure-generated catalytic acids, to spread and degrade.[125,126] Experiments indicate that the diffusion coefficient of acids can be of the order of 10 nm^2/s or more under plausible chemical amplification conditions.[127] Calculations based on Fick's diffusion equation reveal that the latent image of a 0.1-μm-line pattern apparently deforms to a degree that increases with the value of the diffusion coefficient (Figure 6.60). A recent study also indicates that the diffusion of acids depends on the free volume of the resists and influences linewidth control.[128] When the chemical amplification condition is rigorous, the catalytic acids diffuse markedly and the sensitivity of the resist

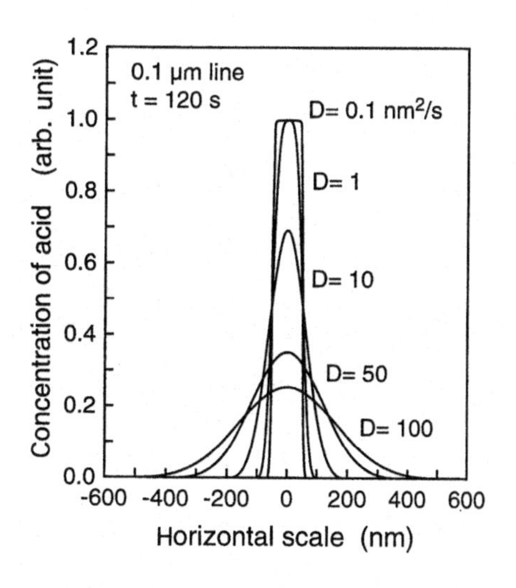

Figure 6.60 Concentration profiles of catalytic acids calculated for various diffusion coefficients. The initial profile is assumed to be a 0.1-μm-wide rectangle. Calculation is based on Fick's diffusion equation.

increases at the cost of resolution. When it is quite mild, however, the resolution can be significantly improved; a PBOCST positive resist accomplished an utmost resolution of 40 nm.[129] Thus, the diffusion of catalytic acids is a prime factor in CA resists.

6.4.6.3 *Pattern collapse*

When pattern size approaches a level below 0.2 µm in X-ray lithography, dense piles of patterns tend to collapse easily, as shown in Figure 6.61. This phenomenon becomes distinct as the aspect ratio of the pattern increases. The cause of this is thought to be the surface tension of the rinse water evaporating during spin-drying. The threshold seems to lie at an aspect ratio of about 5.[130] Since this collapse phenomenon is fundamentally due to a mechanical reason, it seems to be common to all lithographies. The resist thickness should be 0.5 µm or more if the resist is to serve as a dry-etching mask. So, the problem of pattern collapse could be disregarded as long as the pattern size had a design rule of 0.2 µm or more. However, it would become increasingly worrisome as the pattern-size went down to 0.1 µm.

6.4.6.4 *Nanometer-scale roughness*

Pattern-size accuracy is commonly specified to be less than 10% of the coded sizes. This means an accuracy of several nanometers is needed for a 0.1-µm design rule. Such a dimension is almost comparable with the molecular size of resist matrix resins and, thus, an inevitable cause of error with a very narrow pattern size would be the molecular size of the resist matrix

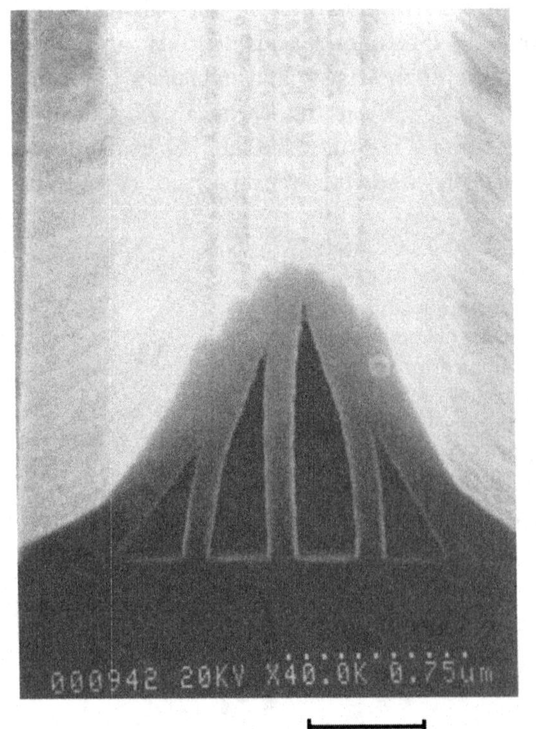

Figure 6.61 SEM photograph of pattern collapse observed for 0.1-µm patterns in 1-µm-thick SAL601 replicated using X-ray lithography.

resin. Several research groups have announced the importance of this issue. A simulation including the statistical molecular size of a resist (PMMA) has been accomplished for EB lithography involving EB radiation, excitation of the resist, latent-image profile, and development processes.[131] The molecular-level non-uniformity has been studied using atomic force microscopy (AFM) for the surface roughness of ZEP nano-patterns.[132] Thus, future resists will need to be intensively investigated from the aspect of molecular size. To minimize the problem of polydispersity of a matrix resin, a resist made of cyclic phenolid oligomer (six-membered calixarene) has been developed.[133] It has a well-defined molecular structure and thus exhibits very high resolution. This proves that the molecular size of the matrix resin is a key factor in resolution. In addition, this field of research involves other fluctuating factors such as the mixing state of resist components, local concentration of sensitizers and reacted sensitizers, and development speed at the resist surface. These factors complicate the controllability of nanometer-scale roughness.

6.4.7 Other processing technologies and related materials

6.4.7.1 *Silicon-containing resists*

The proximity effect caused by the forward-scattering and back-scattering of incident electrons is a crucial obstacle to high resolution in EB lithography. To reduce the severity of this effect, multilayer resist systems are very effective although they are more expensive and their processing requires more steps. Among them, the two-layer resist system using silicon-containing resists was intensively studied in the 1980s because of its simplicity.[134] Lithographed patterns in thin top silicon-containing resists can be transferred to a thick bottom organic layer by one step of O_2 RIE. The compounds SNR, SNP, SPP, and SPR in Figure 6.52 are examples of silicon-containing resists that were studied for EB. Of these, SNR is the first commercialized silicon-containing EB resist and is made of chloromethylated diphenylsilicone. The others are two-component-type resists made of an alkali-soluble silsesquioxane resin and sensitizers. Besides these resists, many other silicon-containing resists have been developed.

Surface imaging by selective silylation of exposed organic resists is an analogue technique,[135] which provides a one-layer resist system, very steep pattern profiles and the advantage of not using developers. Silicon-containing resists and the silylation technique have been explored for their potential usefulness in ArF lithography targeting sub-0.2-μm fabrication generation.[136]

6.4.7.2 *Conductive coating materials*

A local electric charge on substrates, generated by the capture of incident electrons in EB lithography, displaces the position of resist patterns because it repels electron beams. This is serious for insulator substrates but is sometimes observed with silicon wafers. Precise pattern placement will therefore require the use of conductive materials. The conductivity needs less than 10^7 ohm of sheet resistance. The approaches that have been proposed are the use of conductive resist materials, surfactants, ion-implantation, metal deposition, and conductive coating. Among them, the use of a conductive top-coat seems simple and useful.[137] Some of the materials commercialized are organic-soluble TVQ (Nitto Chemical), and water-soluble materials such as ESPACER (Showa Denko) and aquaSAVE (Nitto Chemical).

6.4.8 Summary

The historical development and latest trends of EB and X-ray resists have been described. Although photolithography has been the dominant lithographic tool in LSI production lines, EB and X-ray technologies will be alternatives for the 0.1-μm generation of LSIs. EB and X-ray resists therefore should be further improved. Material designs will be sophisticated, along with CA resist formulations. In addition, fundamental studies into such matters as the nanometer-scale roughness of resist patterns and the probable processes of excitation and sensitization are very important in realizing better resolution than ever before.

6.5 Advanced resist process technology

6.5.1 Introduction

The dry-development process is one of the most important techniques in sub-quarter-micron lithography, and its origin is the tri-layer resist process proposed by J. M. Moran in 1979.[138] Although the tri-layer process had excellent process performance, it required many process steps. Many improvements for simplifying the tri-layer process have therefore been made. The first was to simplify the tri-layer resist to a bi-layer resist by using a silicon-containing resist system. Then, the bi-layer resist was simplified to a single-layer resist using surface silylation techniques. Following this, improvements continued to be made in the resist materials, the silylation process, and the dry-development process. We can now obtain the same performance with single-layer resits as we can with tri-layer resists. This simplification is in direct contrast to the evolution of the conventional wet-developed resists, which have become more complicated with the use of shorter light wavelengths. Progress in dry-development techniques is summarized in Figure 6.62. This section summarizes recent progress and future prospects.

6.5.2 Dry-development techniques

A tri-layer resist consists of an under-planarization layer (hard-baked resist), a middle mask layer (silicon dioxide), and an upper imaging layer (resist). An X-ray resist is used as an imaging resist to form a 1-μm-L&S pattern using dry-development with reactive ion-etching.[138] This tri-layer system is free of problems such as reflection from the substrate and pattern collapse and can achieve very high-resolution and high-aspect-ratio patterns with the help of anisotropic dry-etching process. The lithographic performance of tri-layer resists is, therefore, excellent but the development process has three times as many steps as that of conventional resists, and a dry-etching process has to be performed twice.

After the tri-layer resist process was proposed, many researchers focused on trying to simplify it. The CVD (chemical vapor deposition)-deposited silicon oxide was replaced with spin-coated organic materials. Then, silicon-containing resist materials that act as both an upper imaging layer and an intermediate mask layer during RIE were proposed. This made the development of bi-layer resists possible and reduced the number of process steps needed, but silicon-containing resist materials never became popular because of the difficulty of synthesizing them.

An alternative approach to bi-layer resists is the use of the silicon-added bi-layer resist (SABRE) process, in which a silicon-containing layer is formed through a silylation process after the imaging.[139] In this process, a conventional novolac resist is spin-coated onto a hard-baked resist. Then, after the resist is patterned,

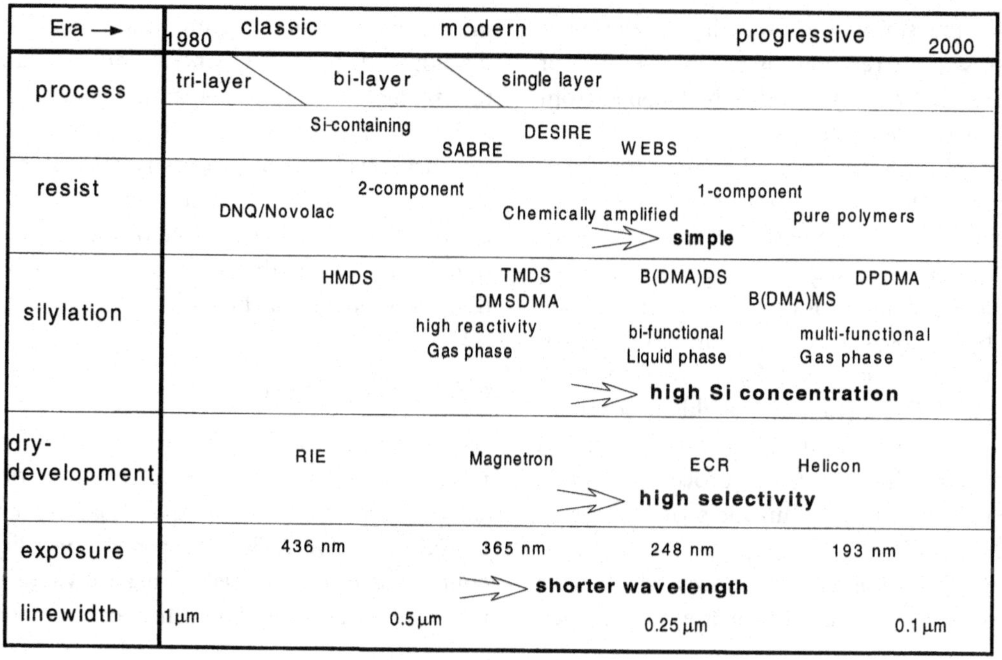

Era →	1980 classic modern progressive 2000
process	tri-layer bi-layer single layer
	Si-containing SABRE DESIRE WEBS
resist	DNQ/Novolac 2-component Chemically amplified 1-component pure polymers
	⇒ simple
silylation	HMDS TMDS DMSDMA high reactivity Gas phase B(DMA)DS bi-functional Liquid phase B(DMA)MS DPDMA multi-functional Gas phase
	⇒ high Si concentration
dry-development	RIE Magnetron ECR Helicon
	⇒ high selectivity
exposure	436 nm 365 nm 248 nm 193 nm
	⇒ shorter wavelength
linewidth	1 μm 0.5 μm 0.25 μm 0.1 μm

Figure 6.62 Progress of dry-development lithography processes.

the wafers are first flood-exposed to change diazonaphthoquinone (DNQ) into indenecarboxylic acid, then silylated with hexamethyldisilazane (HMDS), and finally developed with O_2 RIE. The advantage of this process is that a conventional image resist can be used as an upper imaging layer.

The diffusion-enhanced-silylation resist (DESIRE) process is a single-layer dry-development process in which the functions of the upper imaging layer and the lower layer are combined.[140] In this process, a conventional DQ/N resist is exposed and baked, the exposed area is crosslinked during baking, then silylated with HMDS, and finally a negative pattern is obtained after O_2 dry-development. The advantages of the DESIRE process, which make it possible to simplify tri-layer resists to single-layer resists, are its simplicity and the fact that any commercially available novolac resist can be used as an imaging layer. This means that any exposure wavelengths can be used. As a result, it has been used a lot in research work related to the silylation process. One report, for example, proposed a modified DESIRE process called WEBS, which improves the silylation profile by pre-wet-development.[141] The development of KrF excimer laser lithography has led to a proposal for a single-layer dry-resist process in which positive patterns are formed through the use of chemically amplified negative resists.[142]

Another promising development is the use of an ArF excimer laser to provide shorter-wavelength exposure. At ArF laser wavelength, pure polyhydroxystyrene (PHS) or novolac polymers can be used as a crosslinkable polymer without any sensitizer. For example, D. W. Johnson *et al.* exposed pure PHS polymer at 50 mJ/cm^2 and then silylated with dimethylsilyedimethylamine (DMSDMA) to form a 0.2-μm-L&S.[143] This process, the simplest of the dry-development processes so far reported, is feasible because of the increase in photon energy which comes from shorter-wavelength exposure.

R. R. Kunz *et al.* created a new dry-developed process by combining the use of pure polymer material with an ArF (193 nm) excimer laser light. They used a CVD process to deposit polysilane polymers, formed a silicon dioxide layer by exposing with 193 nm light, then dry-developed it to form 0.2-μm-L&S patterns.[144] The advantages of this process are that the silicon dioxide layer can be formed directly and that the CVD process makes it possible to form very thin films over the step substrate.

The dry-development processes which have been outlined in this section are summarized in Figure 6.63. As the figure indicates, the dry-development process has been simplified down to a single-layer and pure-material process. As a result, the number of process steps in the current dry-development process is comparable to that of conventional wet-development processes. The only disadvantage of the dry-development process is that it requires the use of a dry-development machine.

6.5.3 Resist materials

As mentioned earlier, the first tri-layer resist process could use any type of imaging resist, since that process transfers the image to the silicon dioxide layer. The next-generation silicon-containing resist materials can be divided into two categories, one of which consists of poly-

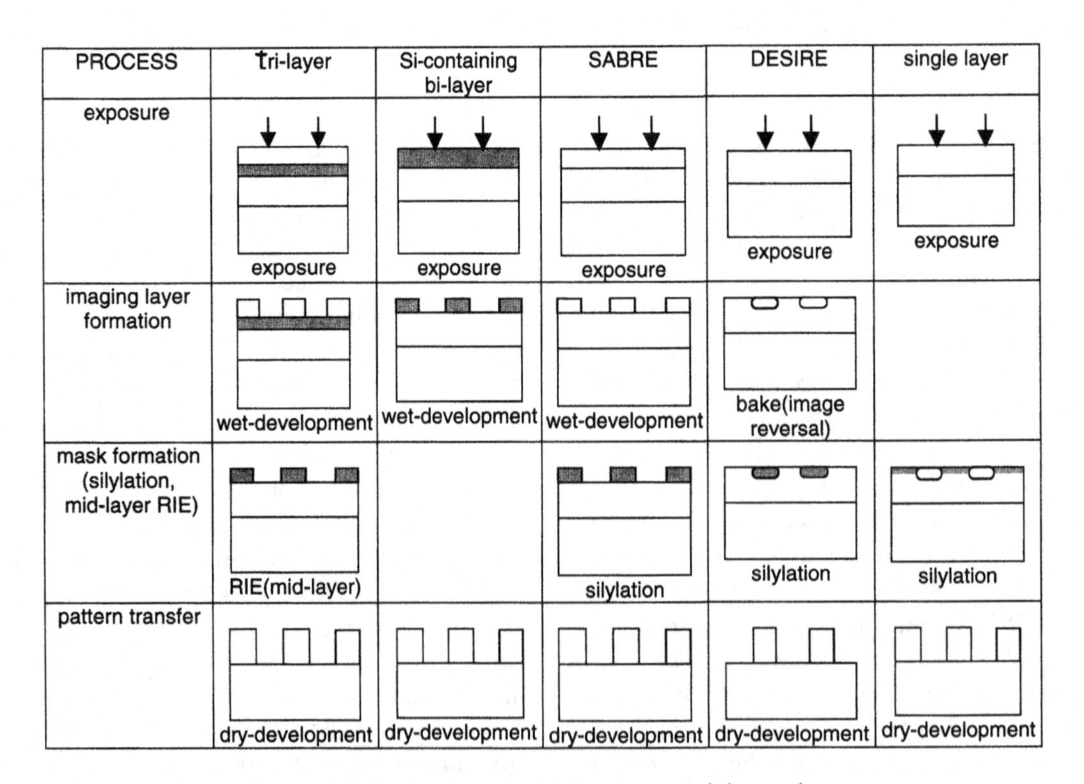

Figure 6.63 Process steps in dry-development lithography.

mers that include the siloxane structure. J. Shaw introduced the phenol structure into a siloxane polymer to obtain O_2 etching resistance and solubility in alkaline developers, and he mixed this polymer with DNQ and used it as a resist (Figure 6.64).[145] The other category consists of a phenol polymer which contain silylation groups. Saotome *et al.* synthesized silicon-containing novolac polymer as an alkaline-developable base polymer, and formed an imaging resist by mixing it with DNQ sensitizer (Figure 6.65).[146] They reported that the amount of silicon was about 10 wt% and etching selectivity at the development process with O_2 ion-beam etching (d.c. bias 50 V) was 67.

In the DESIRE process era, a conventional novolac resist with DNQ sensitizer has been used so far (Figure 6.66). Although the resist used in the DESIRE process is similar to a conventional wet-development resist, the role of the resist is different, since the DNQ sensitizer in the novolac resist is used as a crosslinker in the DESIRE process. Therefore, the best resist material for the DESIRE process may differ from the resist for conventional imaging.

In KrF excimer laser lithography, a chemically amplified resist, originally developed as an imaging resist, has been used for the silylation process. J. P. Schellekens proposed a positive dry-development process with a chemically amplified negative resist by using the crosslinking reaction for selective silylation.[142]

Along with shorter-wavelength exposure using an ArF excimer laser, the direct crosslinking reaction of the pure polymers like novolac or PHS can form positive imaging patterns

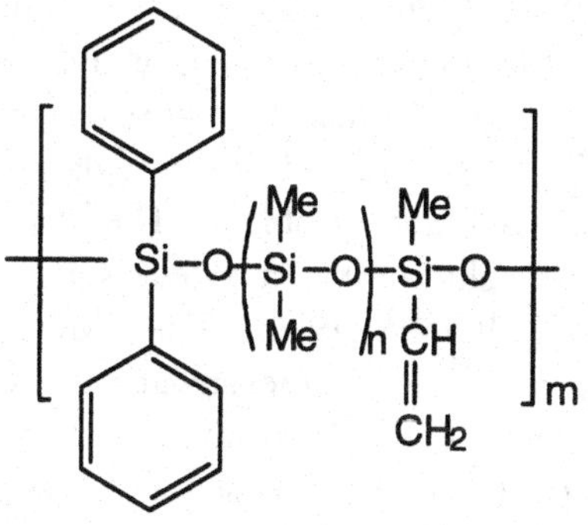

Figure 6.64 Silicon-containing siloxane polymer.

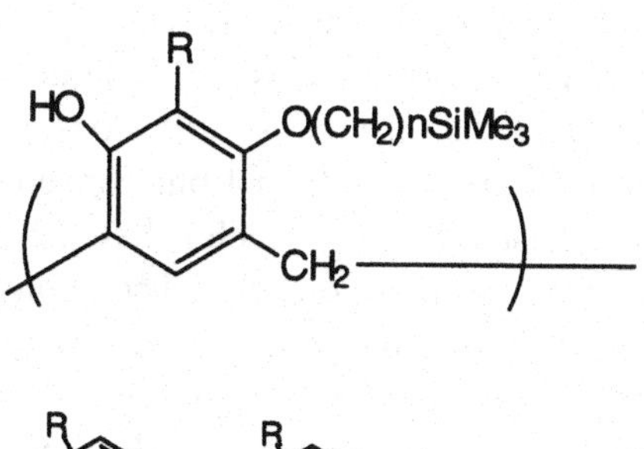

Figure 6.65 Silicon-containing novolac polymer.

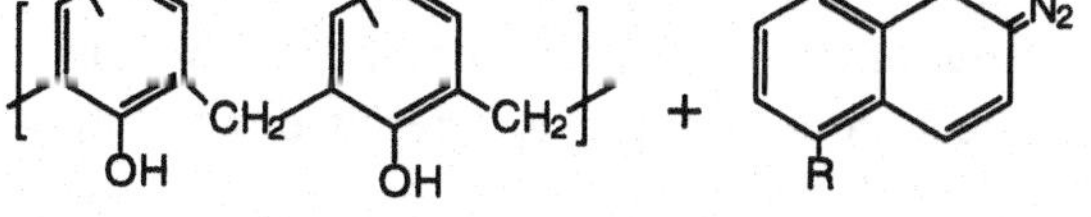

Figure 6.66 Novolac polymer and DNQ sensitizer used in the DESIRE process.

using the silylation reaction. D. W. Johnson reported that the use of a PHS polymer with a higher molecular weight ($M_w = 45\,000$) contributed to good sensitivity (45 mJ/cm^2) with no sensitizer.[143]

Another approach is to use oxide CVD-deposited materials. R. R. Kunz *et al.* deposited polysilane films, like poly-butylsilane and poly-phenylsilane, on to a substrate (0.06-μm), that exposed at 193 nm to a dose of 100 mJ/cm^2 and produced 0.2-μm-L&S patterns. T. W. Weidman deposited a $Me_{1-x}SiH_x$ film (0.1-μm thick) at room temperature, then exposed with 193 nm wavelength in air to a dose of 20 mJ/cm^2 and formed a 0.2-μm-L&S silicon dioxide mask layer.[147]

6.5.4 Silylation reagents

The original silylation reagent was hexamethyldisilazane (HMDS), which had been widely used as an adhesion promoter in lithography processes and was commercially available. Along with its low cost and availability, however, HMDS has several disadvantages. Some of these are its low evaporation pressure, high silylation temperature (~ 150 °C) due to a low diffusion coefficient in the polymer, and extensive swelling due to a relatively high molecular volume.

To overcome these problems, new silylation agents with higher reactivity and smaller molecular volumes have been investigated. The representative silylation agents are trimethyldisilazane (TMDS) and DMSDMA. These silylation reagents achieve high silylation pressure due to high evaporation pressure, a lower silylation temperature due to a higher diffusion coefficient in the polymer, and little swelling due to the smaller molecular volume. TMDS was recommended as the silylation reagent for the optimum DESIRE process.[148] This optimized DESIRE process for i-line exposure (NA = 0.48) with TMDS produced 0.3-μm-L&S patterns.

This silylation process is a vapor-phase process which uses the thermal diffusion property of the silylation reagents. On the other hand, a liquid-phase silylation process was proposed in 1989. In this process, wafers were dipped into a xylene solvent containing 10% hexamethylcyclotrisilazane (HMCTS) as a silylation reagent and 0.5% NMP as a diffusion promoter. This liquid-phase silylation process was first used in the SABRE process, and increased the range of silylation reagents used, because non-volatile silylation reagents could be used. Moreover, K-H. Baik *et al.* found that the use of the bifunctional silylation reagent bis(dimethylamino)dimethylsilane, B(DMA)DS, in the liquid phase enhanced the resolution capability of the DESIRE process.[149] Although B(DMA)DS had been used in the vapor phase before, the amount of silylation in the vapor phase was not sufficient because of the self-crosslinking effect. Liquid-phase silylation solves this crosslinking problem. The optimized silylation solution was a xylene solvent containing 30% B(DMA)DS and 2% NMP as a diffusion promoter, and the silylation was carried out at 40 °C. It was found that the silylation contrast between exposed and unexposed regions is much higher in this process than it is in the process using HMDS in the vapor phase. The reason for this is that liquid-phase silylation is based on case diffusion, so the silylation mechanism differs from conventional vapor-phase diffusion, which is based on Fickian diffusion. Furthermore, the silicon concentration of the resist silylated with B(DMA)DS in liquid-phase silylation was higher than that of the resist silylated with a monofunctional silylation reagent. This higher silicon concentration is a result of the bifunctional silylation reagent being able to form the siloxane structure. As a result, the liquid-phase silylation process produced a very fine pattern of 0.28-μm L&S with i-line exposure (NA = 0.48).

This combination of a multifunctional silylation reagent and liquid-phase silylation has been used in ArF excimer laser lithography.[150] ArF lithography with a high-contrast silylation process produced L&S patterns down to 0.14 μm.

The use of bis(dimethylamino)methylsilane, B(DMA)MS, which has less steric hindrance than does B(DMA)DS, produces twice the silicon concentration. The silicon concentration distribution measured from RBS spectra is shown in Figure 6.67. The silylation reactions of monofunctional DMSDMA and bifunctional B(DMA)MS are shown in Figure 6.68 and Figure 6.69.

Furthermore, B(DMA)MS forms a very shallow silylation layer; so this process is completely free from swelling. This silylation process produced 0.12-μm-L&S patterns when ArF excimer laser lithography (NA = 0.55) was used

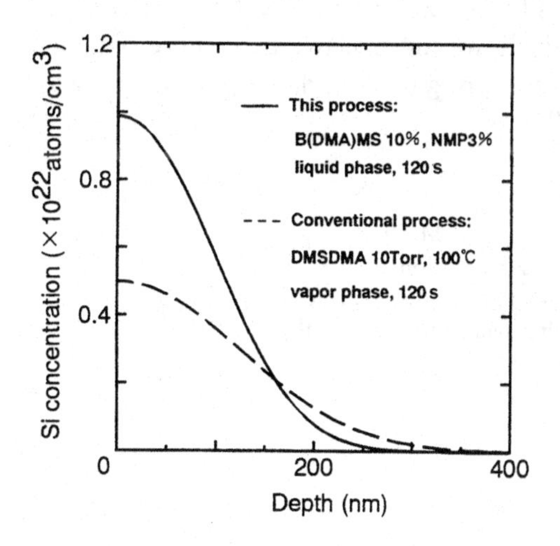

Figure 6.67 Silicon depth profile in silylation reaction measured by Rutherford backscattering spectroscopy (RBS).

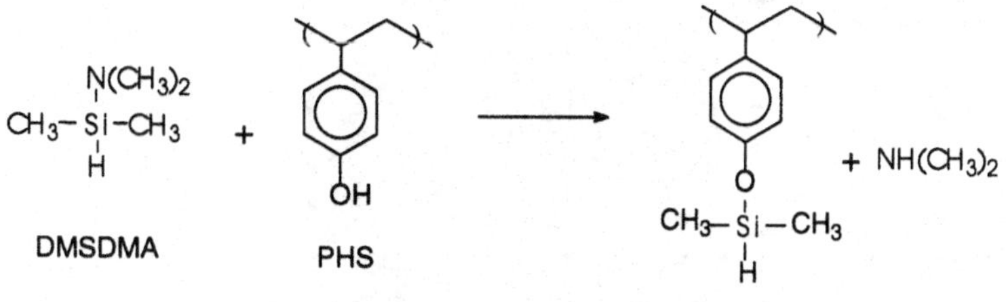

Figure 6.68 Silylation reaction of monofunctional DMSDMA.

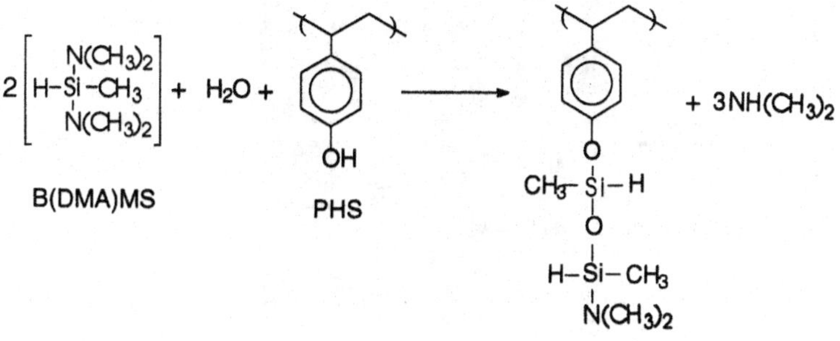

Figure 6.69 Silylation reaction of bifunctional B(DMA)MS.

with a resolution-enhancement technique (Figure 6.70). This resolution corresponds to 0.34 of the k_1 factor, which indicates the resolution capability of the resist process. Moreover, as shown in Figure 6.71, the use of a Levenson-type phase-shifting mask produced 0.075-μm-L&S patterns.[151] The silylation reagents proposed for top-surface imaging are summarized in Table 6.6.

Although multifunctional silylation reagents allow higher silicon concentration and yield higher resolution, one disadvantage of using them is that the liquid-phase silylation process requires a large amount of organic solvents. New disilane silylation reagents that can be used in the vapor phase have therefore been developed.[152] Three of them are dimethylaminopentamethyldisilane (DMAPMDS), dialkylaminodimethyldisilane (DAADMDS), and bis(dimethylamino)tetramethyldisilane, B(DMA)TMDS. All of these silylation reagents have two silicon atoms, quite small molecular volumes, and are free from the swelling problem. Consequently, although not commercially available at present, these materials are expected to be ideal silylation reagents.

Both the materials and processes of the silylation technique has been much improved. The silicon concentration and silylation depth obtained are much better than those obtained by the original silylation process with HMDS.

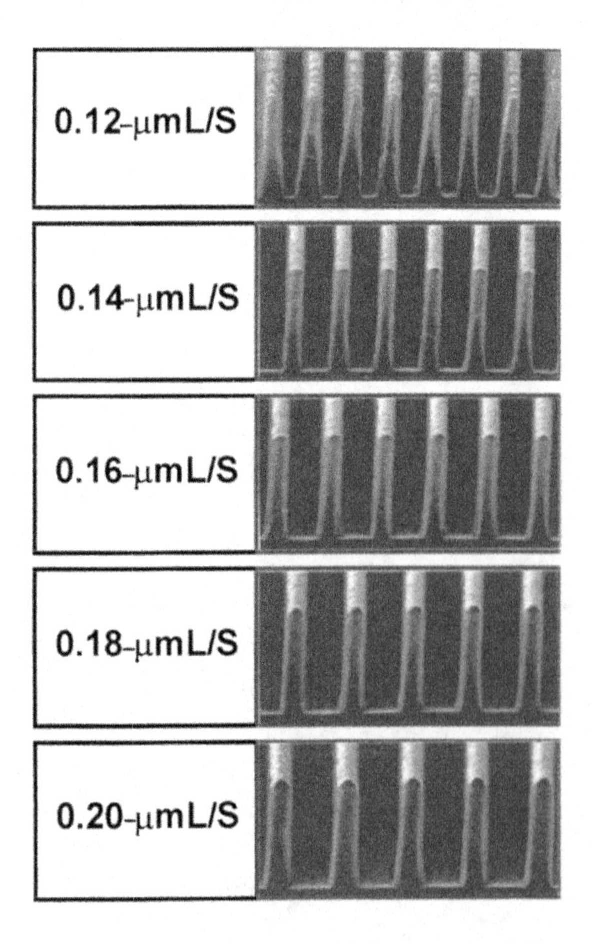

Figure 6.70 SEM photograph showing resolution characteristics of top-surface imaging using silylation with B(DMA)MS.

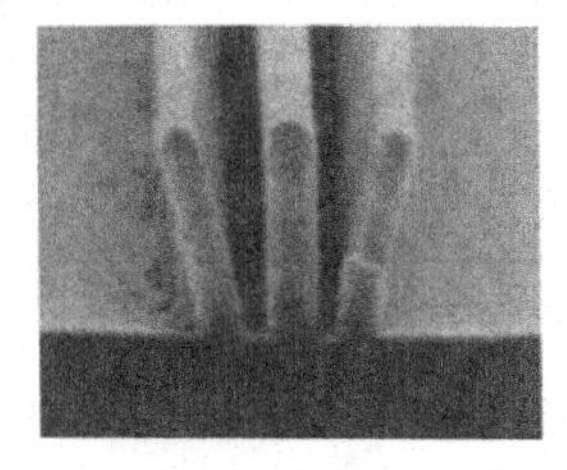

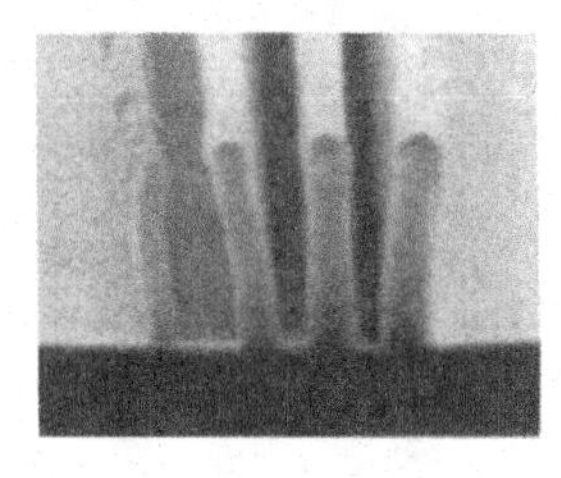

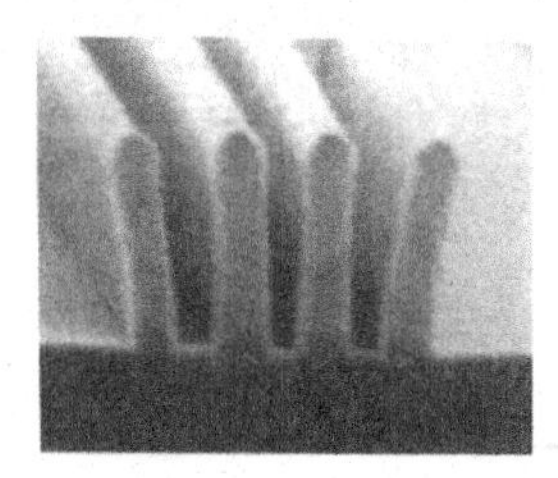

Figure 6.71 SEM photographs showing resolution characteristics of top-surface imaging with a Levenson-type phase-shifting mask.

6.5.5 Dry-development process

Dry-development is one of the most important processes in top-surface imaging. Although the silicon concentration profile in the silylation process is not so abrupt, etching selectivity must be as high as possible to obtain high-resolution capability. Nevertheless, the dry-development process has hardly been investigated compared to silylation. Therefore, further investigation is required in the future.

Reactive ion etching (RIE) was originally used in tri-layer resist development, and the etching selectivity of RIE in a tri-layer system was good enough because the tri-layer process had a good etching mask. But because RIE cannot ensure adequate etching selectivity in the next-generation dry-development process using silylation techniques, magnetron-type etching machines like MERIES have been used for dry-development processes with silylation masks. MERIES can give high plasma density under high-vacuum pressure, so etching selectivity for the silylation mask is improved. Then a more advanced etching machine, with an ECR- or helicon-type plasma source was used as a dry-development machine. The self-bias of these machines is so small that very good etching selectivity was obtained from silylation masks.[153] In addition to the better machine, some other techniques have been developed. Low-temperature dry-etching (−50 °C) can contribute to the formation of more nearly vertical side walls by inhibiting chemical reaction. It is reported that new gas chemistry using SO_2 can also form good pattern profiles.[154]

Table 6.6 *Silylation reagents used in top-surface imaging.*

Compound	Feature	Silylation agent			
disilane	low cost swelling high temperature	$CH_3{-}Si(CH_3)_2{-}NH{-}Si(CH_3)_2{-}CH_3$ HMDS	$H{-}Si(CH_3)_2{-}NH{-}Si(CH_3)_2{-}H$ TMDS		
silane	small swelling low temperature	$CH_3{-}SiH(N(CH_3)_2){-}CH_3$ DMSDMA	$CH_3{-}SiH(N(C_2H_5)_2){-}CH_3$ DMSDEA	$CH_3{-}Si(CH_3)(N(CH_3)_2){-}CH_3$ tri-MSDMA	$CH_3{-}Si(CH_3)(N(C_2H_5)_2){-}CH_3$ tri-MDDEA
bi-functional	liquid phase higher Si conc. high contrast	$[(CH_3)_2Si{-}NH]_3$ (cyclic) HMCTS	$CH_3{-}Si(N(CH_3)_2)_2{-}CH_3$ B(DMA)DS	$H{-}Si(N(CH_3)_2)_2{-}CH_3$ B(DMA)MS	
disilane (Si-Si)	small volume highest Si conc.	$CH_3{-}Si(CH_3)_2{-}Si(CH_3)_2{-}N(CH_3)_2$ DMAPMDS	$N(CH_3)_2{-}Si(CH_3)_2{-}Si(CH_3)_2{-}N(CH_3)_2$ DMADMDS		

6.6 Summary

KrF excimer laser exposure was first tried in 1980 using a contact printer (proximity lithography). Thereafter, the advent of a KrF excimer laser stepper in 1986 definitely positioned KrF excimer laser lithography as the next-generation tool after g- and i-line lithographies. However, KrF excimer laser lithography is now (1999) used for manufacturing some limited devices such as 256-Mb-DRAMs. Thus, a long development time, reaching about 10 years, has been required to put the technology into practical use.

Why was such a long time necessary? This was due to the lack of reliability of the projection lens and the excimer laser light source in the early stages of KrF excimer laser stepper development. However, the most important and definitive reasons were (i) the remarkable improvement of DQ/N resist chemistry, which prolonged the life of i-line lithography, and (ii) the delay of CA resist material development for KrF excimer laser exposure. These historical facts indicate that the key technology for lithography is resist materials and processes.

From this point of view, basic and recent progress in resists were described in this chapter. In both optical and non-optical lithography, the key factor for resist design is the dissolution-rate contrast. A significantly large dissolution-rate contrast is obtainable in DQ/N resist systems by designing the molecular structures of the PAC and novolac to interact with each other efficiently. DQ/N resists have a high performance for g- and i-line exposures. However, their transparency at 248 nm in KrF excimer laser exposure is inadequate to obtain a good resist profile. Thus, a CA system using PHS as a base polymer was developed to overcome this technical problem. As a result, KrF excimer laser resists with high resolution and high sensitivity were successfully obtained.

The issue of transparency again became the critical requirement in the development of ArF excimer laser resists due to the high absorption of the aromatic ring at the 193-nm wavelength. This technical barrier is now being successfully solved by using an aliphatic polymer with alicyclic pendant groups.

The dominant radiochemical reactions in EB and X-ray resists are caused by the ejected electrons which still have keV-level energies. Thus, there is a linear relationship between the sensitivities of EB and X-ray resists. To achieve higher sensitivity, the CA system was also used in EB and X-ray resists. The adoption of the CA system brings a flexibility to resist design, and the CA resist formulations used in deep-UV resists are fundamentally applicable to EB and X-ray resists without any changes.

Advanced resist technologies such as bilayer, tri-layer, and top-surface silylation techniques were considered, as ways of overcoming the resolution limit of single-layer resists on an actual device topology. The key point of these techniques is the dry-development process, which enables pattern formation with a very high aspect ratio. The current dry-development technique, namely top-surface imaging, satisfies both wide process latitude and process simplicity.

6.7 References

1. See for example, *Introduction to Microlithography*, edited by L. F. Thompson, C. G. Willson and M. J. Bowden, *ACS Symposium Series* **219**, (1983), Chapters 3 and 4, American Chemical Society, Washington, D.C.; W. M. Moreau, *Semiconductor Lithography*, Chapters 2 to 5, Plenum Press, New York (1988); *Materials for Microlithography*, edited by L. F. Thompson, C. G. Willson and J. M. J. Fréchet, *ACS Symposium Series,* **266 (1984)**; *Polymers in Microlithography*, edited by E. Reichmanis, S. A. MacDonald and T. Iwayanagi, *ACS Symposium Series,* **412** (1989); *Polymers for Microelectronics, Proc. of PME '98*, edited by Y. Tabata *et al.*, Kodamasha, Tokyo (1990); *Polymers for Microelectronics*, edited by L. F. Thompson, C. G. Willson and S. Tagawa, *ACS Symposium Series,* **537** (1992); *Polymeric Materials for Microlithographic Application*, edited by H. Ito, S. Tagawa and K. Horie, *ACS Symposium Series,* **579** (1994); *Microelectronic Technology*, edited by E. Reichmanis, C. K. Ober, S. A. MacDonald, T. Iwayanagi and T. Nishikubo, *ACS Symposium Series,* **614** (1995)
2. H. Ito *et al., Proc. SPIE,* **1086**, 11 (1989)
3. F. Coopmans, B. Roland and M. K. Templeton, *Proc. SPIE,* **631**, 34 (1986); B. Roland, R. Lombaerts and C. Jakus, *ibid.*, **771**, 69 (1987)
4. B. K. Daniels, C. R. Szmanda and P. Trefonas III, *Proc. SPIE,* **631**, 192 (1986)
5. V. Rao, W. D. Hinsberg, C. W. Frank and R. F. W. Pease, *Proç. SPIE,* **1925**, 538 (1993); *ibid.*, **2195**, 596 (1994); B. T Beauchemin, C. E. Ebersole and L. S. Daraktchiev, *ibid.*, **2195**, 610 (1994)
6. J. B. Lounsbury and M. A. Namashinham, *ACS Conf.: Coating and Plastics,* **37**(2), 125 (1977)
7. E. J. Walker, *IEEE Trans., Electron Devices,* **ED-22**(7), 464 (1975); W. G. Oldham, S. N. Nandgaonkar, A. R. Neureuter and M. O'Toole, *ibid.*, **ED-26**(4), 717 (1979)
8. P. Trefonas III, B. K. Aniels, M. J. Eller and A. Zampini, *Proc. SPIE,* **920**, 203 (1988); H. Ito, *Jpn. J. Appl. Phys.,* **31**, 4273 (1992)
9. H. Gokan, S. Esho and Y. Ohnishi, *J. Electrochem. Soc.,* **130**(1), 143 (1983)
10. L. A. Pederson, *J. Electrochem. Soc.,* **129**(1), 205 (1982)
11. M. Endo *et al., IEDM Tech. Digest,* p. 31 (1992); Y. Kaimoto, K. Nozaki, S. Takeuchi and N. Abe, *Proc.*

SPIE, **1672**, 67 (1992); K. Nakano *et al., ibid.,* **2195**, 194 (1994)

12. S. Fujimura, J. Konno, K. Hikazutani and H. Yano, *Jpn. J. Appl. Phys.,* **28**(10), 2130 (1989); P. Burggraaf, *Semiconductor International,* June, p. 66 (1992)
13. F. H. Dill, *IEEE Trans., Electron Devices,* **ED-22**(7), 440 (1975); F. H. Dill, W. P. Hornberger, P. S. Hauge and J. M. Shaw, *ibid.,* 445 (1975); F. H. Dill, A. R. Neureuther, J. A. Tuttle and E. J. Walker, *ibid.,* 456 (1975)
14. D. J. Kim, W. G. Oldham and A. R. Neureuther, *IEEE Trans., Electron Devices,* **ED-31**(12), 1730 (1984); C. A. Mack, *Proc. SPIE,* **538**, 207 (1987); T. Ohfuji, K. Yamanaka and M. Sakamoto, *ibid.,* **920**, 190 (1988); C. A. Mack, *J. Electrochem. Soc.,* **139**(4), L35 (1992)
15. O. Süs, *Justus Liebigs Ann. Chem.,* **556**, 65 (1944)
16. J. Pakansky and J. R. Lyerla, *IBM J. Res. Develop.,* **23**(1), 42 (1979); *J. Electrochem. Soc.,* **124**, 862 (1977)
17. D. Ilten and R. Sutton, *J. Electrochem. Soc.,* **119**, 539 (1972); B. Broyde, *ibid.,* **117**, 1555 (1970); A. Paramov, *Chem. Abstr.,* **81**, 97720 (1974)
18. M. Hanabata, A. Furuta and Y. Uemura, *Proc. SPIE,* **771**, 85 (1987); M. Hanabata, Y. Uetani and A. Furuta, *J. Vac. Sci. Technol.,* **B7**, 640 (1989)
19. K. Honda *et al., Proc. SPIE,* **1262**, 493 (1990)
20. M. Koshiba, M. Murata, M. Matsui and Y. Harita, *ibid.,* **920**, 364 (1988)
21. K. Honda *et al., Proc. SPIE,* **1672**, 297 (1992)
22. M. Hanabata, Y. Uetani and A. Furuta, *ibid.,* **920**, 349 (1988)
23. T. F. Yeh, H. Y. Shih and A. Reiser, *Proc. SPIE,* **1672**, 204 (1992); C. C. Lin *et al., ibid.,* **1925**, 647 (1993)
24. R. A. Arcus, *Proc. SPIE,* **631**, 124 (1986); M. J. Hanrahan and K. S. Hollis, *ibid.,* **771**, 128 (1987); J. P. Haung, K. Kwei and A. Reiser, *ibid.,* **1086**, 74 (1989); P. Trefonas III, *ibid.,* **1086**, 484 (1989); K. Honda *et al., ibid.,* **1925**, 197 (1993); R. R. Dammel and M. A. Khadim, *ibid.,* **1925**, 205 (1993); H. Y. Shih *et al., ibid.,* **2195**, 514 (1994); P. Trefonas *et al., ibid.,* **707** (1994)
25. K. Uenishi *et al., Proc. SPIE,* **1672**, 262 (1992)
26. H. Miyamoto *et al., Proc. SPIE,* **2438**, 223 (1995); Y. Kawabe *et al., ibid.,* **2724**, 420 (1996)
27. K. Uenishi, Y. Kawabe, T. Kokubo and A. J. Blakeney, *Proc. SPIE,* **1466**, 102 (1991)
28. O. Naramasu *et al., Proc. SPIE,* **1262**, 32 (1990); O. Naramasu *et al., ibid.,* **1466**, 13 (1991)
29. W. deForest, *Photoresist Materials and Processes,* (1975), McGraw-Hill, N.Y., p. 54
30. S. Ogawa, S. Uoya, H. Kimura and H. Nagata, *Proc. 1st Microprocess Conf., Tokyo,* p. 162 (1988)
31. S. Sasago *et al., Proc. SPIE,* **1086**, 300 (1989)
32. J. Pacansky and H. Coufel, *J. Amer. Chem. Soc.,* **102**, 410 (1980); D. W. Johnson, *Proc. SPIE,* **469**, 72 (1984)
33. Examples for novolac resins are references 4–6; M. Hanabata *et al., Proc. SPIE,* **631**, 76 (1986); M. K. Templeton *et al., ibid.,* **771**, 148 (1987); K. Honda *et al., ibid.,* **1466**, 141 (1991); T. Kajiita *et al., ibid.,* 161 (1991); A. T. Jeffries III *et al., ibid.,* **1925**, 235 (1993) Examples for DNQ-PACs are references 10, 12; P. Trefonas III *et al., Proc. SPIE,* **771**, 194 (1987); S. Tan *et al., ibid.,* **1262**, 513 (1990); R. Hanawa *et al., ibid.,* **1672**, 231 (1992); W. Brunsvold *et al., ibid.,* 273 (1992); H. Nemoto *et al., ibid.,* 305 (1992); R. Hanawa *et al., ibid.,* **1925**, 227 (1993); A. J. Blakeney *et al., ibid.,* **2434**, 312 (1995); A. J. Blakeney *et al., ibid.,* 324 (1995)
34. C. A. Mack, *KTI Microlithography Seminar, Interface '90,* 1 (1990); P. Spragg *et al., Proc. SPIE,* **1466**, 283 (1991)
35. T. Ohfuji, K. Yamanaka and M. Sakamoto, *ibid.,* **920**, 190 (1988); R. Hurditch and J. Ferri, *OCG Microlithography Seminar, Interface '94,* 71 (1994)
36. J. B. Lounsbury and M. A. Namashinham, *ACS Conf.: Coating and Plastics,* **37**(2), 125 (1977)
37. M. A. Touky, *Proc. SPIE,* **771**, 264 (1987); M. Furuta, S. Asaumi and A. Yokota, *ibid.,* **1466**, 477 (1991); S. G. Hansen, R. J. Hurditch, D. J. Brzozowy and S. A. Robertson, *ibid.,* **1925**, 626 (1993); B. T. Beauchemine, R. J. Hurditch, M. A. Touky and A. J. Blakeney, *ibid.,* **2438**,

282 (1995); Y. Kawabe *ibid.*, **2724**, 420 (1996)

38. K. Miura *et al.*, *Proc. SPIE*, **920**, 134 (1988)
39. B. J. Lin, *J. Vac. Sci. Technol.*, **12**, 1317 (1975)
40. M. Nakase *et al.*, *Proc. SPIE*, **773**, 226 (1987)
41. N. Abe *et al.*, *J. Photopolym. Sci. Technol.*, **8**, 637 (1995)
42. M. P. Grandpre, D. A. Vidusek and M. W. Legenza, *Proc. SPIE*, **539**, 103 (1985)
43. H. Ito, C. G. Wilson and J. M. J. Fréchet, *Symp. on VLSI Technology, Tech. Digest*, 86 (1982)
44. J. V. Crivello, *Polym. Eng. Sci.*, **23**, 953 (1983)
45. T. Ueno *et al.*, *J. Photopolym. Sci. Technol.*, **7**, 397 (1994)
46. Y. Onishi *et al.*, *Proc. SPIE*, **2724**, 70 (1996)
47. J. W. Thackeray *et al.*, *J. Photopolym. Sci. Technol..*, **7**(3), 619 (1994)
48. T. Kumada *et al.*, *ibid.*, **6**(4), 571 (1993)
49. S. A. MacDonald *et al.*, *Proc. SPIE*, **1466**, 2 (1991)
50. D. J. H. Funhoff, H. Binder and R. Schwalm, *Proc. SPIE*, **1672**, 46 (1992)
51. Y. Kawai *et al.*, *J. Photopolym. Sci. Technol.*, **8**(4), 535 (1995)
52. S. Saito *et al.*, *ibid.*, **9**(4), 677 (1996)
53. T. Hattori *et al.*, *ibid.*, **9**(4), 611 (1996)
54. K. Deguchi, T. Ishiyama, T. Horiuchi and A. Yoshikawa, *Jpn. J. Appl. Phys.*, **29**, 2207 (1990)
55. L. Schlegel, T. Ueno, N. Hayashi and T. Iwayanagi, *J. Vac. Sci. Technol.*, **B9**, 278 (1991)
56. K. Asakawa, *J. Photopolym. Sci. Technol.*, **6**(4), 505 (1993)
57. K. Asakawa, T. Ushirogouchi and M. Nakase, *J. Vac. Sci. Technol.*, **B13**, 833 (1995)
58. J. Nakamura, H. Ban, Y. Kawai and A. Tanaka, *J. Photopolym. Sci. Technol.*, **8**(4), 555 (1995)
59. J. Nakamura, H. Ban, Y. Kawai and A. Tanaka, *ibid.*, **7**(3), 501 (1994)
60. W. D. Hinsberg, S. A. MacDonald, N. J. Clecak and C. D. Snyder, *Proc. SPIE*, **1672**, 24 (1992)
61. W. Hinsberg, S. MacDonald, N. Clecak and C. Snyder, *J. Photopolym. Sci. Technol.*, **6**(4), 535 (1993)
62. H. Ito *et al.*, *ibid.*, **7**(3), 433 (1994)
63. T. Ushirogouchi, K. Asakawa, M. Nakase and A. Hongu, *Proc. SPIE*, **2438**, 609 (1995)
64. H. Roshert *et al.*, *Proc. SPIE*, **1925**, 14 (1993)
65. N. Kihara, S. Saito, T. Ushirogouchi and M. Nakase, *J. Photopolym. Sci. Technol.*, **8**(4), 561 (1995)
66. R. D. Allen *et al.*, *ibid.*, **6**(4), 575 (1993)
67. Y. Kaimoto, K. Nozaki, S. Takechi and N. Abe, *Proc. SPIE*, **1672**, 66 (1992)
68. G. M. Wallraff *et al.*, *J. Vac. Sci. Technol.*, **B11**, 2783 (1993)
69. K. Nakano *et al.*, *Proc. SPIE*, **2195**, 194 (1994)
70. N. Shida, T. Ushirogouchi, K. Asakawa and M. Nakase, *J. Photopolym. Sci. Technol.*, **9**(3), 457 (1996)
71. R. D. Allen *et al.*, *ibid.*, **8**(4), 623 (1995)
72. S. Takechi *et al.*, *J. Photopolym. Sci. Technol.*, **9**(3), 475 (1996)
73. T. Naito *et al.*, *Jpn. J. Appl. Phys.*, **33**, 7028 (1994)
74. I. Haller, M. Hatzakis and R. Srinivasan, *IBM J. Res. Develop.*, **12**, 251 (1968)
75. T. H. P. Chang *et al.*, *IBM J. Res. Develop.*, **20**, 376 (1976)
76. D. L. Spears and H. I. Smith, *Solid State Technol.*, **15**(7), 21 (1972)
77. H. Ito and C. G. Wilson, *ACS Symp. Series*, **242**, *Polymers in Electronics* (1984), ACS, Washington, D.C., p. 11
78. R. Viswanathan *et al.*, *J. Vac. Sci. Technol.*, **B11**, 2910 (1993)
79. K. Deguchi *et al.*, *ibid.*, **B13**, 3040 (1995)
80. Y. Nishioka *et al.*, *Intl. Mtg. on Electron Devices, Tech. Digest*, (1995), IEEE, Washington D.C., pp. 903–6
81. H. Ban *et al.*, *Photon Factory Activity Report*, **8**, 35 (1990)
82. M. Morita *et al.*, *Jpn. J. Appl. Phys.*, **22**, L659 (1983)
83. H. Ban, A. Tanaka, Y. Kawai and K. Deguchi, *Jpn. Appl. Phys.*, **28**, L1863 (1989)
84. A. Tanaka, H. Ban and S. Imamura, *ACS Symp. Series*, **412**, *Polymers in Microlithography* (1989), ACS, Washington, D.C., p. 175.
85. H. Ban, A. Tanaka and S. Imamura, *Jpn. J. Appl. Phys.*, **28**, L1863 (1989)
86. H. Ban, J. Nakamura, K. Deguchi and A. Tanaka, *J. Vac. Sci. Technol.*, **B9**, 3387 (1991)
87. H. Ban, J. Nakamura, K. Deguchi and A. Tanaka, *J. Vac. Sci. Technol.*, **B12**, 3905 (1994)

88. J. Lingnau, R. Dammel and J. Theis, *Solid State Technol.*, Sept., 105 (1989)
89. J. Lingnau, R. Dammel and J. Theis, *Solid State Technol.*, Oct., 107 (1989)
90. K. Mochijji, Y. Soda and T. Kimura, *J. Electrochem. Soc.*, **133**, 147 (1986)
91. B. L. Henke *et al.*, *Atomic Data and Nuclear Data Tables*, **27**, 1 (1982)
92. G. N. Taylor, G.A. Coquin and S. Somekh, *Polym. Eng. Sci.*, **17**, 420 (1977)
93. M. Kaplan, *Polym. Eng. Sci.*, **23** 957 (1977)
94. K. Harada, *J. Appl. Polym. Sci.*, **26**, 3395 (1981)
95. T. N. Bowmer and J. H. O'Donnel, *J. Appl. Polym. Sci., Polym. Chem. Ed.*, **19**, 45 (1981)
96. T. N. Bowmer and M. J. Bowden, *ACS Symp. Series*, **242**, *Polymers in Electronics*, (1984), ACS, Washington, D.C., p. 153
97. S. Imamura and S. Sugawara, *Jpn. J. Appl. Phys.*, **21**, 776 (1982)
98. H. Okamoto and T. Iwai, *Radiat. Phys. Chem.*, **18**, 407 (1981)
99. R. Basheer and M. Dole, *Radiat. Phys. Chem.*, **25**, 389 (1985)
100. J. E. Guillet, *ACS Symp. Series*, **346**, *Polymers for High Technology* (1987), ACS, Washington, D.C., p. 46
101. M. J. Bowden, L. F. Thompson, S. R. Fahrenholtz and E. M. Doerries, *J. Electrochem. Soc.*, **128**, 1304 (1981)
102. S. Tagawa, *ACS Symp. Series*, **346**, *Polymers for High Technology* (1987), ACS, Washington, D.C., p. 37
103. L. F. Thompson, L. E. Stillwagon and E. M. Doerries, *J. Vac. Sci. Technol.*, **15**, 938 (1978)
104. T. Kozawa, Y. Yoshida, M. Uesaka and S. Tagawa, *Jpn. J. Appl. Phys.*, **31**, 4301 (1992)
105. D. R. McKean, Y. Schaedeli and S. A. McDonald, *ACS Symp. Series*, **412**, *Polymers in Microlithography* (1989), ACS, Washington, D.C., p. 27
106. H. Liu, M. P. DeGrandpre and W. E. Feely, *J. Vac. Sci. Technol.*, **B6**, 379 (1988)
107. R. Dammel *et al.*, *Proc. SPIE*, **1262**, *Advances in Resist Technology and Processing VII*, 378 (1990)
108. H. Ito, M. Ueda and W. P. England, *Macromolecules*, **23**, 2589 (1990)
109. H. Ito and C. G. Willson, *Polym. Eng. Sci.*, **23**, 1012 (1983)
110. H. Shiraishi *et al.*, *J. Vac. Sci. Technol.*, **B9**, 3343 (1991)
111. J. Lingnau, R. Dammel and J. Theis, *Polym. Eng. Sci.*, **29**, 874 (1989)
112. J. W. Thackeray *et al.*, *Proc. SPIE*, **1086**, 34 (1989)
113. H. Ito, in *Radiation Curing in Polymer Science and Technology, Vol. IV* (1993), eds. J. P. Fouassier and J. F. Rabek (Elsevier, London), p. 237
114. S. A. MacDonald *et al.*, *Proc. SPIE*, **1466**, *Advances in Resist Technology and Processing VIII*, 2 (1991)
115. W. D. Hinsberg, S. A. MacDonald, N. J. Clecak and C. D. Snyder, *Proc. SPIE*, **1672**, *Advances in Resist Technology and Processing IX*, 24 (1992)
116. K. Kurihara *et al.*, *Jpn. J. Appl. Phys.*, **34**, 6940 (1995)
117. J. Mohr, W. Ehrfeld and W. D. Muenchmeyer, *J. Vac. Sci. Technol.*, **B6**, 2264 (1988)
118. W. Conley *et al.*, *Proc. SPIE*, **1466**, *Advances in Resist Technology and Processing VIII*, 53 (1991)
119. C. P. Babcock *et al.*, *ibid.*, 653 (1991)
120. M. A. McCord, A. Wagner and D. Seeger, *J. Vac. Sci. Technol.*, **B11**, 2881 (1993)
121. K. Deguchi *et al.*, *ibid.*, **B10**, 3040 (1992)
122. W. G. Waldo, A. D. Katnani and H. Sachdev, *Proc. SPIE*, **1671**, 2 (1992)
123. K. Deguchi *et al.*, *J. Vac. Sci. Technol.*, **B13**, 3040 (1995)
124. H. Ito, G. Breyta, D. C. Hofer and R. Sooriyakumaran, *ACS Symp. Series*, **614**, *Microelectronics Technology* (1995), ACS, Washington, D. C., p. 21
125. J. Nakamura, H. Ban, K. Deguchi and K. Tanaka, *Jpn. J. Appl. Phys.*, **30**, 2619 (1991)
126. J. Nakamura, H. Ban and A. Tanaka, *Jpn. J. Appl. Phys.*, **31**, 4294 (1992)
127. L. Schlegel, T. Ueno, N. Hayashi and T. Iwayanagi, *J. Vac. Sci. Technol.*, **B9**, 278 (1991)
128. M. Zuniga *et al.*, *ibid.*, **B11**, 2862 (1993)
129. C. P. Umbach *et al.*, *J. Vac. Sci. Technol.*, **B6**, 319 (1988)

130. K. Deguchi, K. Miyoshi, T. Ishii and T. Matsuda, *Jpn. J. Appl. Phys.*, **31**, 2954 (1992)
131. E. W. Scheckler, S. Shukuri and E. Takeda, *Jpn. J. Appl. Phys.*, **32**, 327 (1993)
132. M. Nagase *et al.*, *Microelectronic Eng.*, **30**, 419 (1996)
133. Y. Ochiai, S. Manako, J. Fujita and E. Nomura, *J. Photopolym. Sci. Technol.*, **10**, 641 (1997)
134. M. Hatzakis, J. Paraszczak and J. Shaw, *Proc. Intl. Conf. Microlithography*, p. 386 (1981)
135. S. A. MacDonald *et al.*, *J. Photopolym. Sci. Technol.*, **4**, p. 487 (1991)
136. D. W. Johnson and M. A. Hartney, *Jpn. J. Appl. Phys.*, **31**, 4321 (1992)
137. T. Fujino *et al.*, *J. Vac. Sci. Technol.*, **B11**, 2773 (1993)
138. J. M. Moran and D. Mayden, *J. Vac. Sci. Technol.*, **16**, 1620 (1979)
139. W. C. McColgin, J. Jech, R. C. Daly and T. B. Brust, *Symp. on VLSI Technology, Tech. Digest*, (1987)
140. F. Coopmans and B. Roland, *Proc. SPIE*, **631**, 34 (1986)
141. W-S. Han *et al.*, *Proc. SPIE*, **1925**, 291 (1993)
142. J. P. W. Schellekens, *Proc. SPIE*, **1086**, 220 (1989)
143. D. W. Johnson and M. A. Hartney, *Jpn. J. Appl. Phys.*, **31**, 4321 (1992)
144. R. R. Kunz, P. A. Bianconi, M. W. Horn and R. R. Paladugu, *Proc. SPIE*, **1466**, 218 (1991)
145. J. Shaw, *Solid State Technol.*, June, 83 (1987)
146. Y. Saotome *et al.*, *Solid-state Sci. and Technol.*, April, 909 (1985)
147. T. W. Weidman, O. Joubert, A. M. Joshi and R. L. Kostelak, *Proc. SPIE*, **2438**, 496 (1995)
148. K-H. Baik *et al.*, *J. Vac. Sci. Technol.*, **B8**(6), 1481 (1990)
149. K-H. Baik, L. Vandenhove and B. Roland, *J. Vac. Sci. Technol.*, **B9**(6), 3399 (1991)
150. T. Ohfuji and N. Aizaki, *Symp. on VLSI Technology, Tech. Digest*, 93 (1994)
151. Maeda *et al.*, *Proc. SPIE*, **2438**, 465 (1995)
152. D. R. Wheeler *et al.*, *J. Vac. Sci. Technol.*, **B11**(6), 2789 (1993)
153. M. W. Horn, M. A. Hartney and R. R. Kunz, *Proc. SPIE*, **1672**, 448 (1992)
154. J. Hutchinson, Y. Melaku, W. Nguyen and S. Das, *Proc. SPIE*, **2724**, 399 (1996)

Metrology, defect inspection, and repair

7

Tadahito Matsuda, Toru Tojo and Seiichi Yabumoto

7.1 Introduction

Metrology, defect inspection, and repair will prove to be extremely useful means of estimation, evaluation, analysis, and yield improvement. Quantitative assessment of exposed pattern size is vital in order to solve a wide variety of practical problems in lithography. Lithographic problems degrading IC device performance relate primarily to the nature of pattern deterioration which in turn, depends on pattern size, shape, positioning error, and defects. Pattern measurement and analysis are thus indispensable to the study of the entire range of lithographic problems.

This chapter covers the current status and projected trends in metrology, defect inspection, and repair. Particular attention is paid to the technologies used for both photomasks and X-ray masks.

7.2 Metrology

7.2.1 Metrology in IC manufacturing

ULSI (ultra-large-scale-integration) devices are made up of three-dimensional structures with more than ten layers. For good IC performance, the dimensions of the patterns (features) created on each layer in the wafer and the placement of the features between the layers should be accurate. Many measurements are used to monitor and to control feature dimensions and feature-placement accuracy. The thickness of films is usually measured by optical methods, such as a multilayer interferometer or ellipsometry. Measurement of critical dimensions (CD) and overlay (feature-placement) accuracy are major methods for monitoring and controling feature size and feature placement. Many measurements still remain off-line. But CD and overlay measurements are carried out in-line. Measurement results obtained using a test sample or a pilot wafer are fed back to the apparatus for calibration.

7.2.2 Principal measurement items in lithography

Optical lithography is now a main-stream technology. In particular, a stepping projection aligner (also called a stepper) is a major technology in 0.25-μm resolution lithography. In the stepping projection aligner, 5× masks (also called reticles) are used widely. For high accuracy with regard to linewidth and overlay accuracy on the wafer, the stepper's performance must be high and the patterns on a mask must be accurate. The CD and overlay required in each LSI 'generation' are listed in Table 7.1. CD uniformity must usually be controlled to be less than 10% of the design rule (minimum feature size) and overlay accuracy must be within 30% of the design rule. Overlay accuracy has been the most critical issue in lithography.

A recent change of mask reduction ratio, from 5× to 4×, and additional means for enhancing optical lithography, such as phase-shift mask technology and proximity-effect correction techniques, have made CD and overlay control more difficult.

7.2.3 Requirement for CD measurement and overlay measurement

As described above, linewidth deviation from the design rule must be less than 10% of the minimum linewidth, and the maximum overlay error on the wafer (total overlay) should be less than 30% of the minimum feature size. It is often said that the 3-sigma (3σ) repeatability of a measurement system must to be less than 10% of the minimum feature size to be measured. The 3-sigma repeatability of the CD and overlay measurement systems for the wafer must, therefore, respectively be less than 1% and less than 3% of the design rule. These wafer metrology requirements are summarized in Table 7.2.

Mask metrology requirements are a little more complicated because the total CD error or total overlay error comes not only from a mask but also from the exposure system (stepper). Errors due to steppers, for example, are caused by magnification variation of the projection lens, projection-lens distortion, machine-to-machine matching error, and alignment error at the time of exposure.

Recent technology requires the CD error on the mask to be less than one-seventh (14%) of the design rule when the reduction ratio is 5× and less than one-ninth (11%) of the design rule when the reduction ratio is 4×. These errors correspond to one-thirty-fifth (3%) of the minimum linewidth on the mask. Pattern-placement error on the mask should be within 20% of the design rule in a system with 5× reduction and less than 16% of the design rule in a system with 4× reduction. These errors are one-twenty-fifth (4%) of the minimum linewidth on the mask. The tolerances for these errors are determined to keep the performance of all devices within a certain range to obtain normally operating semiconductor chips.

Reproducibility requirements for mask metrology systems are about one-fourth (25%) of the CD tolerance or placement accuracy tolerance on the mask. The 3-sigma reproducibility of systems for measuring CD and pattern placement must therefore respectively be less than one-hundred-and-fortieth (0.7%) and one-hundredth (1%) of the minimum linewidth on the mask. These requirements are listed in Table 7.3.

Table 7.1 *CD and overlay requirements in each LSI generation.*

Generation		Requirement	
Design rule (Resolution) (μm)	DRAM capacity (bits)	CD control (nm)	Overlay (nm)
0.35	64M	35	100
0.25	256M	25	75
0.18	1G	18	50
0.13	4G	13	40

Table 7.2 *Wafer metrology requirements in each generation (CD and overlay).*

Generation		Tolerance		Measurement accuracy	
Design rule (Resolution) (μm)	DRAM capacity (bits)	CD (nm)	Overlay (nm)	CD (nm)	Overlay (nm)
0.35	64M	35	100	3.5	10
0.25	256M	25	75	2.5	7.5
0.18	1G	18	50	1.8	5
0.13	4G	13	40	1.3	4

Table 7.3 *Mask metrology requirements in each generation (CD and pattern placement).*

Generation		Mask		Tolerance on a mask		Measurement accuracy	
Design rule on a wafer (μm)	DRAM capacity (bits)	Magnification	Design rule on a mask (μm)	CD (nm)	Pattern placement (nm)	CD (nm)	Pattern placement (nm)
0.35	64M	5×	1.75	50	70	12	18
0.25	256M	4×	1.00	28	40	7	10
0.18	1G	4×	0.72	20	28	5	7
0.13	4G	4×	0.52	15	25	4	6

7.2.4 CD measurement methods

The design rule of a 64-Mb DRAM is about 350 nm and that for a 1-Gb DRAM will be less than 200 nm. Such fine patterns (figures) on a wafer cannot be observed by using an optical probe (optical microscope, laser spot scanning, and so on). Even a confocal method, which can improve optical resolution a little, would not have enough resolution for pattern sizes below 200 nm. EB probes, which have a 5-nm resolution (less than one-tenth of the design rule of a 1-Gb DRAM), have been used for CD measurement. To avoid the charge-up phenomenon, scanning electron microscopes (SEMs) with very low acceleration voltages have been widely used, and SEMs with field-emission guns are the tools most often used in the semiconductor industry to get good images.

Over the past ten years various technologies improving the performance of optical lithography systems have been developed, and some of these technologies, such as phase-shifting masks and proximity-effect correction, have been used in pilot schemes or studies, and in production lines. These technologies reduce the minimum feature size on a mask to a few tenths of a micron, and such patterns are measured using EB-based instruments.

Figure 7.1 shows a typical relationship between actual and measured CD values. For ideal linewidth measurement, this curve should be linear. Here ΔI is repeatability for the same specimen. L_0 is the point where linearity between the actual and the measured linewidth value breaks down, which corresponds to the resolution of the system.

An offset ΔL depends on the algorithm that detects the position of an edge from the signal waveform of the probe (sensor) output. Signal waveforms from an optical probe are usually less sharp than those from an EB probe, so optical metrology usually has a larger ΔL than EB metrology. And ΔL values are affected by the measurement means or the measurement algorithm as well as by the refractivity or film thickness of the specimen.

7.2.5 Overlay measurement methods

Methods for measuring overlay are different for a mask and for a wafer. Pattern placement on a mask is measured in terms of X–Y coordinates from the origin point on the mask. It is an absolute coordinate measurement. Overlay (registration) error on the wafer, on the other hand, is measured from the relative movement of registration target patterns in each lithography layer. In both methods, optical methods are widely used, for the following reasons.

(1) High resolution is not necessary, even in the measurement of wafer overlay, because alignment marks for registration are rather big.
(2) An electron-beam method has the disadvantage of long-term instability.
(3) An electron beam cannot 'see' target marks under an optically transparent layer.

Position data of features (address of pattern center) are measured, but linewidth data of patterns are not measured. The measurement offset, which exists in CD measurement as shown in Figure 7.1, does not occur as long as the specimen has a symmetrical profile.

7.2.6 Measurement accuracy

Each consecutive measured value using the equipment varies slightly. These fluctuations are specified as measurement accuracy and are often described by 3-sigma-variation values.

Repeatability is a term denoting output variation within a short time and without specimen

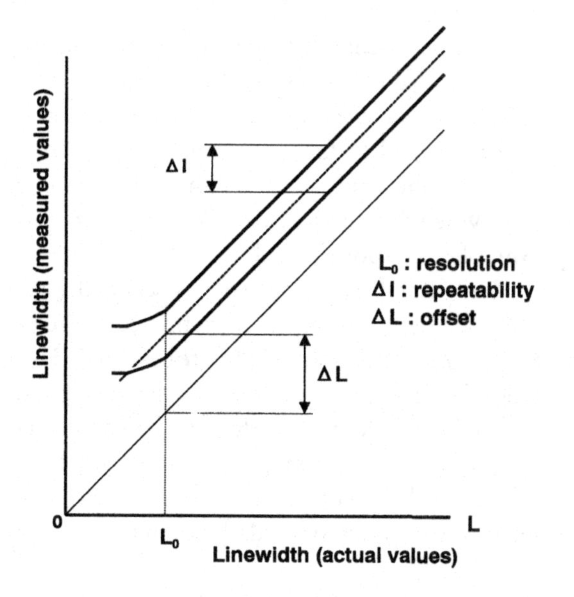

Figure 7.1 Characteristics of a CD measurement system.

resetting. Reproducibility (also called dynamic repeatability), on the other hand, denotes the output variation over a comparatively long time and under conditions with specimen resetting. Even if a measurement probe has good repeatability or excellent reproducibility, its accuracy may not be good enough.

The output of a two-dimensional probe, such as a CCD (charge-coupled device) camera, can also fluctuate because of spatial nonuniformity of signal intensity or because of probe sensitivity. Spatial change of the illumination, aberration (distortion) of an objective lens, and sensitivity variation of the sensor can cause such nonuniformity.

Even in point-to-point probing (point sensing), such as that using a photodiode or a photomultiplier, asymmetry of the target patterns or of specimen deformation causes fluctuation of measurement output. In these cases, measurement output fluctuation is due to movement of the probe relative to the specimen or vice versa. Rotation of the specimen around the sensor axis has an especially large effect on the measurement output.

7.2.7 CD measurement on a mask

Required mask accuracy and mask-measurement accuracy are listed in Table 7.3. Because the design rule for a mask is still larger than the optical resolution, an optical microsocope-based apparatus is widely used in mask metrology.

The optical resolution Δ of a microscope is a function of the wavelength of the illumination light (λ) and the numerical aperture of the objective lens (NA): $\Delta = K \times \lambda/\mathrm{NA}$. The depth of focus (DOF) is proportional to λ/NA^2 (or Δ^2/λ). When the NA is 0.9 and λ is 436 nm for example, the resolution is about 0.3 μm and DOF is about 0.5 μm. It is better to get high resolution by using a shorter wavelength than by using a lens with higher NA. This is because DOF can be made wider with a short wavelength. (DOF is not a major factor in mask measurement but is a key factor in wafer measurement because the surface of a wafer is uneven.)

In an optical-microscope-based system, transmitted illumination has an advantage over reflection illumination because the offset of a measured value (ΔL in Figure 7.1) can be made smaller in transmitted illumination. Figure 7.2 shows one example of such a system. In transmitted illumination, the image intensity profile of a flat specimen (like a mask) produced by an aberration-free lens is well known theoretically. And it is well known that one can determine the edge position accurately if the threshold level is set to 25% of the transmitted light intensity.[1] A CD value without any offset can be obtained under this condition. In edge detection using refracted or scattered light, however, the offset will be influenced by the specimen structure and will not be zero. Generally speaking, one weak point of an optical system is its sensitivity to focusing accuracy: the image intensity profile is severely affected by defocusing.

A laser is a coherent light source and can be easily focused to the resolution-limited size. The minimum spot size is proportional to λ/NA and is nearly equal to the resolution of microscope optics at the same wavelength and the same NA. When NA is 0.75 and λ is 325 nm, for example, the spot size will be 0.4 μm.

When this bright laser spot is scanned on a specimen, either transmitted, refracted, or scattered light can be used as the probe signal. This type of measurement system usually has a good repeatability, but interference phenomena sometimes affect the measurement accuracy.

The resolution of a system with conventional microscope optics or with laser-spot scanning

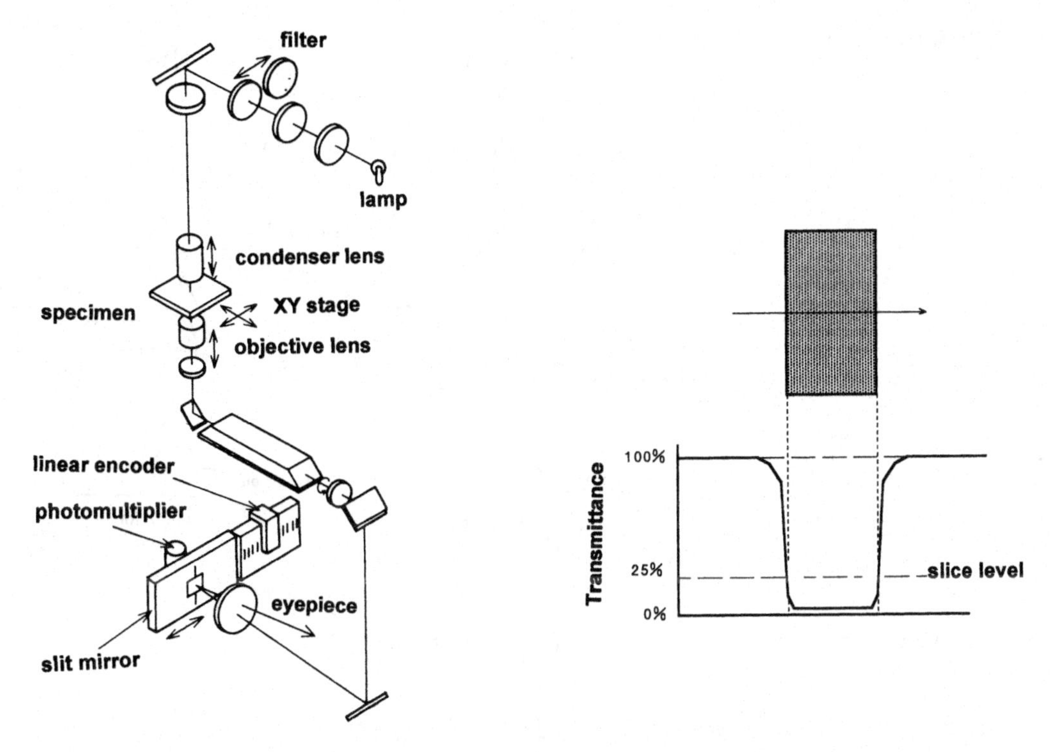

Figure 7.2 Transmission-microscope-based system for mask CD measurement.

can be increased by 30% if a confocal configuration is used.[2] Figure 7.3 shows the principle of a confocal system.

Confocal systems had been available only with laser optics but, in 1991, a real-time confocal system using a rotational disk (Nipkow disk) with many pinholes was invented and made it possible to convert a conventional microscope to a confocal microscope.[3]

In metrology, however, a confocal system is used only in a reflective illumination system or only in combination with a deflective lens system (transparency illumination) to enhance the edge contrast of a pattern image. The minimum linewidth measurable with either of these systems is about 0.3 μm.

The concept of a scanning electron microscope (SEM) is shown in Figure 7.4. This kind of microscope has a higher resolution than a confocal optical microscope, but it is not suitable for observing nonconducting materials (such as mask substrates) because a nonconductive specimen is charged up with electrons. Charging-up degrades measurement accuracy and can even make it impossible to get an image. A very-low-acceleration-voltage (600–1000-V) SEM or low-vacuum SEM[4] has recently been used to avoid the charge-up phenomenon.

SEM-based instruments have recently begun to be used in mask measurement. They are especially useful in observing the edge roughness of

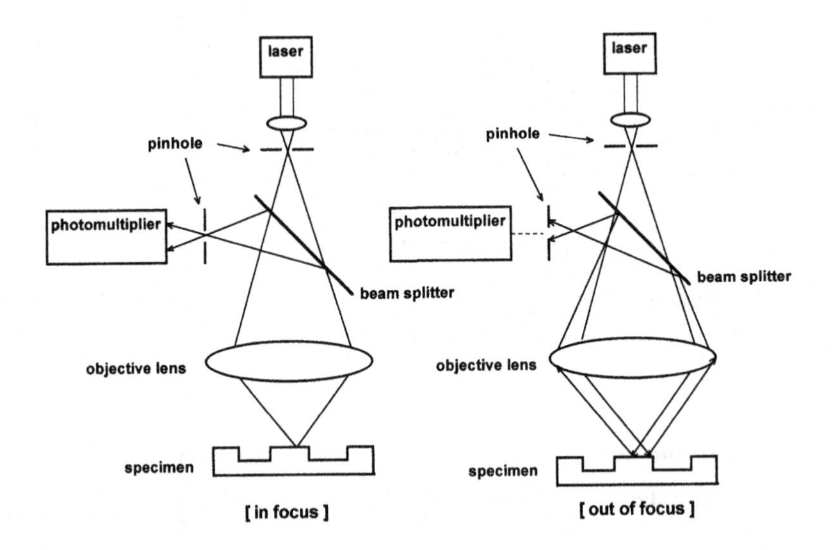

Figure 7.3 Confocal laser optics.

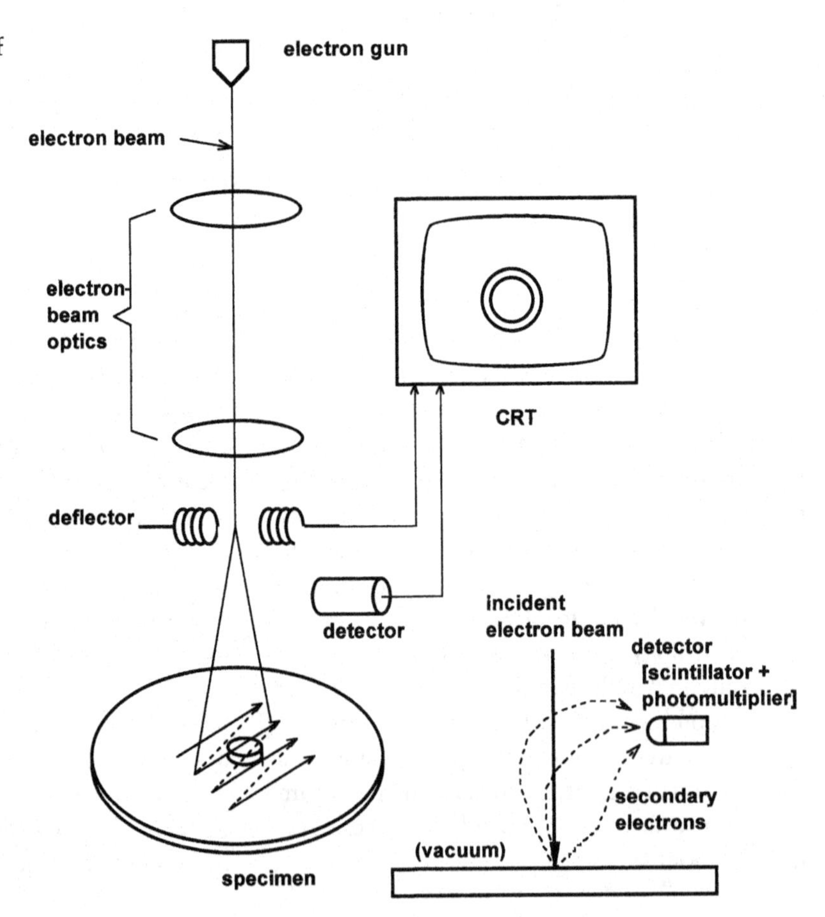

Figure 7.4 Schematic figure of a scanning electron microscope.

line patterns, the shape of contact-hole patterns, and the shifter patterns of phase-shifted masks.

7.2.8 CD measurement on a wafer (current technology)

SEM-based instruments are widely used and may be the only solution in CD measurement on a wafer with a resolution of less than a quarter-micron. In measuring the resist pattern on a wafer, the linewidth at the bottom is the dimension to be measured and controlled. An optical system, however, cannot observe or measure this bottom part. An electron-beam system has an advantage over an optical system for this kind of measurement because of its high resolution and large depth of focus.

In SEM-based systems, electron beams are scanned rasterwise on a specimen and the secondary electron signal from the specimen is usually monitored by an ET (Everhart–Thomley) detector (scintillator + photomultiplier) or a MCP (multi-channel plate). Figure 7.5 shows a popular algorithm for determining the linewidth of the pattern from the electron-beam probe signal.

An EB system has a high potential but there are two major problems using the SEM. One is the charging-up phenomenon (mentioned in the previous section) and the other is contamination of the specimen.

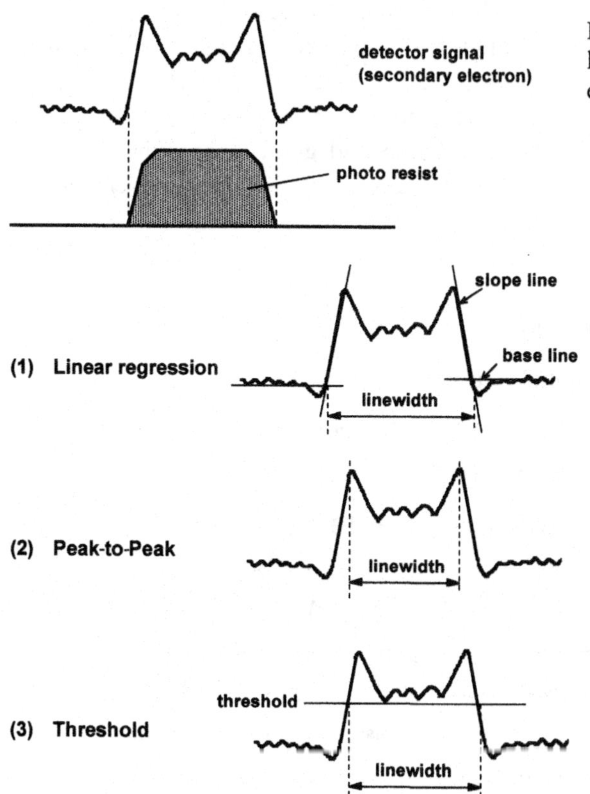

Figure 7.5 Determination of linewidth in a scanning electron microscope (SEM).

7.2.9 Overlay measurement on a mask

Measuring the X–Y coordinates of each pattern center is effective for monitoring pattern-placement accuracy on a mask. Using a probe for edge detection does this measurement by using a laser interferometer equipped with a precise X–Y stage. In transmitted mode, a hollow stage is necessary. Because a hollow stage has poor precision, optical probes in the nontransmitted mode are usually used.

Even though there are some offset problems in CD measurement when the nontransmitted mode is used, the coordinates of the pattern center can be determined precisely because a mask pattern is very thin and the profile of the pattern is symmetrical on both sides. An electron-beam probe is not used widely because it requires a vacuum chamber and it has the disadvantage of poor long-term stability.

Figure 7.6 illustrates a system that uses laser-spot scanning as an edge-detection sensor. A laser beam with a spot size of about 1 μm is focused on a specimen by an objective lens. Scattered light is detected by photodiodes (A and B) at both sides of the objective lens. A pattern edge is determined as a point where the strongest scattering light is detected (as shown in the figure). A pattern position is then determined as the center of the pattern edges on both sides. Moving the X–Y specimen stage under the stationary laser spot scans the laser spot. So this is a kind of point-sensing system. In point sensing there is no sensor-related or lens-related problem causing measurement output to fluctuate (as described in Subsection 7.2.6).

There is another system that uses slit scanning for edge sensing. In this system, a pattern edge is determined from the transmitted light intensity. In both systems, a laser interferometer is used as an x–y scale. To add to X–Y stage position, slit position must also be monitored in a slit-scanning system. The resolution of the laser interferometer is about 1 nm.

High X–Y stage accuracy and stable environmental conditions around the optical path of the laser interferometer are important to get high precision and good measurement repeatability. Figure 7.7 shows measurement errors

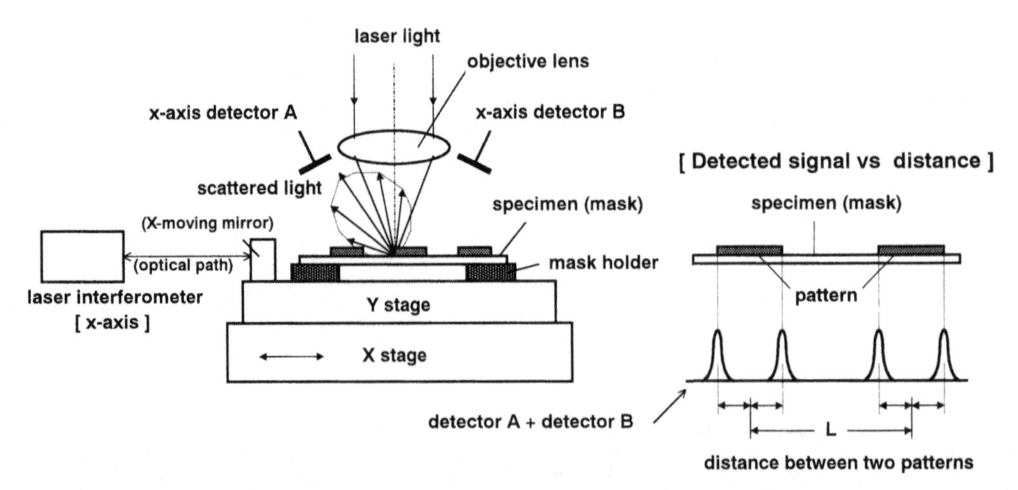

Figure 7.6 (*Left*) Overlay (X–Y coordinates) measurement system for a mask. (*Right*) Detected signal vs distance.

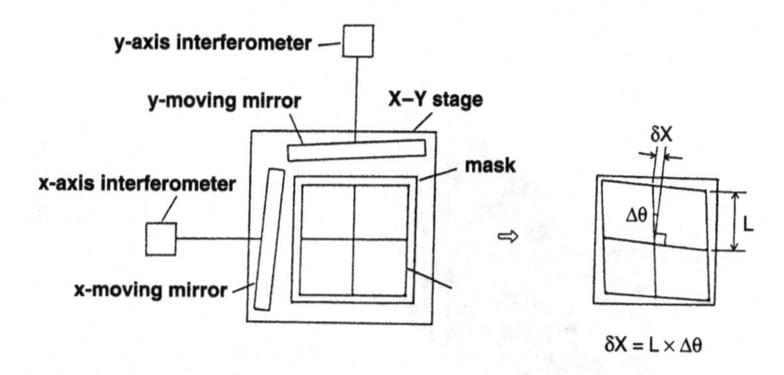

Figure 7.7 Errors caused by orthogonality deviation of laser interferometer in X–Y coordinates measurement.

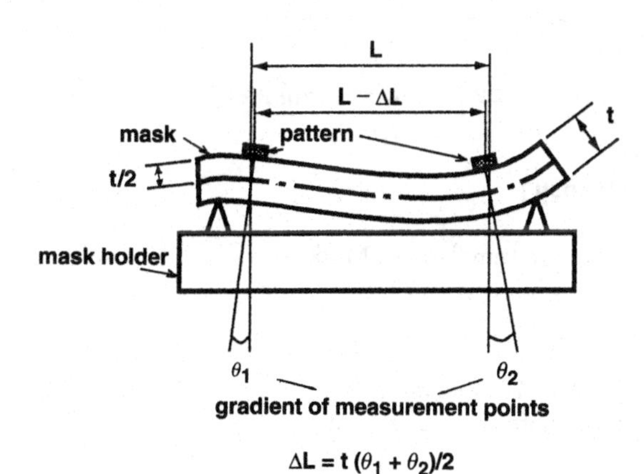

Figure 7.8 Errors caused by mask flexure in X–Y coordinates measurement $\Delta L = t(\theta_1 + \theta_2)/2$.

caused by orthogonal deviation or imperfect flatness of the moving mirrors of an X–Y laser interferometer. To correct errors after measurements, the software usually compensates for static deviations.

The wavelength of a laser in air is changed by changes in air temperature, air pressure, and humidity because these change the refractive index of the air itself. This refractive-index change is monitored by measuring each of these variables (temperature, air pressure, and humidity), and the change is then compensated for using software. A special interferometer called a wavelength tracker can also be used for this compensation.

There is another difficult problem in the overlay measurement. It comes from flexure of the mask substrate, which is quartz glass and is not stiff. The mask substrate is bowed on the X–Y specimen-stage by gravity and also by vacuum chucking. Because of this flexure of the substrate, pattern positions on the mask surface are shifted as shown schematically in Figure 7.8. Figure 7.9 shows an example of the pattern shift of a 5-inch (125-mm) mask. X, Y coordinate values of pattern position on the mask surface are a little fluctuated by the orientation of a mask (90 degrees, 180 degrees, or 270 degrees of rotation around the probe axis), because flexures at each posture are different. This difference is called the posture difference. A recent X–Y coordinate measurement system measures the gradient (grade) coefficient of the measurement point on the specimen, calculates the deviation due to flexure of the substrate, and then corrects the measured X–Y coordinate values.

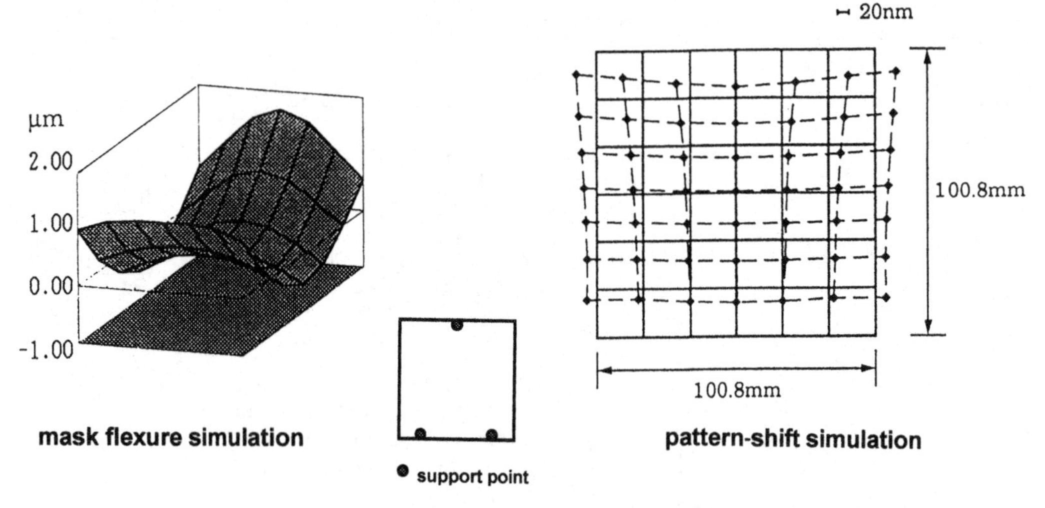

Figure 7.9 Pattern shift caused by gravity (simulation). Mask was 5-inch (125-mm) square and 0.09-inch (2.3–mm) thick.

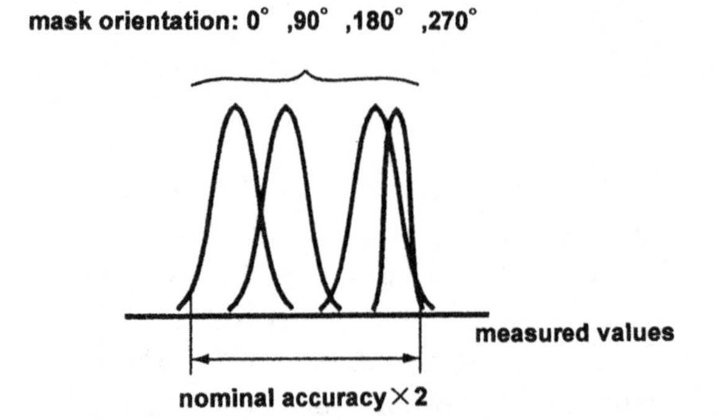

Figure 7.10 Definition of coordinate nominal accuracy.

Measurement repeatability is usually defined as repeatability in one posture, and nominal accuracy (Figure 7.10) is defined as the maximum deviation in four postures (at 0, 90, 180, and 270 degrees of rotation).

7.2.10 Overlay measurement on a wafer

Observing vernier patterns with an optical microscope was formerly used for measuring overlay (registration). As the design rule has been scaled down, however, this observation by human eyes has become difficult. This measurement has recently been done automatically, and target marks have been changed to more simple marks that are suitable for automatic image analysis. Figure 7.11 shows a typical system configuration. Bright-field illumination is widely used and a phase-contrast microscope has begun to be used for a low-contrast target like that made by CMP (chemical–mechanical polishing). A CCD camera is widely used as the

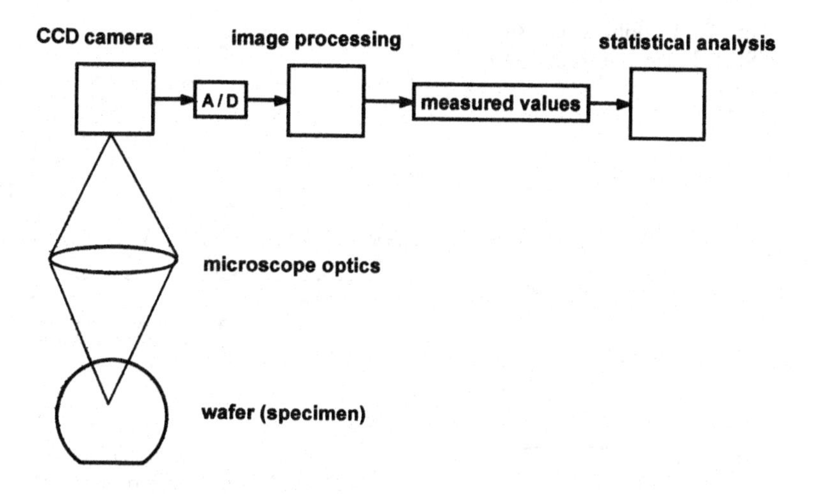

Figure 7.11 Schematic view of wafer overlay measurement system. A/D: analog-to-digital converter.

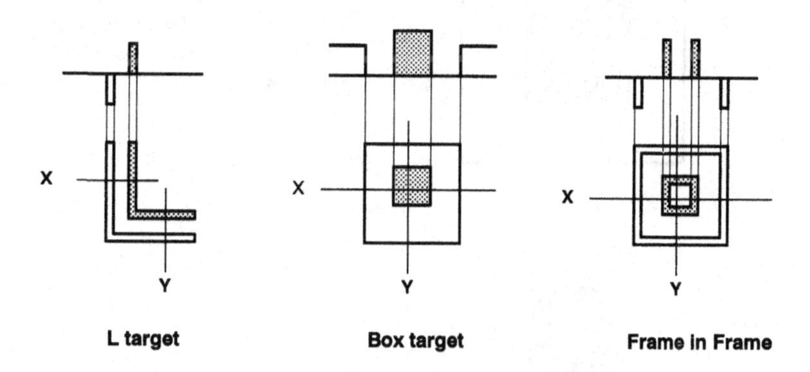

Figure 7.12 Target marks for wafer overlay measurement.

image sensor because of its low distortion. Popular target marks are L target, box target, and frame in frame (Figure 7.12) as well as bar in bar (not shown). In Figure 7.12, plain patterns represent target marks of the under-layer and dotted patterns represent upper-layer targets. Registration errors along the *X*- and *Y*-axes are evaluated by calculating pattern-center positions from measured pattern-edge positions.

A dimension of a target mark is typically about a few tens of microns. The required measurement accuracy is less than 10 nm, so the ratio of measurement accuracy to a target mark dimension is about a few thousandths. Because this ratio is high, wafer overlay measurement is very sensitive to optical quality or the probe characteristics of the system (as described in Subsection 7.2.6). Slight magnification change in an image field of an objective lens (caused by a lens aberration or an inclined optical axis), and asymmetric probe output are major reasons for fluctuating measurement output. TIS (tool-induced shift), WIS (wafer-induced shift) and machine-to-machine difference are terms that specify these measurement output fluctuations. As shown in Figure 7.13, TIS is the measurement output difference between two wafer orientations (0 and 180-degree rotation) for the same wafer. So, focusing offset difference in each machine mainly causes machine-to-machine difference.

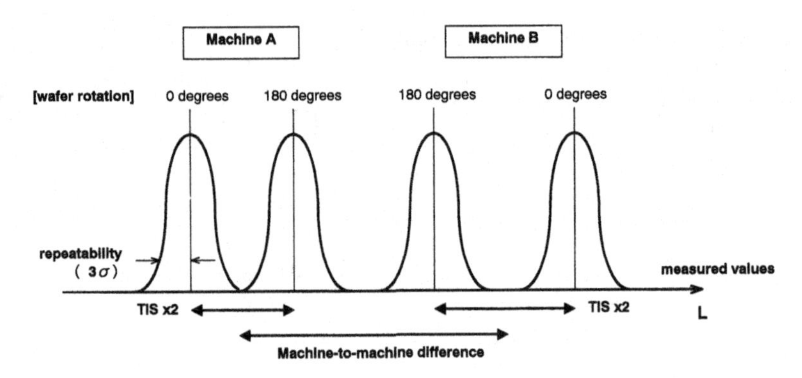

Figure 7.13 Tool-induced shift (TIS) and machine-to-machine differences.

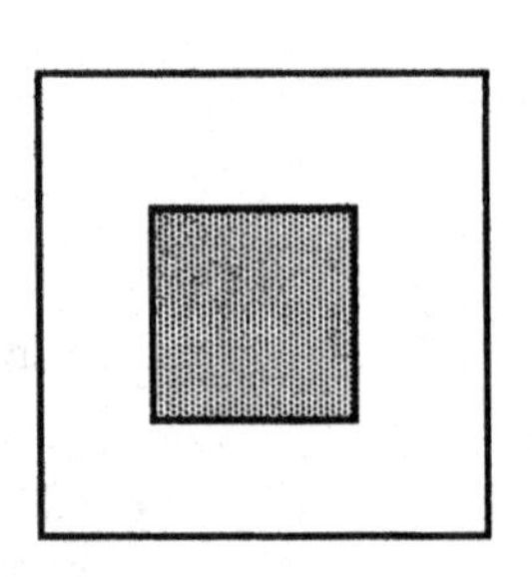

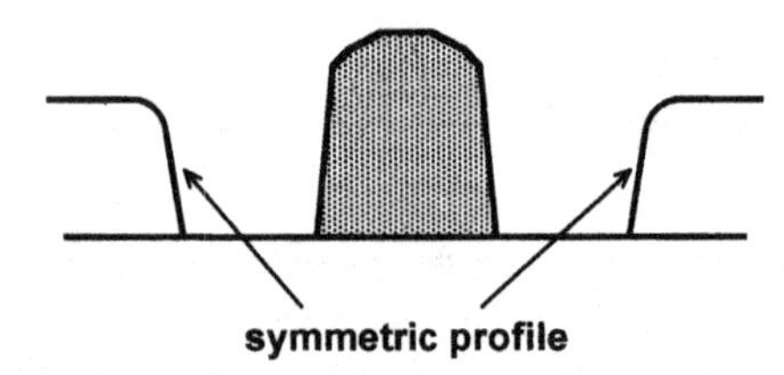

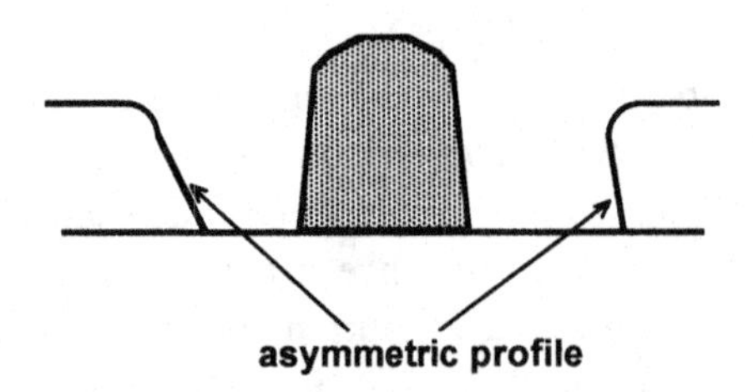

Figure 7.14 Asymmetric profile of a target mark that causes wafer-induced shift (WIS) in wafer overlay measurement.

Figure 7.14 shows an example of asymmetry in the profile of a target mark. This asymmetric profile causes the measurement error called WIS, which is one of the most difficult problems to overcome in wafer metrology. The most effective way to get good accuracy in the overlay measurement is to make target marks with a symmetric profile.

7.2.11 Metrology trends

The progress of microfabrication requires further improvement of CD and overlay measurement. For contact holes with higher aspect ratios, new probes with higher resolution, such as SPM (scanning probe microscope), STM (scanning tunneling microscope), or AFM (atomic force microscope), are needed. Sensors must be noncontact or nondestructive sensors, and automatic measurement is essential.

Not only must resolution be improved, but machine-to-machine difference and method-to-method difference must also be eliminated. This will require a standard for length and calibration methods in the sub-half-micron range.

At present, measurements are done in one or two dimensions, but the structure of ULSIs such as the capacitor fins of a memory cell has become more complicated. Three-dimensional measurements will be necessary for such complicated structures. Adoption of in-line *in-situ* control will also be accelerated in IC manufacturing.

7.3 **Inspection**

7.3.1 Introduction

Photomasks used in LSI circuit fabrication must be of high quality as well as free of dust and pattern defects. Pattern-defect inspection is thus increasingly important in a mask inspection process.[5,6]

In the early-1970s, operators used optical microscopes to inspect masks for defects. Entire masks were inspected for defects such as registration error, dimension variations, and random visual defects. This took from 4 to 20 hours for a single 4-inch (100-mm) mask. In contact printing, statistical sampling was generally used for inspecting working masks because this was more economical. In the late-1970s, 64–256-kb DRAM mass production began, and manual inspection was no longer reliable or economical. When projection printing was introduced, the increased mask lifetime made more extensive mask testing and repair worthwhile, and die-to-die inspection was introduced.[5,7] Die-to-die inspection was used for comparing two masks or two adjacent dies for differences because masks are uniquely repetitive, that is, two masks or all dies should have the same pattern. Common defects at the same position in each die are, however, not detected in die-to-die inspection. Mask inspection comparing pattern images to their database definition (die-to-database inspection) has been under development since 1980. Die-to-database inspection provides an absolute comparison with design-pattern data[8,9] and has been increasingly used. These developments in inspection technology were associated with increased industrialization in the late-1980s. Since 1990, optical lithography engineers have held that patterns as small as 0.1–0.2 μm may be formed by pattern image contrast enhancement using phase-shift masks and special stepper illumination.[10,11] For this reason, mask inspection engineers have increasingly emphasized the necessity of phase-shift mask inspection.

7.3.2 Mask manufacturing process and defects

7.3.2.1 *Mask manufacturing and inspection*

The mask substrate (Figure 7.15) consists of a glass plate coated with films, such as an optical shield film (e.g. Cr) and an anti-reflection film. First, the mask substrate is optically checked for dust particles, then the substrate is resist-coated. The condition of the resist coating is checked and the coating is inspected optically for dust

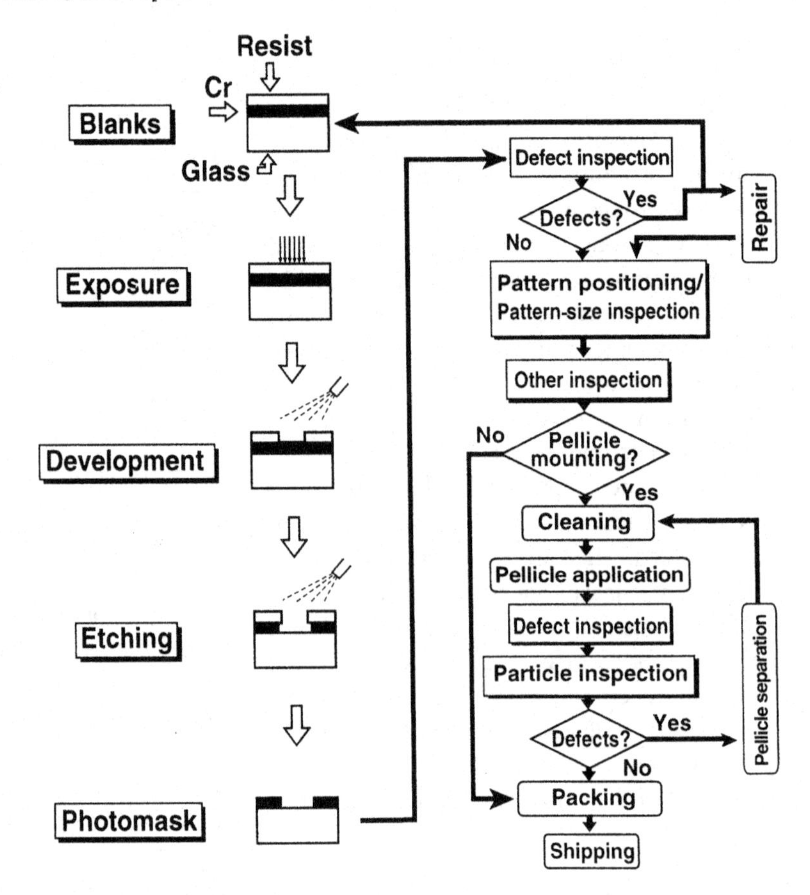

Figure 7.15 Example of photomask manufacturing and inspection.

particles. Then, patterns in the resist are delineated by exposure with an electron beam or a laser beam. Next, the resist is developed and then the mask substrate undergoes etching of the optical shield film.

Final inspection to verify mask quality before shipment involves measuring pattern-position accuracy as well as the size of pattern defects and particles. Substrates with repairable defects are sent back for retouching or cleaning. A fatal defect requires complete remanufacturing of the mask.

7.3.2.2 *Defects*

As shown in Figure 7.16, many kinds of defects may be found on a photomask. The inspection methods for various types of defects in the mask manufacturing process are as follows:[12]

(a) *Contamination and dust particles*: A photomask is often contaminated by products of the chemical reactions in mask processes and by substances in the molecular-film deposition environment. The photomask is also contaminated by dust particles during mask handling. So, the photomask must be inspected for contamination and dust particles after each process.

(b) *Pattern-positioning error*: A pattern-positioning error is defined as a position deviation in the pattern's micro- or macro-geometrical features. This error strongly depends on the positioning accuracy of the pattern exposure system and is usually evaluated using laser-interferometric coordinate measurement systems.

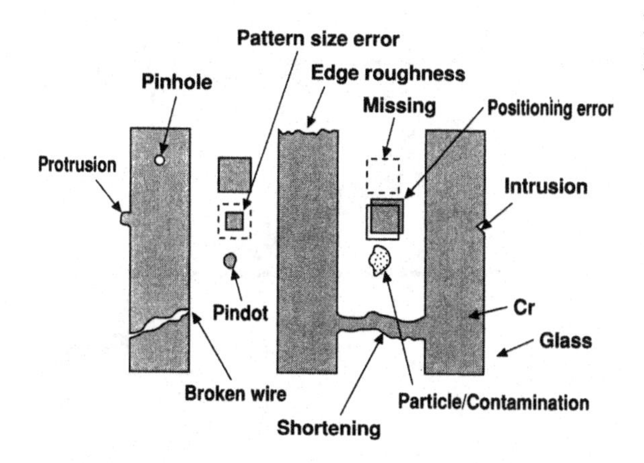

Figure 7.16 Examples of pattern defects on a mask.

(c) *Pattern-size error*: Pattern-size error is defined as deviation from the designed pattern. This kind of error occurs in the development and etching processes and in the pattern exposure process. This error strongly reduces the yield and must be evaluated before mask shipping.

(d) *Pattern defects*: Pattern defects include all kinds of pattern irregularities, such as pattern-edge roughness error, pattern missing, shortening, intrusion, protrusion, pindot, pinhole, and soft defects (see Figure 7.16). These defects occur in all mask processes, including pattern writing, etching, mask handling, and cleaning. In some cases, a device pattern is destroyed by electrostatic discharge (ESD).

Evaluations for (b)- and (c)-type defects involve inspection based on a small sample of patterns on a mask. This sampling, however, cannot be used in evaluating (a)- and (d)-type defects.

7.3.3 Device development trends and defect-detection size

ULSI devices are continuously getting more complex, and defects that must be detected on masks are becoming smaller (Figure 7.17). Generally, the size of defects that are to be detected is from one-third to one-quarter of the minimum pattern feature size on a mask. In the case of 256-Mb-DRAM devices, defects smaller than 0.25 μm must be detected.[13] The detection of defects smaller than 0.15 μm will be required to satisfy 1-Gb-DRAM mask manufacturing.

It is becoming extremely important to inspect attenuated and alternative phase-shift masks (PSMs), because PSMs can improve the resolution of the optical lithographic tool and the depth of focus. Unfortunately, defects in PSM cause high printability on the wafer. The detection of 0.08-μm defects, for example, is required in manufacturing devices with a 0.3-μm design rule.[14] Measuring the phase angle of PSM patterns is also an important issue.

Pattern formation using X-ray lithography is expected to become the key technique for fabricating future devices beyond 4-Gb DRAMs. X-ray lithography requires increasingly stringent detection sensitivity for X-ray mask patterns since it uses 1 : 1 pattern transfer. As shown in Figure 7.17, defects with a size of 0.035 μm or less must be detected when manufacturing 4-Gb DRAMs. Mask pattern inspection technology will thus enter a new phase in the 4-Gb-DRAM era.

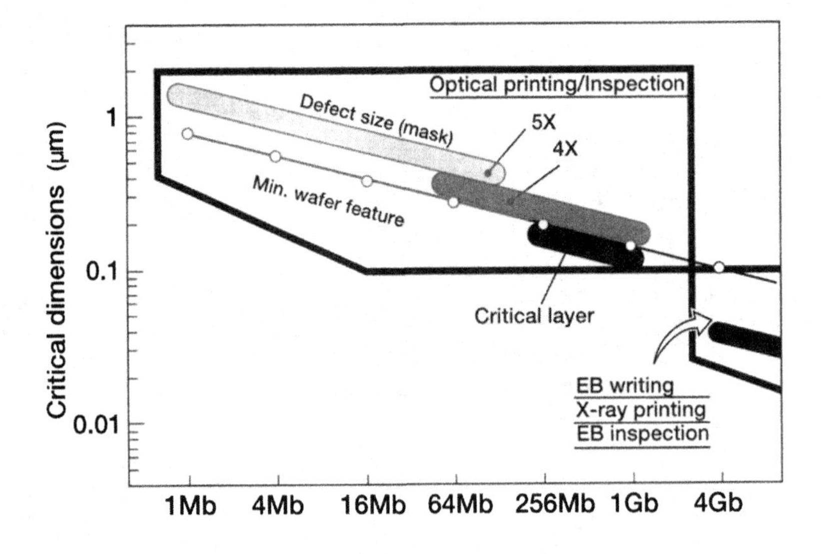

Figure 7.17 Trends in DRAM wafer and mask feature size, and required defect-detection size in Cr mask patterns. Shading shows required defect-detection size. Open circles indicate minimum pattern feature size in wafer pattern.

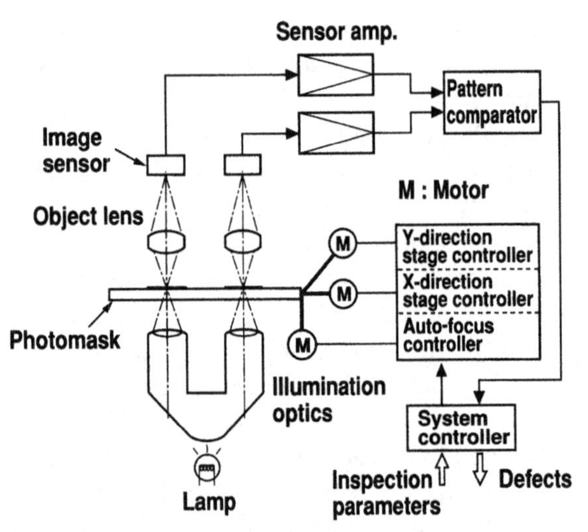

Figure 7.18 Die-to-die inspection system architecture.

7.3.4 Die-to-die and die-to-database comparison and current inspection performance

One method currently used for detecting mask defects is die-to-die comparison, which compares patterns in adjacent dies on the same mask. Studies were conducted by Ciarlo *et al.*[15] and Dyer[16] in order to determine inspection system requirements, and Figure 7.18 shows a fundamental system configuration for die-to-die inspection. The die-to-die comparison makes use of the fact that IC masks contain a regular array of nominally identical die images, so defects can be detected by comparing adjacent images. An appropriate comparison algorithm compares optical pattern images from two microscopes, and difference between these images is recognized as a defect. If it is not possible to prepare the same dies on a mask because of increased die size, die-to-die comparison is done by preparing two identical masks. A new method that compares measured image data for the die pattern with previous die-pattern image data stored in computer memory has recently been developed.[13] This method requires the use of only one microscope.

Generally, the die-to-die comparison method has high defect-detection sensitivity and high inspection speed because it intrinsically compares the same pattern signals. The commercially available KLA301 (KLA Instruments Co.) die-to-die inspection system, for example, detects Cr pattern defects as small as 0.2 μm and requires only 70 min to inspect at 100×100 mm^2 area (0.7 min/cm^2).[17] The 9MD83SR (Lasertech Co.)[18] also detects Cr pattern defects of 0.2–0.25 μm at 86 min per 100×100 mm^2 area (0.86 min/cm^2). These die-to-die inspection systems are thus sufficient for development of 256-Mb-DRAM device masks and first-stage production. KLA Instruments Co. has developed a die-to-die inspection system using an electron-beam probe for X-ray mask inspection.[19] This system detects defects of less than 0.05 μ m at an inspection speed of 27 min/cm^2 by scanning a spot beam with a diameter of approximately 50-nm full-width at half maximum (Figure 7.19). These die-to-die inspection methods do not, however, detect defects that are at the same position in each die. Additional inspection is therefore required.

Inspection of mask pattern images against their database definition is used increasingly for accurate defect recognition because die-to-database inspection provides an absolute comparison with design pattern data (Figure 7.20). The die-to-database inspection system is basically made up of five components: an optical system, a mechanical system, image-acquisition circuits, data processing circuits, and a system controller. Image data from the image-acquisition circuits is compared with image data from an appropriate comparison algorithm, and differences are recognized as defects.[20,21,22] Die-to-database mask-inspection systems are commercially available from KLA Instruments Co., Orbot, and Toshiba Machine. Although die-to-database inspection is slower and less sensitive than die-to-die inspection, because of the difficulty of reconstructing the geometric representation of the computer-aided design (CAD) pattern images being inspected, many technical improvements are possible to overcome this problem, as discussed in Subsection 7.3.5.2. The die-to-database inspection system is sensitive enough for inspection of 256-Mb-DRAM device masks. The most advanced commercial inspection system detects Cr pattern defects as small as 0.2 μm at 70 min per 100×100 mm^2 area (0.7 min/cm^2).

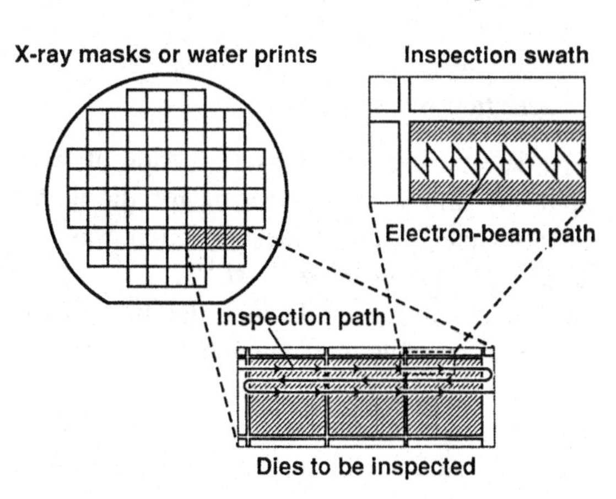

Figure 7.19 SEM die-to-die inspection strategy. As the stage follows the inspection path, the electron beams scan regions to be inspected, acquiring nominally identical 'inspection swath' images that are compared by the system.

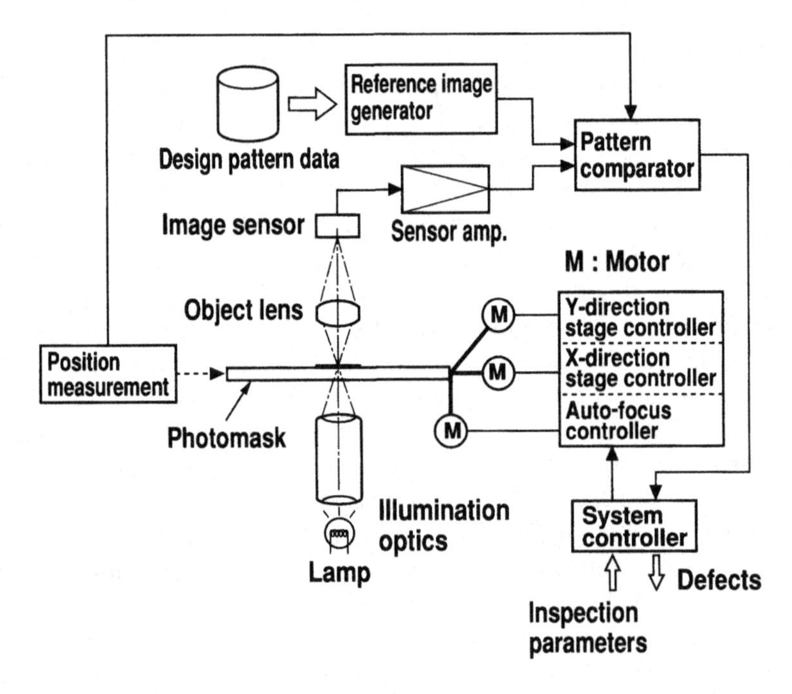

Figure 7.20 Die-to-database inspection system architecture.

7.3.5 Defect signal acquisition and image processing

7.3.5.1 *Defect signal acquisition*

Many inspection systems use combinations of optical microscopes and linear photodiode arrays to measure the light transmitted through the mask and to obtain pattern images (Figures 7.18 and 7.20). The whole field of view of the object lens is illuminated continuously by transmitted light and the mask pattern image is focused on the linear photodiode array by the object lens. The linear photodiode array measures intensity variation of the light through the mask pattern.

In another method, which involves a flying-spot scanner, two die patterns are compared using two small spots of light to simultaneously scan two corresponding positions on different die patterns (Figure 7.21).[23,24] Automatic mask inspection systems (AMIS) using this method were developed for measuring the defects on different types of IC masks at Bell Laboratories in 1975 and at RCA Laboratories in 1977.[25,26] Recent automatic mask inspection systems use a flying-spot scanner to measure image data for the die pattern. These data are then compared with image data stored in a computer memory.[17] This system uses only one microscope.

7.3.5.2 *Defect detection algorithms and detection sensitivity*

The detection sensitivity as a function of laser-spot size and scan-line spacing for the flying-spot scanner in die-to-die inspection was studied by J. H. Bruning *et al.*[24] It is assumed that light transmitted through the mask is collected at an angle wide enough to intercept all relevant diffraction components. The scanning spot illuminates defects and the transmitted light is modulated in proportion to total

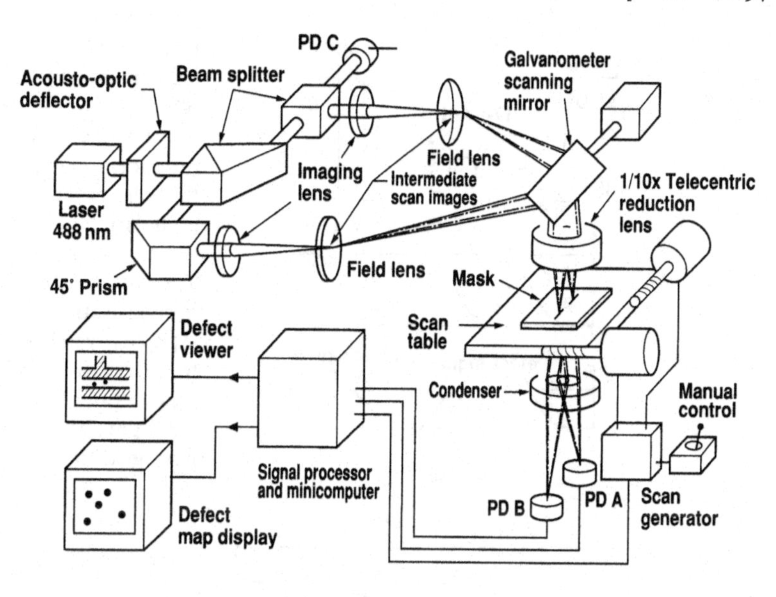

Figure 7.21 Basic automated mask inspection system (AMIS). An argon laser with blue light at 488 nm was used. A beam splitter and movable prisms produce two beams of variable separation to accommodate different pattern spacing on the mask.

power. If the light spot is Gaussian with a diameter of $2W_0$, total power P_0 is given by:

$$P_0 = I_0 \exp\left(-\frac{2r^2}{W_0}\right), \tag{7.1}$$

where I_0 is a constant and r is the distance from the light-spot center. If this light spot encounters an opaque square with sides $2a$, parallel to the X- and Y-axes, and offset from the spot center by a distance δ, as shown in Figure 7.22(a), the power transmitted P_s is:

$$P_s = P_0 - I_0 \int_{-a}^{a} \int_{\delta-a}^{\delta+a} \exp\left[\frac{-2(x^2+y^2)}{W_0^2}\right] dxdy. \tag{7.2}$$

Using the tabulated integral ($\mathrm{erf}(t)$), we can write the signal modulation as:

$$1 - \frac{P_s}{P_0} = \frac{1}{2}\left(\frac{\sqrt{2}a}{W_0}\right)\left\{\mathrm{erf}\left(\frac{\sqrt{2}(\delta+a)}{W_0}\right) - \mathrm{erf}\left(\frac{\sqrt{2}(\delta-a)}{W_0}\right)\right\}. \tag{7.3}$$

Solid curves in Figure 7.22(b) show the signal modulation, $1-(P_s/P_0)$, vs $2a/W_0$ for various values of δ/W_0. This modulation decreases with δ/W_0 and increases with $2a/W_0$. It is pointed out that the lowest usable signal modulation is not generally limited by the system's electrical noise, which is overcome by using an adequately high beam power P_0. A more serious limitation results from residual signals due to edge misregistration. This occurs if the two light spots are not located at the proper corresponding elements of adjacent chips. Assuming that the light-spot center is placed a distance x_0 from the edge of the opaque half-plane $x \leq 0$, we can write the power transmitted as:

$$\frac{P_s}{P_0} = \frac{1}{2}\left\{1 - \mathrm{erf}\left(\frac{\sqrt{2}x_0}{W_0}\right)\right\}. \tag{7.4}$$

If the spacing of the scanning spot and that of the corresponding feature edges in the two channels differ by misregistration D, as shown in Figure 7.22(a), subtraction of the two output signals leaves the misregistration signal:

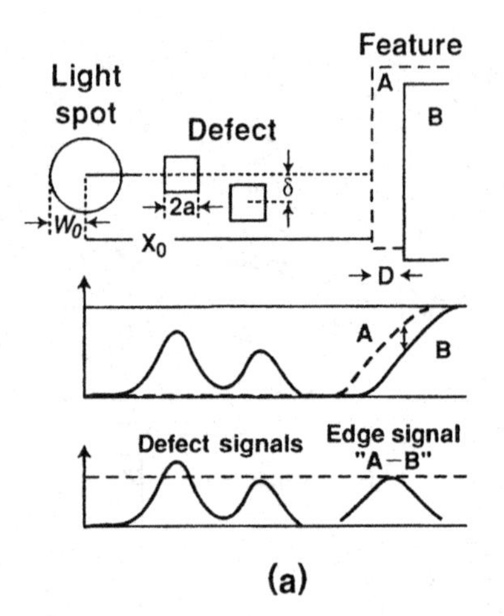

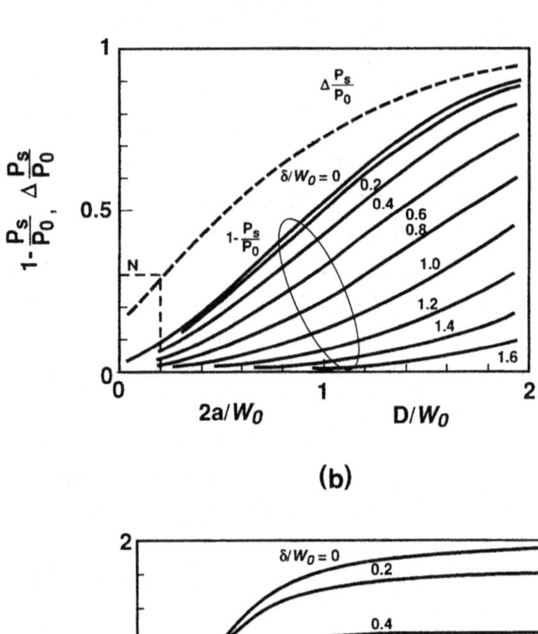

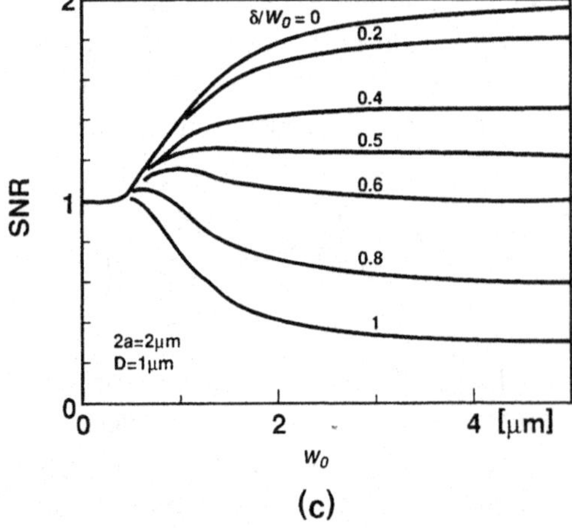

Figure 7.22 (a) Two Gaussian light spots scan defects with offset δ from the spot pass. Feature edge misregistration D produces an edge signal. (b) Solid curves show signal modulation, $1 - (P_s/P_0)$, as a function of $2a/W_0$ and δ/W_0. The dashed lines show misregistration signal, $\Delta(P_s/P_0)$ vs D/W_0. (c) Signal-to-noise ratio, SNR, as a function of scan overlap δ/W_0 for 2-μm defect and 1-μm misregistration.

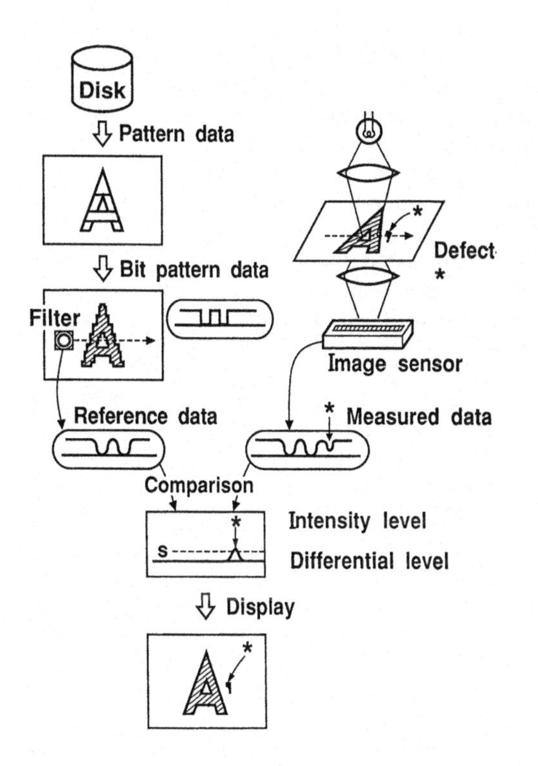

Figure 7.23 Defect inspection using multilevel signal comparison.

$$\Delta \frac{P_s}{P_0} = 1 + 2 \operatorname{erf}\left(\frac{D}{W_0\sqrt{2}}\right), \qquad (7.5)$$

which has a maximum at $x_0 = D/2$.

The dashed line in Figure 7.22(b) show the misregistration signal vs D/W_0. It was found that a misregistration of $D/W_0 = 0.2$ suffices to generate a signal modulation of 0.3. Bruning *et al.* concluded that smaller defects are more readily detected when a spot with a smaller radius is used. The scan overlap δ/W_0 is adjusted to obtain the desired signal-to-noise ratio (SNR) for the smallest defect and the largest expected misregistration (Figure 7.22(c)).

Many comparison methods in die-to-database inspection have been developed in order to improve sensitivity. In the MC-100 and MC-2000 die-to-database inspection systems,[20,27,28] for example, the multilevel comparison shown in Figure 7.23 is used. Output signals from the CCD image sensor are multilevel. Signals from the design pattern data, on the other hand, are binary. To generate reference image signals, the design pattern data are converted to multilevel signals by window scanning. Window scanning works as special filtering corresponding to an optical lens and sensor characteristics. Calculation of the reference image signals from the design pattern data is based on the principles outlined below.

Assuming a light flux-density on the image plane Σ_i through the optical system from the point-source irradiance on the object plane Σ_o; that is, the point-spread function is $\xi(x, y; X, Y)$. As shown in Figure 7.24(a), the irradiance distribution on the object plane (that is, pattern data) is $I_o(x, y)$. Under the circumstances of space invariance and incoherence, the flux-density distribution on the image plane is:

$$I_i(x, y) = \int_{-\infty}^{\infty}\int_{-\infty}^{\infty} I_o(x, y)\xi(X - x, Y - y)\, dxdy. \qquad (7.6)$$

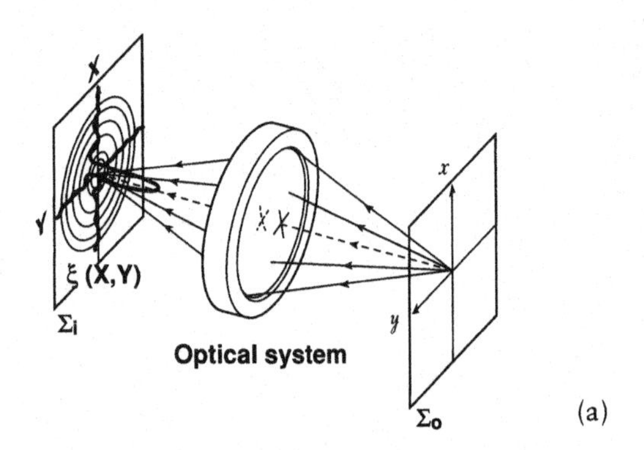

Figure 7.24 Light flux-density on the image plane is obtained by using (a) a point-spread function and designed pattern data. (b) Reference image signal at the center of pattern data is the sum of the value of each cell, calculated by the product for the point-spread function and designed pattern data.

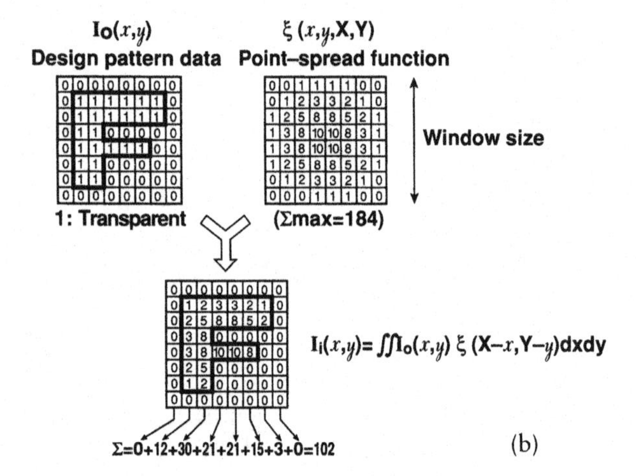

This equation is a two-dimensional convolution integral. To solve it quickly and easily, we define the windows in the *n*th row, the *n*th column for pattern data, and the point-spread function in the finite region. We obtain the light flux-density (that is, the reference image signal) on the center point of the pattern data by using the product and sum of each cell in the windows as shown in Figure 7.24(b). By scanning the same operations over the whole pattern area, we generate reference image signals corresponding to multilevel signals measured by the sensor.

The multilevel signals from the CCD image sensor and the reference image generator are compared with an intensity level and a differential-calculus level. First, the shape of the pattern inspected is recognized from the design data, and then suitable comparison algorithms and comparison levels corresponding to these shapes are selected. In differential-calculus level comparison, the decision of the direction of the differential calculus is very important for reducing 'false defects'. The most important adjustment in a die-to-database inspection system is that needed to obtain a high coincidence of sensor and reference signals. The defect-detection sensitivity of the MC-100 system for a Cr mask is listed in Table 7.4.[27] Die-to-database and die-to-die inspection-system performances show almost the same defect-detection sensitivity (Table 7.5) and have sufficient sensitivity for 256-Mb-DRAM device mask manufacturing and for 1-Gb-DRAM device mask development and first-stage device manufacturing.

Table 7.4 *Detected defect size (in μm) for Cr patterns when 0.33-μm optical pixel size is used in inspection.*

Pixel: 0.33 μm

PD	Protrusion defect							
0.24	0.22	0.26	0.24	0.21	0.25	0.21	0.22	0.27
PH	**Intrusion defect**							
0.30	0.24	0.19	0.22	0.16	0.19	0.27	0.21	0.20

Table 7.5 *Inspection performance for state-of-the-art mask inspection systems.*

Manufacturer	KLA	Orbot	Toshiba	Lasertech
Type	KLA-353UV	ARIS-1	MC-3000	MD2000
Die-to-die (DD) or Die-to-database (DB)	DB	DB	DB	DD
DRAM capacity (bit)	256 M	1G	1G	1G
Pixel size (μm)	0.23	0.15	0.10/0.15	0.25
Detection sensitivity (μm)	0.23	0.15	0.12/0.16	0.12
Inspection time (min/100 mm × 100 mm)	138	133	280/133	100

7.3.5.3 *Defect image enhancement and image processing*

Measured signals from pinhole defects and pindot defects show a big difference in optical properties because of the partially coherent optics of the inspection system. In calculated signals from pinhole and pindot defects that are smaller than the inspection wavelength, the output signal from a pinhole defect is one-half that of the pindot defect (Figure 7.25). To obtain excellent sensitivity when detecting defects smaller than 0.2 μm, it is therefore important to improve the signals or the SNR of images passed to the defect comparator. Image-enhancement techniques must be introduced into the defect-detection circuit in the inspection systems.

In a technique reported by S. Takeuchi *et al.*,[22] the image-enhancement circuit contains finite-impulse-response (FIR) filters used to either smooth or sharpen geometry edges. Figure 7.26(a) shows a sample frequency response of the FIR filter when used as a low-pass filter. The filter is used to attenuate various forms of edge roughness. Figure 7.26(b) shows simulated input modulation along a line with 0.1-μm edge roughness and a 0.3-μm defect. The resulting modulation after passing through the filter is shown in Figure 7.26(c). Input noise peaks are softened in the output and are distributed along the ideal edge.

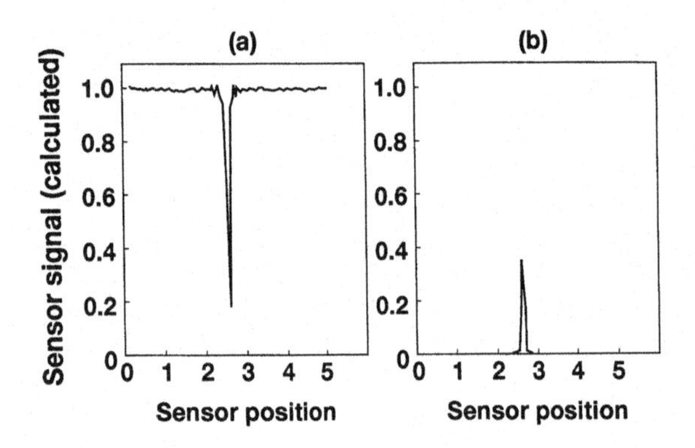

Figure 7.25 Calculated sensor signals from (a) pindot defect and (b) pinhole defect. Defect size is 0.5 μm and the inspection wavelength is 550 nm.

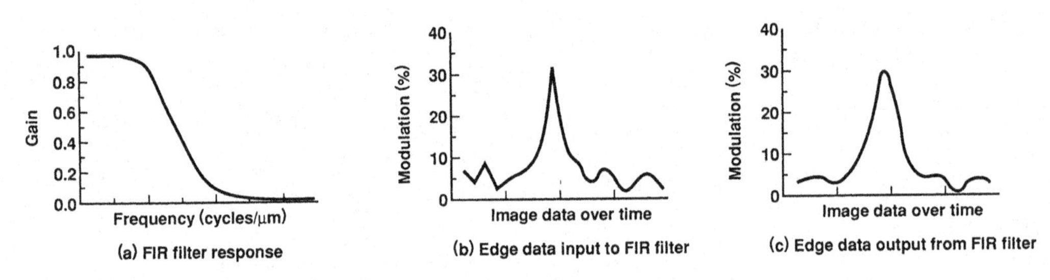

Figure 7.26 (a) Example of FIR filter frequency response as a low-pass filter. (b) Simulated input modulation to the FIR filter showing the signal along an edge with 0.1-μm edge roughness and a 0.3-μm defect. (c) Resulting modulation smoothed after passing through the FIR filter.

Figure 7.27 shows the improved inspection performance of an advanced inspection system equipped with an FIR filter and a 16-gray-level detection circuitry. These curves demonstrate that the sensitivity improvements are directly attributable to the new technologies.

By adding the defect signal from light transmitted through the mask, we can considerably improve inspection performance. The simultaneous transmitted and reflected (STAR) light system[28,29] (developed by KLA Instruments Co.) detects (i) particles on the Cr parts of the mask, and (ii) soft defects, such as contamination and thin-film defoliation. The STAR light system is illustrated in Figure 7.28. Detection of 0.4-μm particles is possible.

Another approach of this transmitted light and reflected light application is applied to automatic classification of photomask defects.[30] The effectiveness of inspection using transmitted and reflected light is most clearly seen by considering the correlation function between images from these kinds of light. Figure 7.29 shows these correlation curves for (a) illumination variance, (b) lateral misalignment of images illuminated by transmitted and reflected light and (c) defocus. These curves can be obtained from a pattern in the mask. In this figure, dashed lines show correlation curves without any perturbation and solid lines show the range of correlation curve depending on fluctuations. Pindot defects, for example, disappear

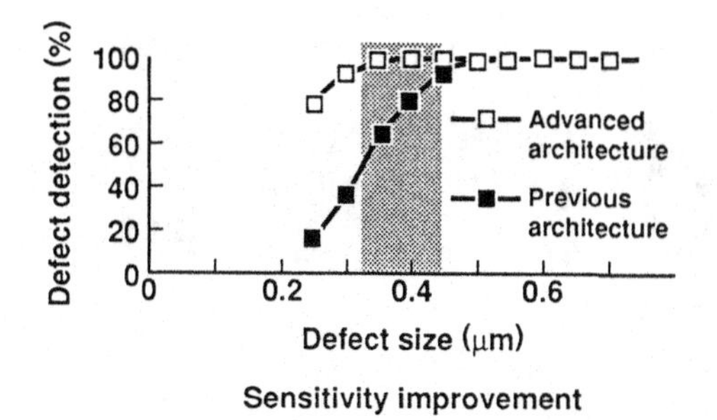

Figure 7.27 Sensitivity comparison of previous and advanced architecture with FIR filter and 16-gray-level detection circuitry. Pixel size is 0.36 μm.

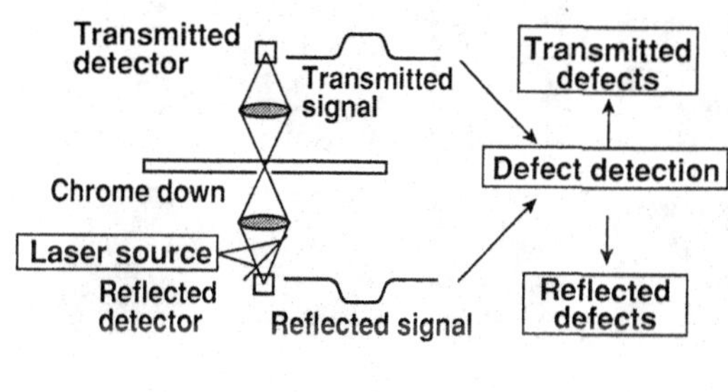

Figure 7.28 Simultaneous transmitted and reflected (STAR) light system detects contamination defect with patterns cancelled by the sum of transmitted gray-level signals and reflected gray-level signals.

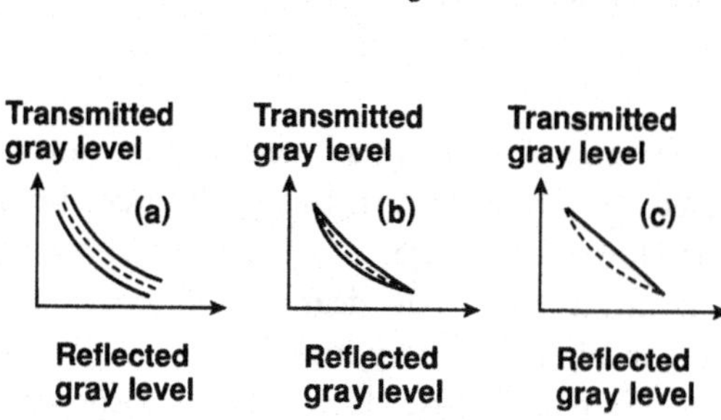

Figure 7.29 Reference correlation curves induced by transmitted and reflected light. Effects of (a) illumination variance, (b) lateral misalignment, and (c) defocus variation.

under transmitted light and glitter under reflected light. For contamination defects, both transmitted light and reflected light become very weak as shown in Figure 7.29. For contamination defects, signals from both transmitted light and reflected light become very weak and are thus automatically classified by recognizing their position in the correlation curve. Figure 7.30 shows the inspection results obtained using 0.3- to 0.5-μm standard particle defects. (T), (R) indicate the measured images on glass (left side) and on chromium (right side) of the mask using transmitted and reflected light, respectively. (Qz) and (Cr) show defect images after image processing using the correlation-curve comparison method. In this experiment, a correlation curve was obtained from a line-and-space pattern indicated in the 'ref' images. This method enables contamination defects on the mask to be discriminated from those on the pattern.

7.3.6 Inspection of phase-shift masks

It is becoming increasingly important for inspection systems to detect defects in phase-shift masks such as attenuated masks[31] and alternative masks.[14,32] Phase-shift-mask defects smaller than the defects on Cr masks must also be detected because they show higher printability on the wafer pattern. The main difficulty in detecting defects on phase-shift masks is caused

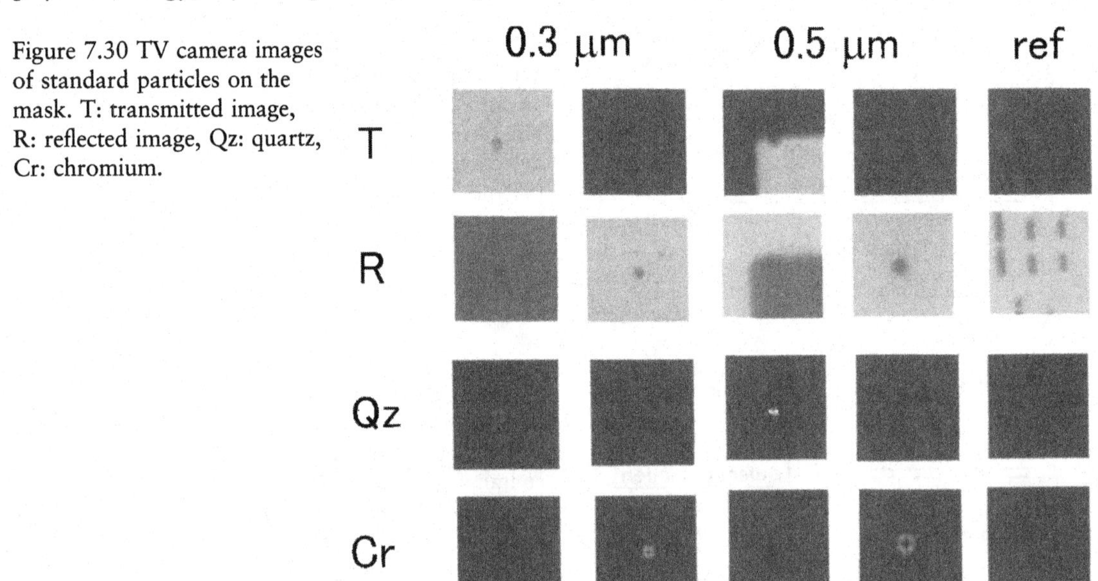

Figure 7.30 TV camera images of standard particles on the mask. T: transmitted image, R: reflected image, Qz: quartz, Cr: chromium.

by the difference between the wavelengths of the inspection and pattern-transfer systems.

There are some typical differences between sensor signals from Cr-mask defects and from attenuated-phase-shift-mask defects (Figure 7.31).[27] One major difference is caused by the increased transmittance of attenuated phase-shift materials in the inspection wavelength. For attenuated phase-shift materials, the longer the inspection wavelength, the higher the transmittance and vice versa (Figure 7.32). The amplitude of the signals from the sensor will therefore be reduced, and this lifts the ground level to T, as shown in Figure 7.31(b) if, for example, we use a 550-nm wavelength when inspecting a phase-shift mask used with a 365-nm exposure wavelength. The SNR of defect signals decreases more than that of defect signals for Cr pattern masks. The other major difference is that sensor signals at the edge of the phase-shift material undershoot the T level as shown in Figure 7.31(b). Sensor signals for large defects are similar to those for Cr defects; that is, a defect signal can be obtained at a level higher than T. For small defects, no signals exceed the T level; that is, all signals undershoot T as shown in Figure 7.31(b). Undershooting phenomena and SNR degradation pose problems in high-sensitivity inspection.

Alternative phase-shift mask inspection encounters different difficulties if we use wavelengths different from the exposure wavelength. Figure 7.33 shows measured sensor signals from a 0.75-μm L/S alternative phase-shifting mask. The amplitude of the sensor signal alternately changes because the phase angle deviates from the 180-degree phase difference for the inspection wavelength. Thus phase-shift mask inspection differs in several ways from the conventional Cr pattern inspection.

Ichioka and Suzuki investigated the combined effects of amplitude- and phase variation in the irradiation on the image of a complex object such as attenuated and alternative phase-shift masks.[33] Image deformation due to amplitude- and phase variation was derived using an optical system with partially coherent object illumination. Examples of numerical calculations of

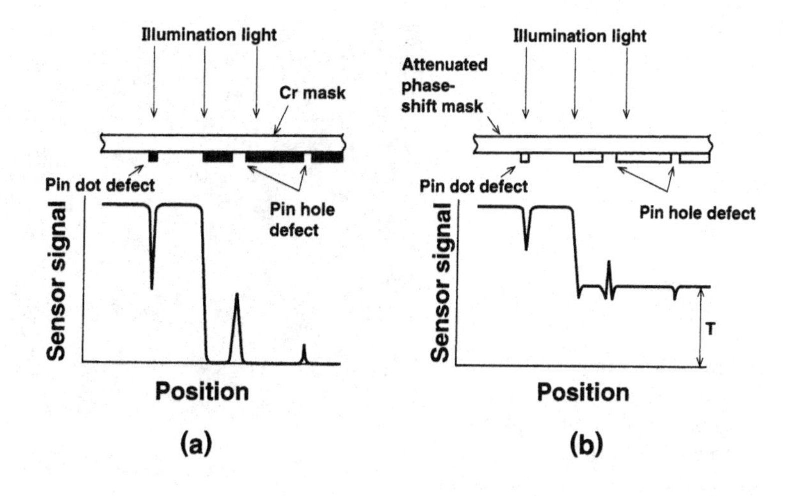

Figure 7.31 Differences in inspection wavelength between (a) Cr and (b) attenuated-mask defect signals for pindot and pinhole defects.

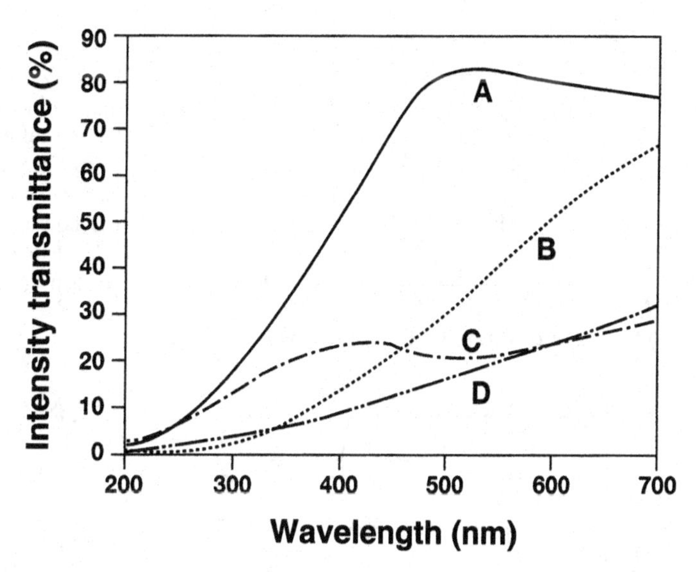

Figure 7.32 Example of transmittance curves for various attenuated phase-shift materials (A, B, C, D). Transmittance increases with the inspection wavelength.

image amplitude and object phase are shown in Figure 7.34. Figure 7.34(i) shows a model of a grating-like complex object, (ii) and (iii) show calculated images for different amplitude transmittance and different phase illuminated with coherent conditions of 1.0 and 0.5, respectively. In this figure, images are drawn for one period and for changes of amplitude and phase parameters. In image formation under coherent and near-coherent conditions, the phase change greatly influences the appearance of the image for pure phase objects and low-amplitude-contrast objects that have sharp phase boundaries. The most important phenomena are the generation of heavy ringing and sharp notches or peaks at pattern boundaries. Increasing amplitude- and phase contrast in the complex object increases the overall image contrast. Increasing phase variation sharpens the edge appearance at partially coherent illumination.

These phenomena degrade defect-detection sensitivity in phase-shift mask inspection. Die-

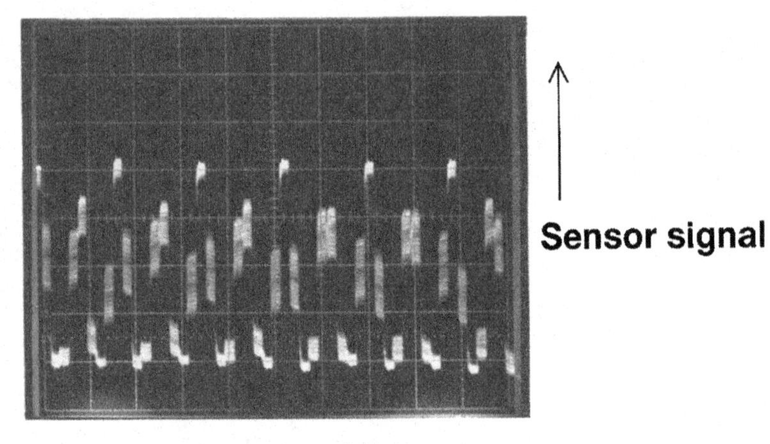

Figure 7.33 Example of measured sensor signal from a 0.75-μm L/S alternative phase-shift mask with 0.33-μm optical pixel size.

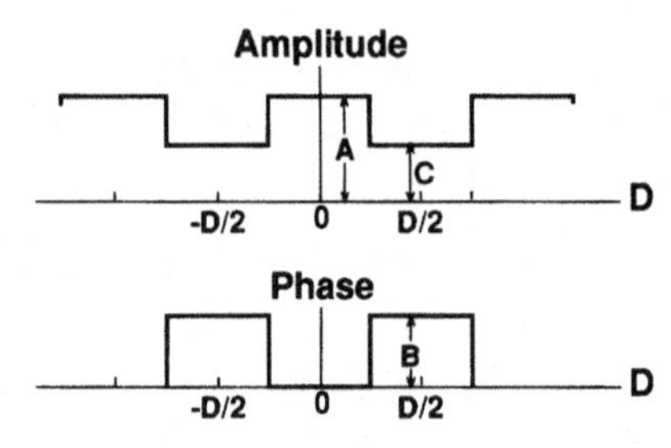

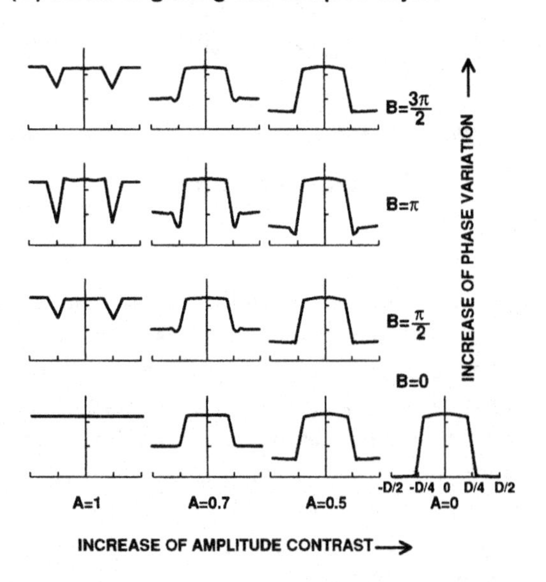

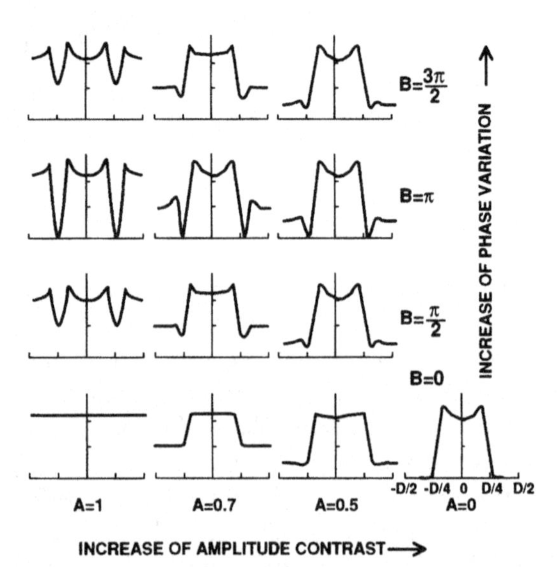

Figure 7.34 Images of square-grating-like complex objects having different amplitude transmittance and phase variations illuminated by partially coherent sources ($C = 1.0$; R: ratio of NA for condenser and objective lenses).

to-die inspection, however, is hardly influenced by sensor signals changing due to optical phenomena. This is because the image of one die pattern is compared to another on the same mask. This is, however, not possible with die-to-database inspection. The comparison of designed-pattern data to measured-pattern images deformed by optical phenomena must thus be improved. Inspections using either design-data conversion in which design-pattern data images are corrected to images to be measured in partially coherent illumination, or measured-image data conversion in which measured image data are corrected to images that can use a conventional comparison algorithm, are under development. Figure 7.35 shows an example of defect-detection sensitivity of a die-to-database inspection system for attenuated phase-shift masks. Although detection sensitivity for pinhole defects is inferior to that for Cr mask defects, detection sensitivity for pindot defects is nearly equal to that for Cr mask defects. However, in order to detect defects more sensitively in phase-shift masks, a new defect-detection method and a new algorithm for comparing measured-image data and the design pattern must be developed.

7.3.7 Projected trends and problems in defect inspection

During the last decade, inspection has become an indispensable part of mask manufacturing. State-of-the-art systems for automated defect inspection satisfy the performance required for inspecting masks for manufacturing 1-Gb-DRAM devices and for developing 4-Gb-DRAM devices. Current inspection systems usually use 436 nm wavelength for illumination. Using a shorter wavelength, such as 365 or 248 nm, and improving comparison algorithms will improve detection sensitivity to the 0.1-μm levels required for manufacturing 1-Gb-DRAM masks and for developing 4-Gb-DRAM masks.[34] Inspection technologies for phase-shift masks will become an increasingly important issue. For later-generation DRAMs, inspection using electron-beam technologies will be required for verifying pattern defects.

Despite the best efforts of engineers, many fundamental inspection problems remain. Three main problems define the limits of performance in future inspection technologies.

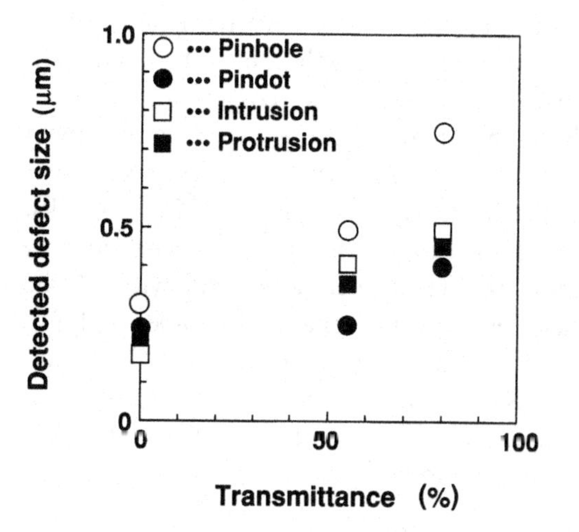

Figure 7.35 Defect-detection sensitivity for attenuated phase-shift mask vs transmittance in inspection wavelength. Transmittance 0% indicates inspection results for conventional Cr mask.

First, from the mask-production manager's standpoint, the limitation of inspection using optical technologies is a big issue. Research and development engineers have to answer that question. Although defect detection remains difficult, it is generally accepted that we can detect defects smaller than the patterns formed using optical lithography. At the present stage, we believe that defects as small as 0.1 μm can be detected using optical technology, as shown in Figure 7.17. Many reports indicate the importance of good optical systems and excellent detection algorithms and the need to maintain ongoing research and development.

The second main problem is the difficulty in classifying defects in the defect-review process and supplying the program with standardized defect-containing masks to verify the inspection system. Visual observation in review mode is reaching a limit. For 0.1-μm defects, we need other techniques, such as image recognition using computer graphics.

The third problem is that it is necessary to consider the effective application of defect information gained by inspecting masks. To improve mask pattern quality, inspection results based on defect classification should be sent to each mask process and should be used as feedback data. Next-generation mask-production processes for controlling 0.1-μm defects will require systems that use such inspection information effectively.

7.4 Repair

7.4.1 Introduction

Lithography is the key patterning step in LSI fabrication. In the projection exposure method wafers are exposed by using masks that contain LSI patterns. From the point of view of cost efficiency, the masks should not contain any defects. Repair technology is necessary in order to make defect-free masks. Mask defects are roughly classified into two groups: opaque defects, which are regions of extra, unwanted absorber; and clear defects, which are regions in which the absorber is missing. These defects mainly result from mask-fabrication processes: patterning errors, resist adhesion problems, resist residue, poor adhesion of absorber, and pinholes in the resist and absorber material.

From the late-1970s to the 1980s, the lift-off repair technique was one of the most commonly used repair methods.[35] Lift-off repair is a multistep, time-consuming technique and often introduces additional defects. Since the 1980s, several new methods have been proposed. These techniques involve laser-based repair[36] and focused-ion-beam (FIB)-based repair.[37] Optical masks generally have been repaired using laser-based tools. Opaque defects are repaired by thermal evaporation using short laser pulses, whereas clear defects are repaired by laser-induced chemical vapor deposition (CVD) of optically opaque material. More recently, the thermal processes effected by laser beam have been replaced by inherently higher-resolution FIB, and FIB repair tools are now commonly used in photomask manufacturing when conventional laser-based tools cannot meet resolution requirements. In the case of X-ray masks, the small feature size, high-aspect-ratio features, and large thermal conductivity of the masks make laser repair unsuitable, and FIB repair is the most feasible choice.[38] Defect-free X-ray masks have consequently been obtained in the laboratory by using the FIB repair technique.[39,40]

7.4.2 Repair system description

7.4.2.1 *FIB system*

A block diagram of a FIB repair system is shown in Figure 7.36. The system consists of a FIB column containing a liquid-metal ion source, electrostatic lenses, electrostatic deflectors and blankers, a secondary-electron ion detector, a pattern generator, image-acquisition electronics, a gas-injection controller, an *X–Y* stage, and a computer. The ion beam from the liquid-metal ion source is focused using an electrostatic lens. The beam is typically 20- to 30-keV Ga ions focused to a diameter of 25–100 nm at a current of 10–100 pA. The FIB is scanned on the mask substrate for imaging and repairing defects automatically by computer control.

Opaque defects are removed by physical sputtering and clear defects are repaired by ion-induced deposition. Defect location, size, and type are determined from the images obtained by scanning ion microscopy. The repair system receives information about the locations of defects in the mask directly from the inspection tool. When the substrate is nonconductive, a charge neutralizer consisting of an electron flood gun and electrostatic shield is required. The main specifications of commercial FIB repair systems are listed in Table 7.6.

7.4.2.2 *Laser system*

The laser-based repair system resembles the FIB repair system except for the incorporation of laser-beam optics, the basic configuration of which is shown in Figure 7.37. A variable-shaped aperture image of a laser beam, magnified by a beam expander, is projected onto a mask through a projection lens. Defects are repaired by directing the laser beam through the aperture to the defect on the mask. To do this easily, the variable-shaped aperture image and a defect image on the mask can be superimposed by adjusting the *X–Y* stage position while watching both of the images displayed

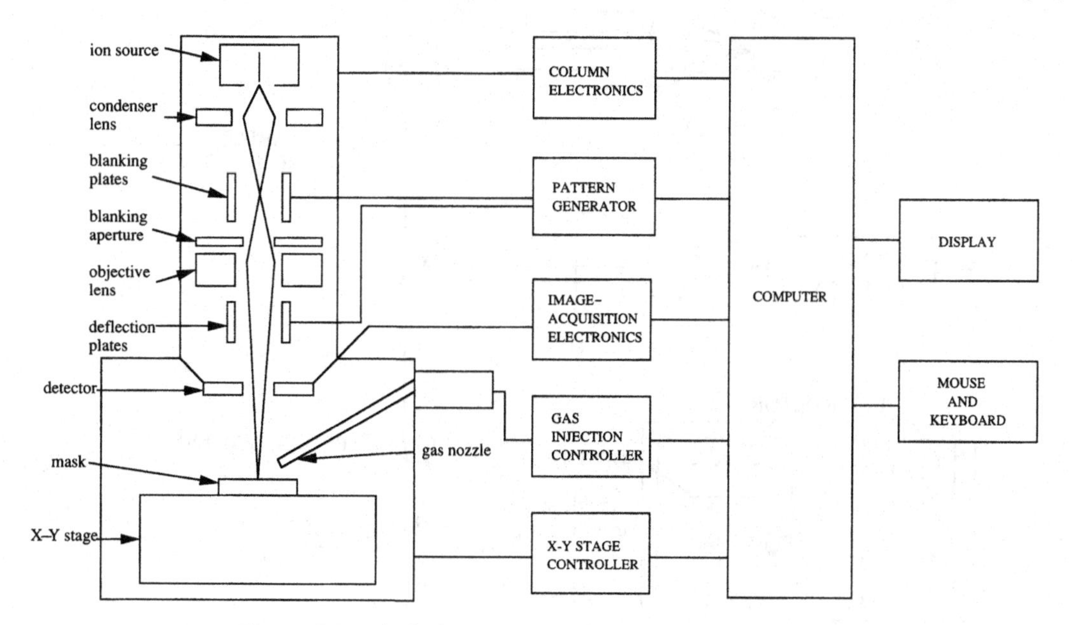

Figure 7.36 Block diagram of a focused-ion-beam repair system.

Table 7.6 *Main specifications of commercial focused-ion-beam repair systems.*

Item	SIR-1500α (SEIKO Instruments Inc.)	MICRION-8000 (MICRION)
Ion-beam source	Liquid Ga^+ ion	Liquid Ga^+ ion
Beam size (nm)	< 100	20–95
Ion energy (keV)	200 (typical value)	30 (typical value)
Repair accuracy	$\pm 0.09\ \mu m$ (2σ)	$\pm 0.1\ \mu m$ (99%) for photomask ± 25 nm (99%) for X-ray mask
Repair function	Both sputter and deposition (carbon)	Both sputter and deposition (carbon for photomask; gold, tungsten or tantalum for X-ray mask)

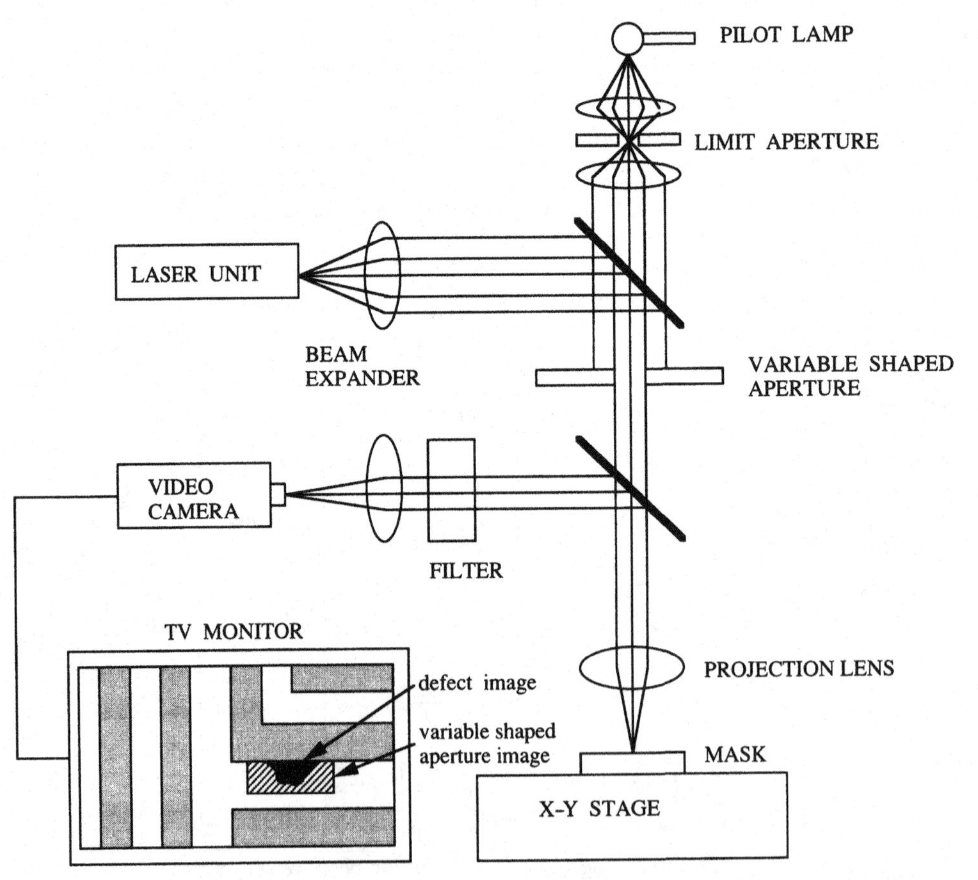

Figure 7.37 Basic configuration of laser-beam optics for a laser-based repair system.

Table 7.7 *Main specifications of commercial laser-based repair systems.*

Item	SL-453C (NEC)[a]	DRS-III (Excel/Quantronix)
Laser source	Nd:YAG (2nd harmonic: 532 nm)	Ti:sapphire (3rd harmonic: 248 nm)
Spot size (μm)	0.7	0.17 (calculated value)
Repair accuracy (μm)	±0.15	±0.1
Repair function	Only ablation	Both ablation and deposition (tungsten)

[a] Nippon Electric Corp.
Nd:YAG: Neodymium-doped yttrium aluminum garnet

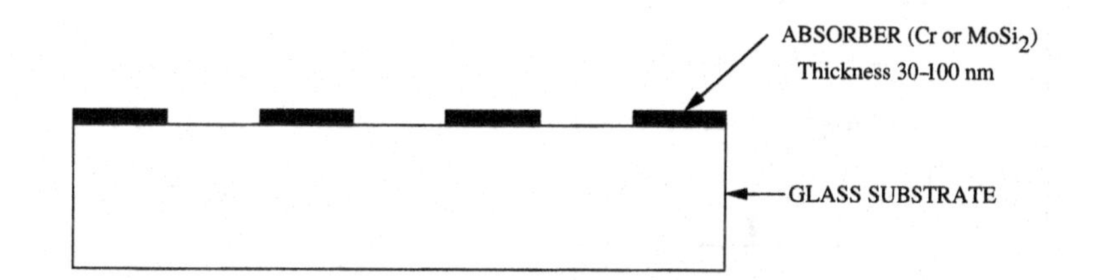

Figure 7.38 Schematic sectional view of a photomask.

on a TV monitor. The accuracy to which the two images can be superimposed is limited by the magnification of the microscope system, including the projection lens, a video camera, and the TV monitor.

Opaque defects are removed by laser ablation and clear defects are repaired by laser-induced photolytic or pyrolytic deposition. The main specifications of commercial laser-based repair systems are listed in Table 7.7. A second-harmonic Nd:YAG (Q-switched) laser[36] and a third-harmonic Ti:sapphire (gain-switched) laser[41] are used commercially as light sources in laser-based repair tools.

7.4.3 Photomask repair

Photomasks and reticles are made of transparent glass (substrate) on which a thin film of metal, such as Cr or $MoSi_2$, is patterned (Figure 7.38). The metal film thickness is typically less than 100 nm. The sizes of the defects which need repair are usually greater than $\frac{1}{3}$–$\frac{1}{4}$ of the minimum feature size formed on the mask. A flowchart of the basic repair sequence is shown in Figure 7.39. The defective region of the mask is first positioned under the beam axis using location information read from the inspection tool. A magnified image of the area is then taken, stored in the computer, and displayed on the TV monitor. Then an operator outlines the defect in the displayed image, and the computer controls the repair of the area. For an opaque defect, the excess absorber is removed; for a clear defect, a thin film is deposited as an absorber. Finally, a new image of the area is obtained to confirm the repair.

7.4.3.1 *Opaque defect*

Opaque defects are repaired by removing excess absorber material such as Cr or $MoSi_2$. These repairs constitute more than 75% of all mask repairs,[36] so throughput is an important factor to consider. It is also important that the underlying glass substrate is not damaged during the

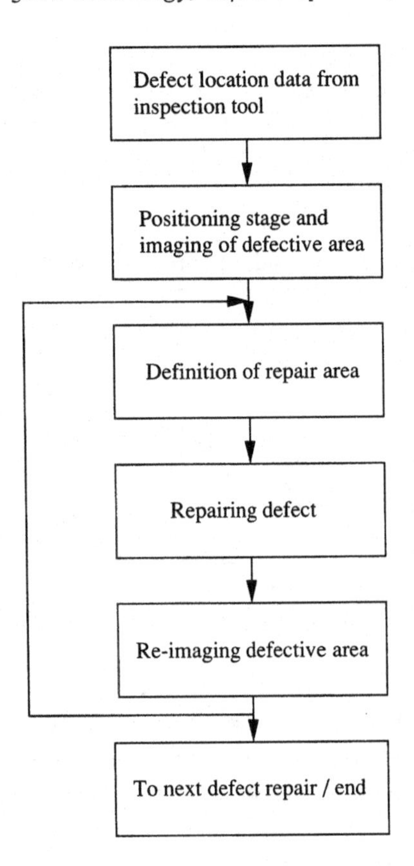

Figure 7.39 Flowchart of the basic repair sequence.

repair process. Laser-based repair tools are commonly used for opaque defects because laser repair is a high-throughput process and does not damage either the adjacent absorber material or the underlying glass substrate. When a laser beam irradiates an opaque defect area, the absorber material instantaneously evaporates. The laser firing time is only a few nanoseconds and the underlying glass substrate is not damaged because it absorbs little of the energy.

There is no problem in using laser repair for opaque defects larger than 1 μm, but the resolution and positioning accuracy of the laser-based repair system become problems when the defect size is less than a half micron. Although the throughput of FIB is lower than that of laser repair, FIB system has higher resolution and position accuracy. Physical sputtering by Ga ions is used to repair opaque defects. The ion beam is scanned repeatedly over the defect until it is sputtered away.

FIB repair inevitably damages the glass under the defect: the incident Ga ions strike the glass substrate and produce a post-repair stain.[42] The stain is normally visible under optical inspection and is considered a mask defect because it may print during photolithography. The transmission of the glass substrate is reduced by the Ga implanted during repair, and the reduction is much greater for shorter wavelengths: the transmission values for the g-line (436 nm), i-line (365 nm), and DUV (deep-ultraviolet light: 248 nm) are respectively about 75%, 60%, and

25%.[43] There are several strategies for minimizing this effect: (1) etching the damaged layer by CF_4 plasma[44] or $CHF_3 + O_2$ reactive ions[45] after the opaque defect is removed, (2) FIB-assisted etching with XeF_2 gas,[46] and (3) detecting the endpoint during repair.[47] The transmission of a repaired glass substrate recovers by more than 97% when these methods are used.

7.4.3.2 *Clear defects*

Clear defects are repaired by depositing an opaque film over the area where the absorber material is missing. The film must be: (1) opaque, conformal, free from pinholes and must adhere to both metal and glass surfaces; (2) confined to the beam profile of the repair tool; and (3) chemically and physically resistant to mask cleaning procedures.

There are two types of methods for using a FIB to repair clear defects. One is to cover the defect with an opaque film such as carbon,[48] and the other is to mill the defect by ion-beam sputtering so that it becomes opaque.[49] The former is the film-deposition process illustrated in Figure 7.40(a). A nozzle in close proximity to the mask surface delivers a gas to the area around the defect. This builds up a high pressure close to the nozzle while a good vacuum is maintained in the rest of the system. Gas molecules adsorb to the substrate surface where the ion beam induces chemical reactions. A hydrocarbon gas such as styrene (C_8H_8) or pyrene ($C_{16}H_{10}$) is commonly used and a carbon film is formed by ion-beam-induced deposition. The film thickness of the deposited carbon is directly proportional to the ion dose and the deposition rate for styrene corresponds to 1–2 carbon atoms per incident Ga ion.[50] About 150 nm of carbon film is required to ensure optical opacity, and the carbon film is remarkably resistant to a variety of acids and solvents used in cleaning processes.[51]

The FIB milling method is used to form optical microstructures in the mask as illustrated in Figure 7.40. An optical microstructure is a prism-like, grating-like, or lens-like structure etched into the glass. By total internal reflection, it blocks the incident light as effectively as an opaque film does. It is difficult, however, to reproduce a microstructure having optical opacity.

There are two laser-based processes for repairing clear defects: pyrolytic and photolytic. These processes utilize an organometallic compound such as $Al(CH_3)_3$, $Cr(CO)_6$ or $W(CO)_6$.

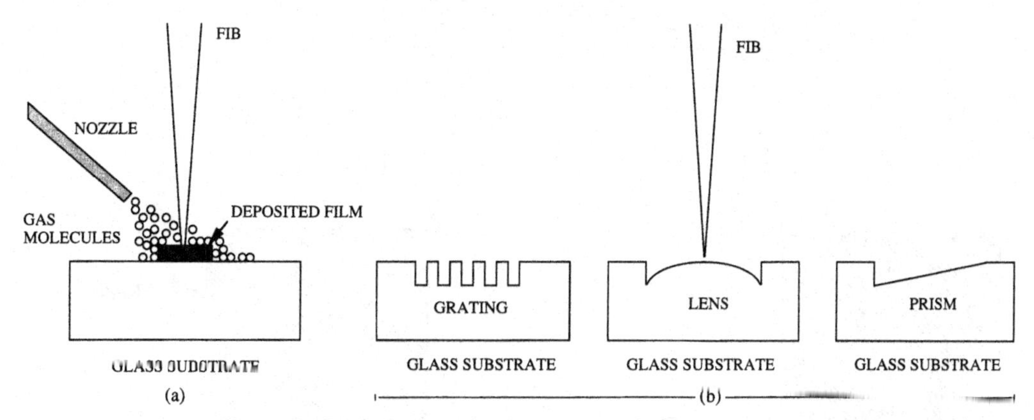

Figure 7.40 Focused-ion-beam processes for repairing clear defects: (a) by deposition and (b) by milling grating, lens or prism of optical microstructures.

The pyrolytic deposition process for optical (photomask) repair is difficult to perform because the transparent substrate does not absorb sufficient laser photons to increase the temperature enough to activate thermal decomposition. However, laser-induced deposition takes place by nucleating a substrate.[52] The nucleation layer can be created from the parent organometallic molecule by flood exposure with deep-ultraviolet (DUV) light. Such pyrolytic deposition is difficult to control and results in nonuniform surface qualities.[41] Photolytic deposition can also be performed using a DUV laser. This process is similar to the FIB-induced deposition process. An organometallic gas fully dissociates in a multiphoton process with a DUV laser. The molecules partially dissociate in the gas phase and are deposited on a photomask surface where they continue to absorb photons. Only those partially dissociated molecules within the area exposed to the laser continue to absorb photons and fully dissociate to form metal atoms. This process proceeds at a much lower laser intensity, thus making it easier to control deposition and produce films with high-quality surfaces.[41]

7.4.4 X-ray mask repair

A schematic of the sectional configuration of an X-ray mask is illustrated in Figure 7.41. The mask consists of a membrane, an absorber, and a frame. The membrane material must be transparent to X-rays and the absorber material must be opaque to X-rays. Currently, silicon carbide or diamond are commonly used for the membrane, and tantalum, tungsten, or their compounds are used for the absorber. X-ray mask repair also requires the etching and deposition of heavy metals. Compared with photomasks, X-ray masks have extremely small feature sizes, high-aspect-ratio patterns, and large thermal conductivity. These features make laser repair unsuitable, so FIB repair is commonly used.

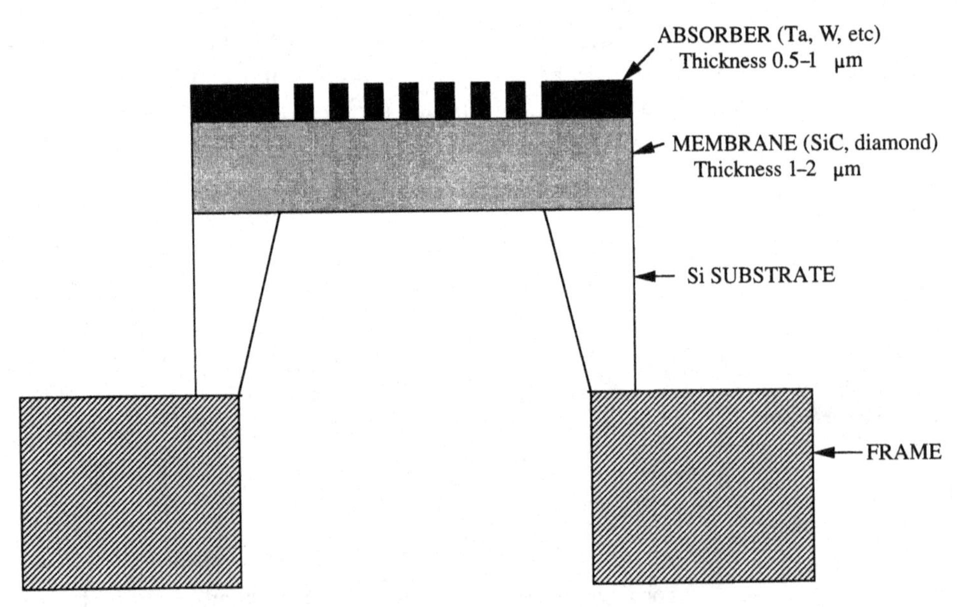

Figure 7.41 Schematic sectional configuration of an X-ray mask.

7.4.4.1 *Opaque defects*

Opaque defects are removed by physical sputtering. The ion beam is raster-scanned over the defect to erode it. While this technique is similar to that used for photomask repair, there are several factors that are especially important for repairing X-ray masks. One factor is the redeposition of sputtered absorber material. Some of the absorber atoms ejected during the sputtering process may not be permanently removed from the mask because they can strike existing features on the mask and stick to nearby features in the pattern. Redeposition of sputtered atoms is likely to occur during opaque repair because of the density and high aspect ratio of the patterns on an X-ray mask. A photograph showing a top view of a repaired X-ray mask is shown in Figure 7.42. The absorber material has been redeposited on neighboring lines 0.2 μm from the repaired line. When features are less than 0.5 μm away, the redeposition occurs not only on nearby features but also on the feature being sputtered.[53] There are several ways to minimize this effect: (1) multiple scanning of the ion beam to reduce redeposition within the feature; (2) trimming the areas surrounding the defect to remove redeposited absorber; (3) milling the center of the defect before milling the edge in order to reduce the volume of absorber available for redeposition on adjacent features;[53] and (4) chemically enhancing the physical sputtering process by using a reactive gas that forms a volatile compound when it reacts with the absorber.[54] This last way is the most practical because it is very simple and the volatile by-products are not likely to stick to adjacent features.

Another factor especially important in X-ray mask repair is the non-channeling effect, in

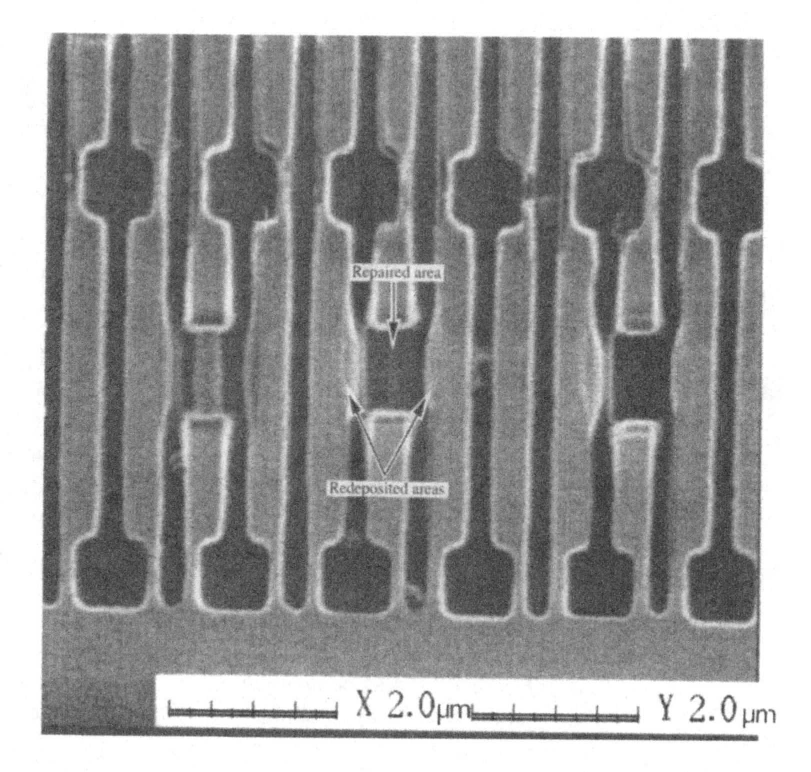

Figure 7.42 A scanning-ion-beam photograph showing the effect of the redeposition of sputtered material during opaque repair. The Ta absorber material has been redeposited on neighboring lines 0.2 μm from the repaired line.

which the ability of grains to stop ions is relatively small when the grains are aligned with the ion beam. Since the channeled ion has a much smaller stopping power than a non-channeled ion, the sputter and the secondary-electron yield of the grains becomes very small. This results in large variations of sputter yield in opaque repair and in the brightness of the defects showing up in the images. Sputter yield variations of 5 to more than 25 atoms per incident ion have been observed.[55] Some solutions to this variation problem are (1) to minimize the grain size of the absorber material; (2) to deposit a thin film of amorphous or fine-grained, randomly oriented, crystalline material on the absorber; or (3) to increase the etching rate by using a chemically assisted etching process.[54]

7.4.4.2 *Clear defects*

Clear defects are repaired by ion-beam-induced deposition. An organometallic gas is supplied directly over the mask and a FIB is scanned on the area over the defect. Nonvolatile metal atoms are left on the surface while volatile components desorb. The ion beam also sputters several atoms from the surface. In clear-defect repair, since sputtering always occurs during deposition, redeposition of sputtered material affects resolution. Net deposition of metal occurs when more atoms are added through decomposition than are removed by sputtering. Increasing the net number of atoms deposited per incident ion increases the writing rate and decreases the relative contribution of redeposition. The deposition yield (net number of atoms

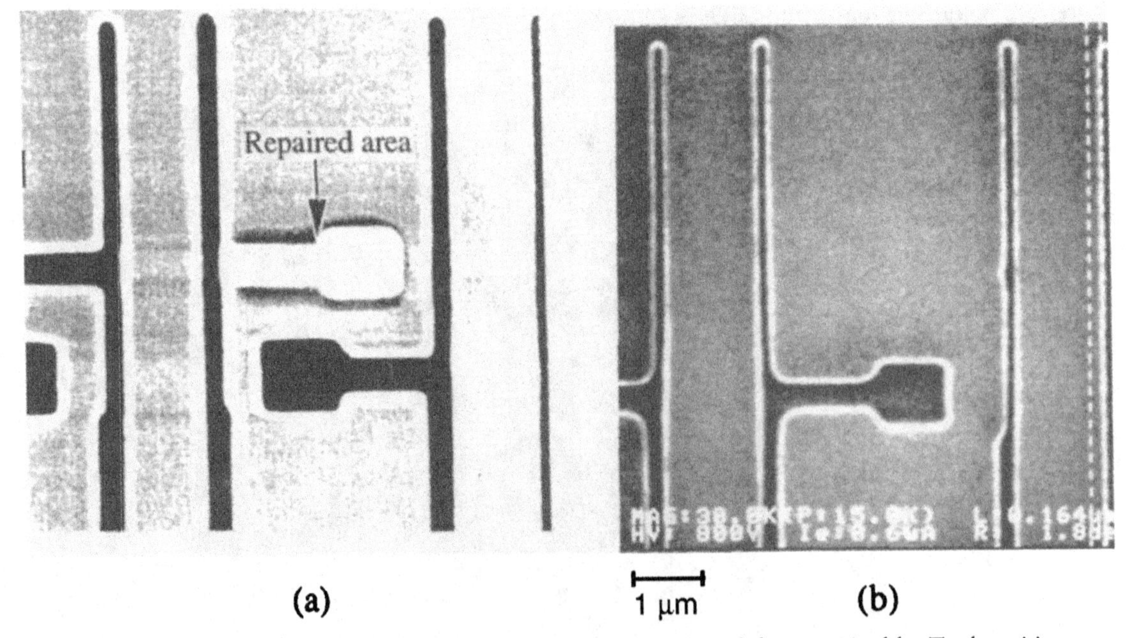

Figure 7.43 SEM photographs showing clear defects repaired by Ta deposition: (a) repaired X-ray mask and (b) resist image replicated by the repaired mask.

deposited per incident ion) should therefore be large in order to reduce redeposition and pattern-size error.

Increasing the number of organometallic molecules absorbed on the surface is an efficient way to improve the deposition yield because the deposition yield is directly proportional to the number of molecules absorbed on the surface.[56] At the gas pressures typically used during ion-beam deposition, however, the arrival rate of organometallic molecules at the surface of the mask is an order of magnitude less than the ion arrival rate in a focused-ion-beam tool. So the ion beam depletes the gas absorbed on the surface rapidly and then ion sputtering proceeds. This makes the deposition yield low. The effect can be reduced by selecting organometallic compounds and controlling deposition process parameters[57] Since films deposited for X-ray masks must be made of a heavy metal in order to absorb X-rays efficiently, an organometallic gas containing gold, tantalum, or tungsten is often used.[58] The deposition yield becomes larger as the vapor pressure of the gas is increased. The best deposition yields are obtained for gold (75 atoms/ion), tantalum, (7–10 atoms/ion) and tungsten (1–2 atoms/ion).[59] Gold is excellent in terms of deposition yield, but tantalum and tungsten are more suitable materials for use in LSI fabrication lines. An example of an X-ray mask with a clear repair is shown in Figure 7.43(a). The resist pattern produced by the repaired mask is shown in Figure 7.43(b). Tantalum tetraethoxyacetyl acetonate was used as the organometallic gas, and the deposited film was about 30% tantalum and 70% carbon.[60] To improve exposure contrast, the thickness of the repaired absorber film was made three times that of the original absorber. The thicker the deposited film, however, the worse the resolution. It is important to increase the density of the deposited film by reducing the amount of carbon in the film.

7.5 Summary

Current critical pattern dimensions on a wafer for mass-production IC devices are about 0.25 μm. These critical pattern dimensions will become 0.15 μm for 1-Gb DRAMs and 0.13 μm for 4-Gb DRAMs. The attainment of these critical dimensions (CD) in the manufacturing environment will require improvements in current lithographic capabilities. Progress in this field already owes much to metrology, defect inspection, and repair, and an investigation into these technologies could significantly influence estimation, verification, analysis, study, development, and yield-improvement in lithographic tools vital to the future of the semiconductor industry.

As described in this chapter, these technologies are extremely useful for design rules below a half micron. New probes with resolution better than 1 nm, such as a scanning probe microscope (STM, AFM) are expected to be widely used in measurement. The sensors in metrology tools must be noninvasive, noncontact and nondestructive; automatic and in-line measurement will be essential for mass production. Two-dimensional monitoring is no longer sufficient, and three-dimensional measurement will be required to measure a complex pattern structure such as memory-cell capacitor fins. The most important requirement for the next generation of metrology tools is to develop measurement algorithms that eliminate measurement error.

Inspection systems have, over the last decade, become indispensable to photomask manufacture. Shorter-illumination-wavelength (such as 248- and 193-nm) optics and improved comparison algorithms will boost defect detection sensitivity to 0.1 μm, the level required in manufacturing 1-Gb-DRAM masks and developing 4-Gb-DRAM masks. Inspection techniques for phase-shift mask defects, particularly the detection of alternative phase-shift mask defects, will become an important issue as we move beyond the 1-Gb DRAM. For later generations of 4-Gb DRAMs, electron beams will be used to detect pattern defects because of their sensitivity. Defect detection is absolutely essential to mask fabrication and development and, as research in this area expands, even greater progress is expected.

Both laser and focused-ion-beam (FIB) repair techniques are now widely used for repairing photomasks and X-ray masks. Laser repair is commonly used in mass-production lines, while FIB repair is used primarily when resolution requirements cannot be met by laser repair. FIB repair is commonly used in X-ray mask repair because of the X-ray masks' distinctive structure: extremely tiny features, high aspect ratio of features, and large thermal conductivity of the substrate. Improving resolution and repair accuracy, however, is essential for satisfying the requirements of next-generation masks.

The effective application of metrology, defect inspection, and repair information from each process must also be considered in device fabrication. Effective application and the strategic use of information from these processes will eliminate many metrological problems and will provide good cost-performance and elegant solutions to problems in semiconductor manufacturing. By using this information, we will be able to set the guidelines needed in choosing the most appropriate systematic solutions to some problems that may arise in device fabrication.

7.6 References

1. D. Nyyssonen, *Appl. Opt.*, **16**, 2223 (1997)
2. M. Minsky, US Patent 3,013,346 (1961)
3. G. S. Kino and G. Q. Xiao, US Patent 5,022,743 (1991)
4. G. D. Danilatos, *Scanning*, **4**, 9–20 (1981)
5. P. Sandland, *Proc. SPIE*, **100**, *Semiconductor Microlithography II*, 26 (1977)
6. D. B. Novotny and D. R. Ciarlo, *Solid State Technol.*, May, 51 (1978)
7. I. Tanabe, *Production of High Quality Photomasks for MOS LSIs*, Hitachi Ltd, Kodaira-shi, Tokyo, Japan, Interface 75, Kodak Pub. G45, 91 (1975)
8. H. Yang, *Proc. SPIE*, **334**, 216 (1982)
9. I. A. Cruttwell, *Proc. SPIE*, **394**, 223 (1983)
10. R. F. Pease, *Jpn. J. Appl. Phys.*, **31**, 4103–9 (1991); M. Shibuya, T. Ozawa, M. Komatsu and H. Ooki, *Proc. 7th Intl. Microprocess Conf.*, July 11–14, Hsinchu, Taiwan, 6874–7 (1994)
11. A. Imai *et al.*, *Jpn. J. Appl. Phys.*, **33**, 6816–22 (1994)
12. T. Furukawa *et al.*, *J. Jpn. Soc. Precision Engg.*, **58**, (2), 215 (1992)
13. J. N. Wiley and J. A. Reynolds, *Solid State Technol.*, July, 65 (1993)
14. Y. M. Ham *et al.*, *Jpn. J. Appl. Phys.*, **31**, 4155 (1992)
15. D. R. Ciarlo, P. A. Schultz and D. B. Novotny, *Proc. SPIE*, **55**, *Photofabrication Imagery*, 84–9 (1975).
16. D. L. Dyer, *Research/Development*, **24**, 40–4 (1973)
17. *KLA301 Photomask/Reticle Inspection System*, KLA Instruments Co. (1995)
18. *9MD83SR Mask Inspection System*, Laser Tech. Co. (1995)
19. P. Sandland *et al.*, *J. Vac. Sci. Technol.*, **B9**(6), 3005 (1991)
20. R. Yoshikawa and S. Sasaki, *Toshiba Review*, No. 147, 44 (1984)
21. S. Yabumot, T. Arai, Y. Fujimori and T. Azuma, *Proc. SPIE*, **633**, *Optical Microlithography V*, 138 (1986)
22. S. Takeuchi *et al.*, *Proc. SPIE*, **1261**, *Integrated Circuit Metrology, Inspection and Process Control IV*, 195 (1990)
23. E. K. Sittig and M. Feldman, *Kodak Microelectron*

Seminar, Atlanta, GA, Oct. (1973), p. 49
24. J. H. Bruning *et al., IEEE Trans. Electron Devices,* **ED-22**(7), 487 (1975)
25. J. D. Knox, P. V. Goedertier, D. W. Fairbanks and F. Caprari, *Solid State Technol.,* **20**, 48 (1977)
26. *5MD44 Mask Inspection System*, Laser Tech Co. (1995)
27. T. Tojo *et al., Jpn J. Appl. Phys.,* **33**, 7156 (1994)
28. J. N. Wiley, *Proc. SPIE,* **2196**, 219–33 (1994)
29. Y. Suzuki *et al., Tech. Proc. SEMICON/Japan*, 255 (1993)
30. K. Amashita, K. Matsuki and K. Akeno, *Proc. SPIE,* **2793**, *Photomask and X-ray Mask Technology III*, 279–87 (1996)
31. Y. Eran and G. Greenberg, *Proc. SPIE,* **2512**, *Photomask and X-ray Mask Technology II*, 453–6 (1995)
32. H. Watanabe, Y. Todokoro and M. Inoue, *Jpn. J. Appl. Phys.,* **30**, 3016 (1991)
33. Y. Ichioka and T. Suzuki, *J. Opt. Soc. Amer.,* **66**(9), Sept., 921 (1976)
34. M. Tabata *et al., Proc. SPIE,* **3096**, 415 (1997)
35. A. J. Serafino *et al., Proc. SPIE,* **1088**, 74 (1989)
36. J. K. Tison and M. G. Cohen, *Solid State Technol.,* Feb., 113, (1987)
37. N. P. Economou, D. C. Shaver and B. Ward, *Proc. SEMI Tech. Symp. '86,* Abst. B-4-1 (1986)
38. D. K. Atwood, G. J. Fisanick, W. A. Johnson and A. Wagner, *Proc. SPIE,* **471**, 127 (1984)
39. R. Viswanathan *et al., J. Vac. Sci. Technol.,* **B11**, 2910 (1993)
40. I. Okada *et al., Proc. SPIE,* **2437**, 253 (1995)
41. J. W. Herman *et al., Proc. SPIE,* **2437**, 264 (1995)
42. A. Wagner and J. P. Levin, *Nucl. Instrum. and Meth.,* **B37/38**, 224 (1989)
43. P. D. Prewett, B. Martin, A. W. Eastwood and J. G. Watson, *J. Vac. Sci. Technol.,* **B11**(6), 2427 (1993)
44. T. Cambria, STEP/SEMI Technical Education Programs (1987)
45. K. Saitoh *et al., J. Vac. Sci. Technol.,* **B6**(3), 1032 (1988)
46. H. Nakamura, H. Komano and M. Ogasawara, *Jpn. J. Appl. Phys.,* **31**, 4465 (1992)
47. Y. Nakagawa *et al., Proc. SPIE,* **923**, 114 (1988)
48. M. Yamamoto *et al., Proc. SPIE*, **632**, 97 (1986)
49. A. Wagner, *Nucl. Instrum. and Meth. Phys. Res.,* **218**, 355 (1985)
50. L. R. Harriott and M. J. Vasile, *J. Vac. Sci. Technol.,* **B6**(3), 1035 (1988)
51. T. D. Cambria and N. P. Economou, *Solid State Technol.,* Sept., 133 (1987)
52. M. M. Oprysko and M. W. Eranek, *J. Vac. Sci. Technol.,* **B5**(2), 496 (1987)
53. D. Stewart, T. Olson and B. Ward, *Proc. SPIE,* **1924**, 98 (1993)
54. L. R. Harriott, R. R. Kola and G. K. Celler, *Proc. SPIE,* **1924**, 76 (1993)
55. A. Wagner *et al., J. Vac. Sci. Technol.,* **B8**(6), 1557 (1990)
56. A. D. Dubner and A. Wagner, *J. Appl. Phys.,* **66**, 870 (1989)
57. J. P. Levin, P. G. Baulner and A. Wagner, *Proc. SPIE,* **1263**, 2 (1990)
58. K. Gamo and S. Namba, *Microelectronic Eng.,* **11**, 403 (1990)
59. D. K. Stewart, J. A. Doherty, A. F. Doyle and J. C. Morgan, *Proc. SPIE,* **2512**, 398 (1995)
60. I. Okada *et al., Proc. SPIE,* **2512**, 173 (1995)

Index

Bold page numbers indicate the main entries.